Textbook of Human Reproductive Genetics

A basic understanding of human genetics is vital for all those working in the field of assisted human reproduction. Genetic makeup can hamper reproduction, and insight into this is making genetic diagnosis and counselling increasingly important. This fully updated textbook continues the clear structure of the original edition, beginning with a chapter on the basics of genetics and cytogenetics. Genetic causes of infertility and the effect of epigenetics and transposons on fertility are discussed in detail. Several new chapters are included in this edition, reflecting the advances of the field, including preconception genetic analysis and screening in IVF and mitochondrial genetics. Combining genetics, reproductive biology, and medicine, this is an essential text for practitioners in reproductive medicine and geneticists involved in the field looking to improve their knowledge of the subject and provide outstanding patient care.

Stéphane Viville is a professor at the University of Strasbourg. He leads a laboratory of fundamental research at the Laboratoire de génétique médicale (UMRS-1112) Strasbourg. He is a former chair of the ESHRE Reproductive Genetics Special Interest Group.

Karen D. Sermon is Head of the Reproduction & Genetics Research Group, Vrije Universiteit Brussel, Brussels. She is a former chair of the ESHRE Reproductive Genetics Special Interest Group and is currently chair-elect for ESHRE (2021–2023) and chair (2023–2025).

Textbook of Human Reproductive Genetics

SECOND EDITION

Edited by
Stéphane Viville
Institut de Génétique et de Biologie Moléculaire et Cellulaire, Strasbourg
Karen D. Sermon
Research Group Reproduction and Genetics, Vrije Universiteit Brussel

Shaftesbury Road, Cambridge CB2 8EA, United Kingdom

One Liberty Plaza, 20th Floor, New York, NY 10006, USA

477 Williamstown Road, Port Melbourne, VIC 3207, Australia

314–321, 3rd Floor, Plot 3, Splendor Forum, Jasola District Centre, New Delhi – 110025, India

103 Penang Road, #05–06/07, Visioncrest Commercial, Singapore 238467

Cambridge University Press is part of Cambridge University Press & Assessment, a department of the University of Cambridge.

We share the University's mission to contribute to society through the pursuit of education, learning and research at the highest international levels of excellence.

www.cambridge.org
Information on this title: www.cambridge.org/9781009197724
DOI: 10.1017/9781009197700

First published 2023

Printed in the United Kingdom by TJ Books Ltd, Padstow Cornwall

A catalogue record for this publication is available from the British Library.

Library of Congress Cataloging-in-Publication Data
Names: Viville, Stéphane, editor. | Sermon, Karen D., 1964– editor.
Title: Textbook of human reproductive genetics / edited by Stéphane Viville, Karen D. Sermon.
Description: 2. | Cambridge, United Kingdom ; New York, NY : Cambridge University Press, 2022. | Includes bibliographical references and index.
Identifiers: LCCN 2022011824 (print) | LCCN 2022011825 (ebook) | ISBN 9781009197724 (paperback) | ISBN 9781009197700 (epub)
Subjects: MESH: Reproduction–genetics | Embryonic Development | Genetic Services
Classification: LCC QP251 (print) | LCC QP251 (ebook) | NLM WQ 205 | DDC 573.6–dc23/eng/20220404
LC record available at https://lccn.loc.gov/2022011824
LC ebook record available at https://lccn.loc.gov/2022011825

ISBN 978-1-009-19772-4 Paperback

Contents

Contributors

Esther B. Baart
Department of Obstetrics and Gynaecology, Erasmus MC, University Medical Center, Rotterdam, The Netherlands

Ashwini Balakrishnan
Magee-Womens Research Institute, Department of Microbiology and Molecular Genetics, University of Pittsburgh School of Medicine, Pittsburgh, PA, USA

Déborah Bourc'his
Institut Curie, CNRS UMR3215/Inserm U934, Paris, France

J. Richard Chaillet
Magee-Womens Research Institute, Department of Microbiology and Molecular Genetics, University of Pittsburgh School of Medicine, Pittsburgh, PA, USA

Effrosyni A. Chavli
Department of Obstetrics and Gynaecology, and Department of Clinical Genetics, Erasmus MC, University Medical Center, Rotterdam, The Netherlands

Christine de Die-Smulders
Department of Clinical Genetics, Maastricht University Medical Centre, Maastricht, The Netherlands

Wybo Dondorp
Department of Health, Ethics and Society and GROW School for Oncology and Developmental Biology, Maastricht University, Maastricht, The Netherlands

Ursula Eichenlaub-Ritter
Faculty of Biology, University of Bielefeld, Bielefeld, Germany

Patricia Fauque
Hôpital de Dijon, Université de Bourgogne, Laboratoire de Biologie de la Reproduction; and Université Bourgogne Franche-Comté – INSERM UMR1231, Dijon, France

Elia Fernandez Gallardo
Laboratory of Reproductive Genomics, Centre for Human Genetics, and KU Leuven Institute for Single Cell Omics (LISCO), KU Leuven, Leuven, Belgium

Juan José Guillén
Clinica Eugin, Barcelona, Spain

Thomas Lefevre
Laboratory of Reproductive Genomics, Centre for Human Genetics, and KU Leuven Institute for Single Cell Omics (LISCO), KU Leuven, Leuven, Belgium

Özlem Okutman
Laboratoire de Diagnostic Génétique, UF3472-Génétique de l'Infertilité, Hôpitaux Universitaires de Strasbourg; and UMR 1112 Laboratoire de Génétique Médicale, Strasbourg, France

Aleksandar Rajkovic
Departments of Pathology and Obstetrics, Gynecology and Reproductive Sciences, University of California San Francisco, San Francisco, CA, USA

Carmen Rubio
PGT-A Research, Igenomix, Valencia, Spain

Karen D. Sermon
Research Group Reproduction and Genetics, Vrije Universiteit Brussel, Brussels, Belgium

Claudia Spits
Vrije Universiteit Brussel, Jette, Belgium

Koen Theunis
Laboratory of Reproductive Genomics, Centre for Human Genetics, and KU Leuven Institute for Single Cell Omics (LISCO), KU Leuven, Leuven, Belgium

Jan Traeger-Synodinos
Laboratory of Medical Genetics, National and Kapodistrian University of Athens, Choremeio Research Laboratory, St. Sophia's Children's Hospital, Athens, Greece

Ron van Golde
Department of Obstetrics and Gynaecology, Maastricht University Medical Centre, Maastricht, The Netherlands

Aafke P.A. van Montfoort
Department of Clinical Genetics, GROW School for Oncology and Developmental Biology, Maastricht University Medical Centre, Maastricht, The Netherlands

Diane Van Opstal
Department of Clinical Genetics, Erasmus MC, University Medical Center, Rotterdam, The Netherlands

Rita Vassena
Clinica Eugin, Barcelona, Spain

Stéphane Viville
Laboratoire de Diagnostic Génétique, UF3472-Génétique de l'Infertilité, Hôpitaux Universitaires de Strasbourg; and UMR 1112 Laboratoire de Génétique Médicale, Strasbourg, France

Thierry Voet
Laboratory of Reproductive Genomics, Centre for Human Genetics, and KU Leuven Institute for Single Cell Omics (LISCO), KU Leuven, Leuven, Belgium

Guido de Wert
Department of Health, Ethics and Society and GROW School for Oncology and Developmental Biology, Maastricht University, Maastricht, The Netherlands

Svetlana A. Yatsenko
Departments of Pathology, Obstetrics, Gynecology and Reproductive Sciences, and Human Genetics, University of Pittsburgh; and Magee-Womens Research Institute, Pittsburgh, PA, USA

Filippo Zambelli
Clinica Eugin, Barcelona, Spain

Basic Genetics and Cytogenetics: A Brief Reminder

Karen D. Sermon

1.1 Introduction

This brief reminder chapter aims to freshen up what professionals in reproduction may have learned a while ago at university, and will also serve the reader as a source of information to comprehend the following, more complex chapters. At the end of this chapter, basic study books or broad reviews are recommended for further reading rather than regular scientific references, to help the reader in the further understanding of this textbook.

Human reproduction and genetics are intimately intertwined and indeed often confused and rolled into one. Understanding reproduction is impossible without a firm basis in genetics, and the readiness to acquire more knowledge when needed. However, human genetics is much broader than just reproduction – think of, for instance, oncogenetics – so in this chapter I will summarize those aspects of basic genetics that are indispensable for specialists in reproduction.

In this chapter, I introduce the general organization of our genome, how this genome behaves when it goes through a reproductive cycle (meiosis), how our genome is used as a template for making proteins and how this is broadly regulated, describe major genetic and hereditary abnormalities at both the chromosome and the monogenic level, and conclude with a brief overview of current genetic diagnostic techniques.

1.2 The Organization of the Human Genome

1.2.1 The Basic Building Material: DNA

Deoxyribonucleic acid (DNA) consists of four different nucleotides [1,2]. Each nucleotide consists of a sugar, which is deoxyribose in DNA, a phosphate group and a base. Four different bases are present in the nucleotides of DNA: adenine (A), cytosine (C), guanine (G), and thymine (T) (Figure 1.1a) [3]. These four nucleotides are strung together in long strands of DNA, alternating a sugar (with a base attached) and a phosphate group, and where the order of the different bases defines the genetic code (Figure 1.1b). The phosphate group can be bound either to the 5′ carbon of the sugar or to the 3′ carbon of the sugar, while the base is bound to the 1′ carbon. When strung together, the first sugar in the DNA strand has a free 5′ carbon, while its 3′ carbon is covalently bound to a phosphate group. This phosphate group is bound to the 5′ carbon of the next sugar further down the strand. This is why in a DNA strand, the 5′ carbon of the first sugar is free and, at the end of the strand, the 3′ carbon of the last sugar is free. This is why DNA base pairs are always read from 5′ to 3′.

Moreover, cellular DNA is usually found in a double helix form. The bases A and T on the one hand and C and G on the other can form hydrogen bonds, thus stabilizing the double helix (Figure 1.1c). According to the Ensembl database [4], our nuclear DNA contains 4 537 931 177 base pairs (bp). An estimated half of the human genome consists of transposons, also known as mobile DNA elements or "jumping genes," and are extensively discussed in Chapter 6.

1.2.2 DNA Is Organized in Chromosomes

The DNA of our whole genome is not ordered in one long strand, nor does it lie naked and unprotected in the nuclei of our cells. Human nuclear DNA is organized in 46 separate strands, so each chromosome contains one long DNA strand. Twenty-two of these chromosomes are paired, one inherited from the mother and one from the father. These are called the autosomes. The two remaining chromosomes are the sex chromosomes: females have two X chromosomes, while males have one X and one smaller Y chromosome.

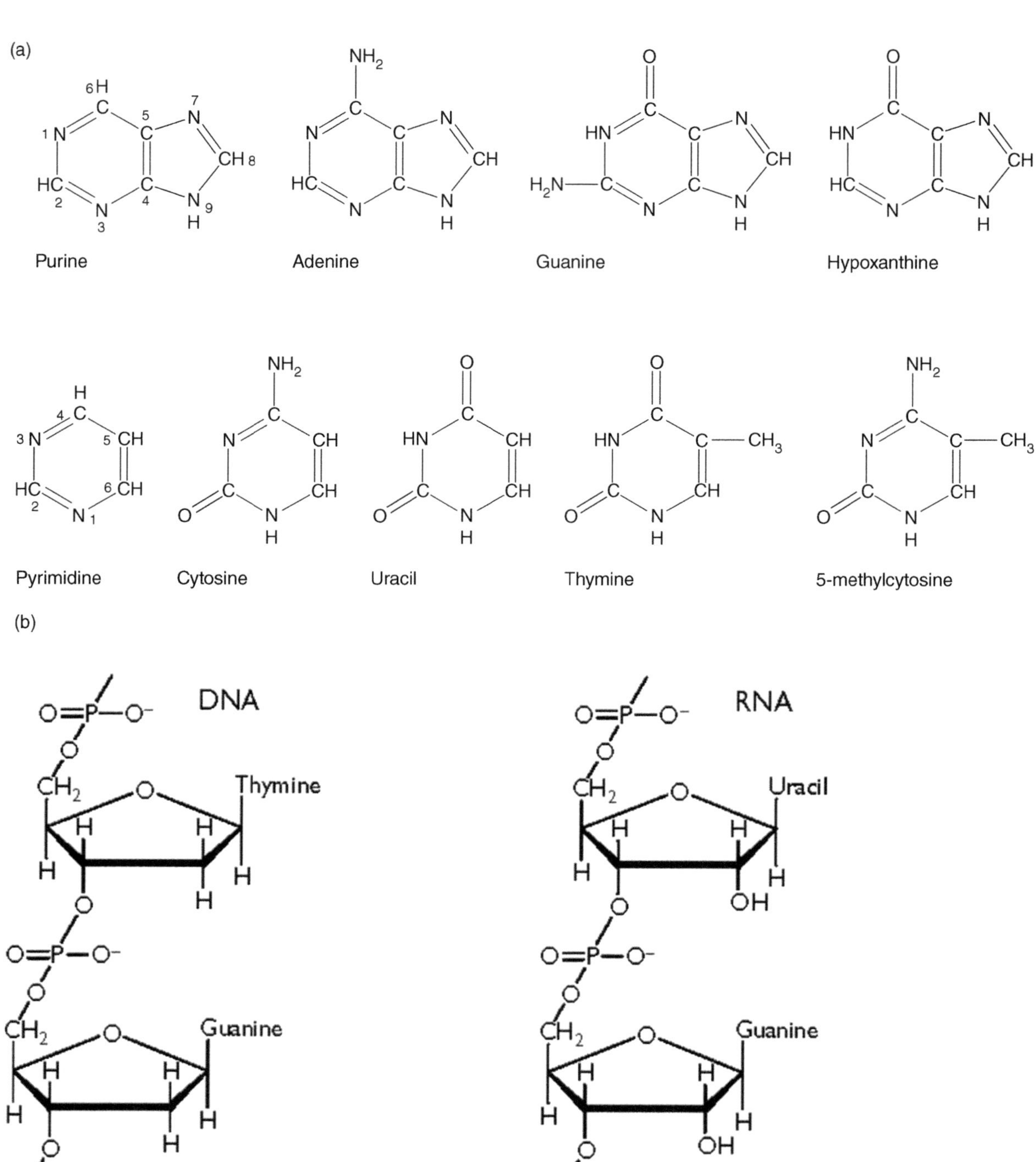

Figure 1.1 (a) The four bases in DNA and RNA. (b) Sugar–phosphate backbone of nucleotides. DNA has 2′-deoxyribose as sugar, RNA has ribose as sugar. (c) A and T form two hydrogen bonds while G and C form three hydrogen bonds. Together with the sugar–phosphate backbone this forms the double helix of DNA. Source: Figures 2.4 and 2.7–2.9 from Ringo [3].

(c)

A:T base pair (MW = 615 Da)

G:C base pair (MW = 616 Da)

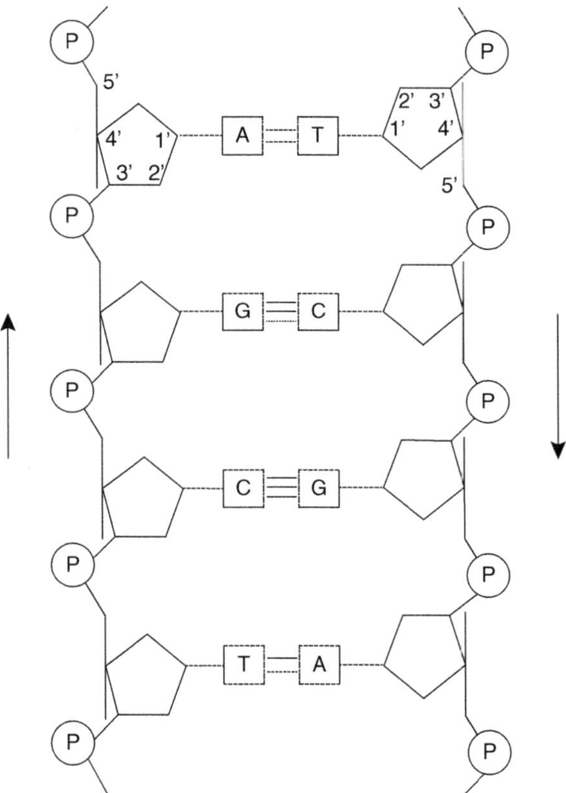

Figure 1.1 *(cont.)*

When the cell is not dividing, the DNA is organized in chromatin. The double helix is wound around protein structures called histone octamers, which in their turn are further coiled into structures called solenoids. These solenoids are attached to a protein scaffold within the nucleus to form loops (Figure 1.2). Important features of the chromosome are the centromere and the telomeres. Each chromosome has one centromere that holds the sister chromatids together until anaphase. They are most prominent at metaphase, where they can for instance be readily recognized after G-banding. The telomeres consist of highly conserved repeats (TTAGGG) and have an important function in stabilizing the chromosome; a chromosome that has lost a telomere due to breakage is unstable. The way the chromatin is organized in a specific cell, and the way some histones are chemically modified, define which genes can or cannot be transcribed and expressed. For instance, some genes will be so tightly packed in the chromatin that the transcription machinery cannot reach them and thus these genes are silenced in this particular cell. The chromatin structure is thus an important determinant of the cell's expression pattern and therefore its function.

A particular type of DNA is found in mitochondria. Mitochondria are cell organelles that are essential for the respiration and energy production of the cell. They carry their own circular DNA of about 16 000 bp which replicates independently of the DNA in the nucleus. The genes on the mitochondrial DNA (mtDNA) encode for their own ribosomal RNA (rRNA), transfer RNA (tRNA), and ribosomal proteins, as well as a handful of aerobic metabolism enzymes. However, many mitochondrial proteins are encoded by the nuclear DNA and are later imported into the mitochondrion to contribute to mitochondrial function. How mtDNA impacts reproduction is discussed in Chapter 12.

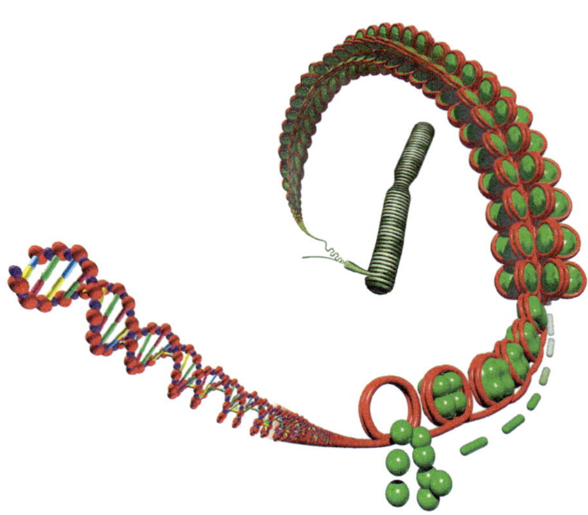

Figure 1.2 DNA is organized in chromatin. Source: QC Science Photo Library.

1.2.3 The Functional Genetic Entity: The Gene

A gene can be defined as "a sequence of DNA in the genome that is required for the production of a functional product which can be a polypeptide or a functional RNA molecule" [2]. The number of coding genes in our genome is estimated to be around 20 000 [4]. These are not equally scattered along the chromosomes: some parts of chromosomes are very gene-rich while other stretches of more than a million base pairs contain no genes at all and are called gene deserts. Some genes are quite small and comprise only a few kilobase pairs, while others span a million or more base pairs. One such large gene is the dystrophin gene, which spans more than 2 million base pairs. The two gene copies on the autosomes are usually both transcribed (or both silenced), except for a number of developmentally important genes where only one copy is transcribed. In these so-called imprinted genes, either the paternal or the maternal copy is exclusively expressed. Some of these genes have been implicated in the congenital defects that occur more frequently after assisted reproductive technology (ART); this topic is discussed in Chapter 14.

Genes that typically code for a polypeptide have recurring structural features (Figure 1.3). Not all base pairs in a gene will be translated into a protein: genes typically contain exons that are the translated parts interspersed with introns that are not translated. In many genes, the introns represent a significantly larger proportion of the gene than the exons. Other recurring structural features include sequences conserved among many different genes that provide the transcription machinery of the cell with appropriate

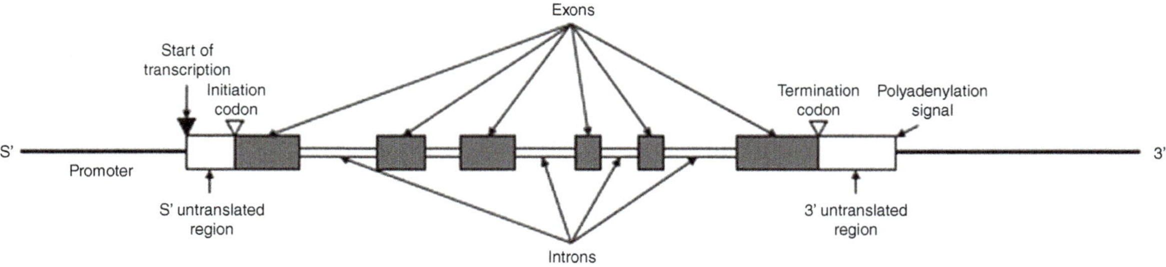

Figure 1.3 The typical anatomy of a human gene. The promoter contains the signal for transcription, and the transcription is started downstream. The initiation codon gives the signal for the start of translation. Only exons are translated; the 5′ and 3′ untranslated regions and the introns are transcribed into mRNA but not translated. The termination codon gives the signal to end the transcription, and the polyadenylation signal causes a polyA tail to be attached to the mRNA for stabilization.

signals of when and where to transcribe specific genes. Every gene has a start and a stop signal, as well as a promoter sequence in the 5′ end. This promoter specifies the pattern as well as the level of expression of a gene. At the end of the 3′ untranslated region is a signal for the polyadenylation of the transcribed messenger RNA (mRNA), which will carry a polyA tail that confers stability to the molecule. Other regulatory elements include enhancers and silencers, and can be found in either the 5′ or 3′ untranslated region, or in intronic sequences of the gene. Some can even lie far away from the coding sequence of a gene: for instance, enhancers can bind gene regulatory proteins, which then interact with proteins on promoters to upregulate gene expression. The DNA between the enhancer and the promoter is typically looped out. Silencers work in a similar way to enhancers, but instead inhibit transcription of a particular gene. Alleles are alternative forms of the same gene: changes in the DNA sequence will lead to a different protein with different function. A well-known example of multiple alleles is the ABO blood group system.

1.3 Making More Copies: The Cell Cycle

1.3.1 The Different Parts of the Cell Cycle

In order to get from the fertilized zygote to the estimated 100 trillion cells in the human body, cells need to undergo continuous divisions. Cell division and mitosis are also crucial for differentiation.

When a cell is not in mitosis, it is said to be in interphase (Figure 1.4a). The first part of interphase, immediately after mitosis, is called the G1 phase; in this phase, the chromosomes each contain only one copy of the DNA strand. This G1 phase typically lasts for several hours in rapidly dividing cells until the S-phase is reached, although some terminally differentiated cells (neurons or white blood cells) may withdraw from the cell cycle altogether and are said to be in G0, while other cells such as blastomeres have a truncated G1 phase, allowing them to divide every 13–17 h.

When the cell starts to replicate its DNA, it is said to enter the S-phase. The two DNA strands are

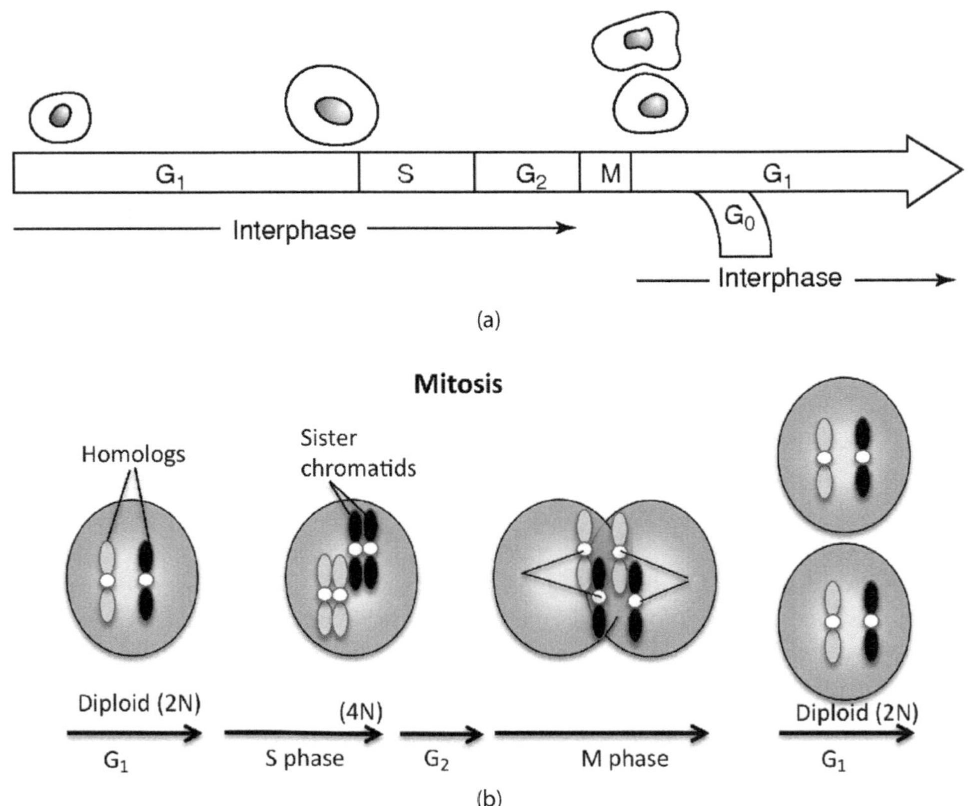

Figure 1.4 (a) The cell cycle of a somatic cell and (b) the different steps in mitosis. Source: Figure 19.1 from Ringo [3] and Chapter 3.

separated, and a complementary strand is made of each strand. The chromosomes at the end of the S-phase consist of two sister chromatids, each containing an identical DNA strand. The chromatids are held together at the centromere that associated with specialized proteins forms the kinetochore, by which the chromatids will be attached to the mitotic spindle. Before the cell enters mitosis, it goes through a brief control G2 phase in which the cell visibly grows after the accumulation of synthesized proteins during the whole cell cycle. The whole interphase of a typical cell lasts between 16 and 24 h but may extend to months, whereas mitosis is completed in a couple of hours.

1.3.2 Passing on the Information: Mitosis

At the end of G2, every chromosome consists of two chromatids and one of each of these chromatids needs to end up in one of the daughter cells in orderly chromosome segregation. The first step in mitosis is prophase, in which chromosomes start to condense and a mitotic spindle starts to form (Figure 1.4b). The formation of the mitotic spindle is organized through two centrosomes from which microtubules will radiate to form the spindle. During prometaphase, the nuclear membrane breaks up and the chromosomes are attached to the mitotic spindle by their kinetochore. Led by the microtubules of the spindle, the chromosomes move to the metaphase plate in a process called congression. During metaphase, the chromosomes have reached their maximal condensation at the equatorial plane and are easiest to visualize. Anaphase starts when the chromosomes separate into two chromatids and each chromosome moves to the different poles. Mitosis concludes with telophase, during which the chromosomes decondense and the nuclear membrane is built up again. Concomitantly to telophase, the cytoplasm is divided between the two daughter cells and cytokinesis is completed.

1.4 Making More Humans: Meiosis

During the process of mitosis, two daughter cells are produced that carry exactly the same genetic information both in content and volume. Mitosis could thus not be used to form gametes, since with every generation the amount of DNA would double. Meiosis is a specialized form of cell division that solves this problem by resulting in cells (gametes) with only half of the DNA content (i.e. one of each chromosome) of the somatic cells (Figure 1.4c). An added bonus is that during meiosis, the DNA from the two parental chromosomes is exchanged to form new chromosomes built with genetic material from the two parents. This process of recombination is important for generating genetic diversity in a species and thus for securing its evolution. Meiosis is the most important step for the survival and evolution of sexually reproducing species.

The different steps in male and female meiosis are discussed in depth in Chapter 3.

1.5 From DNA to Protein: The Transcription and Translation Machinery

1.5.1 RNA Comes in Many Forms and Functions

If DNA is important as the keeper of genetic information, the role of RNA is at least as important – some say more important – because of the versatility in form and function of RNA. RNA differs from DNA in three aspects: (1) the sugar moiety is different, ribose instead of deoxyribose; (2) the thymine base is replaced by a uracil base; and (3) RNA can be found in many different three-dimensional structures, usually as a single strand.

RNA is transcribed from the DNA template by RNA polymerases that first unwind the DNA and then synthesize an RNA strand that forms a temporary double helix with the DNA. The RNA is synthesized on the 3′–5′ DNA strand in the 5′ to 3′ direction. This 3′–5′ template DNA strand is often called the antisense strand because it is in the opposite sense from the RNA, while the 5′–3′ DNA strand which does not serve as a template has the same nucleotide sequence as the synthesized RNA strand (except that thymine is replaced by uracil) and is often called the sense strand.

Different examples of RNA forms, and functions, include mRNA that is transcribed from polypeptide-encoding genes, rRNA that will make up the ribosomes, and tRNA that will ultimately translate the information in the mRNA sequence to an amino acid sequence. When RNA is transcribed from the DNA strand, it will need to undergo a great deal of processing before it can fulfill its function. Introns will have to be spliced out of the mRNA so that only the exons are translated into protein, and a polyA tail will be added to the mRNA to ensure its stability. The rRNA and tRNA also undergo extensive changes before they are functional.

A more recently discovered type of RNA is transcribed from noncoding RNA genes. These RNAs have an important function in the regulation of other genes. Well-known examples of this are the microRNAs that can regulate the amount of mRNA available for translation. Other examples are other small noncoding RNAs such as piwi-protein interacting RNA (piRNA) and short interfering RNA (siRNA) as well as long noncoding RNAs. After the chromatin structure, this is the second example of how gene expression is regulated and explains the increasing importance that scientists are giving to RNA.

1.5.2 Different RNAs Work Together to Translate DNA into Protein

1.5.2.1 Genetic Code

In the DNA sequence of a polypeptide gene, a set of three base pairs constitutes a codon. One codon codes for one amino acid. Looking at all the possible codons using combinations of three of the four base pairs, we get 4^3 or 64 possibilities. These 64 codons are the genetic code. Because there are only 20 amino acids, one amino acid can be represented by more than one codon. There is only one codon for the amino acid methionine, and this codon also represents the start of the polypeptide. Finally, three codons represent a stop and instruct the translation machinery that the end of the polypeptide has been reached.

1.5.2.2 Translation Machinery

Once the mRNA, rRNA, and tRNA have been transformed into their mature and functional forms, they are transported to the cytoplasm of the cell where the translation into proteins takes place. The large ribosomes, built of several rRNA molecules, will start to read the mRNA at the start codon, which is always AUG coding for methionine (Figure 1.5a). Within the ribosome, a tRNA that carries an anticodon that is complementary to the methionine codon on the mRNA will then bind to the methionine codon. The other end of the tRNA carries a methionine amino acid. The ribosome then moves one codon along the mRNA to the next codon. In Figure 1.5b, this codon is GGG coding for glycine. A tRNA that has CCC as anticodon binds to the mRNA codon and carries a glycine amino acid. This glycine amino acid is then bound to the first methionine amino acid, and the two first amino acids of the polypeptide are formed (Figure 1.5c). The whole mRNA is read in this fashion until the ribosome reaches one of the three stop codons. The synthesis of the polypeptide is then terminated, and the polypeptide is released from the ribosome to be further processed into a functional protein by other cytoplasmic organelles such as the endoplasmic reticulum before reaching its cellular destination.

1.6 Behold the Genome: Tools in Human Genetics

We discuss here only a handful of tools as they appear in the following chapters. This should help the reader to understand the contents, but is by no means comprehensive. We refer the interested reader to the study books listed in the References for additional information.

1.6.1 The Big View: Cytogenetics

1.6.1.1 Know Your Classics: G-Banding and Karyotypes

Cytogenetics is the study of chromosomes, their structure, and their inheritance. At the end of the 1950s, the exact number of chromosomes was known, and at the first Chromosome Conference in Denver, the chromosomes were classified according to their size: the largest chromosome is chromosome 1 and the smallest is chromosome 22; in addition, chromosome X and chromosome Y were also included. Later, banding techniques became available and the current classification was established in Paris in 1971 (Figure 1.6). The complete picture, with every chromosome of an individual arranged from the largest to the smallest, is called a karyotype.

Most commonly, chromosomes are visualized using lymphocytes obtained from peripheral blood. These are placed in culture and forced to divide and then arrested when they go into metaphase. Once in metaphase, the microtubules in the spindle are stopped, and the cells lysed and spread on a glass slide. The spread metaphases can be dyed with Giemsa stain, which produces the classical G-banding found on most karyotype protocols. According to the position of the centromere, three types of chromosomes can be distinguished: metacentric chromosomes, with a centromere approximately in the

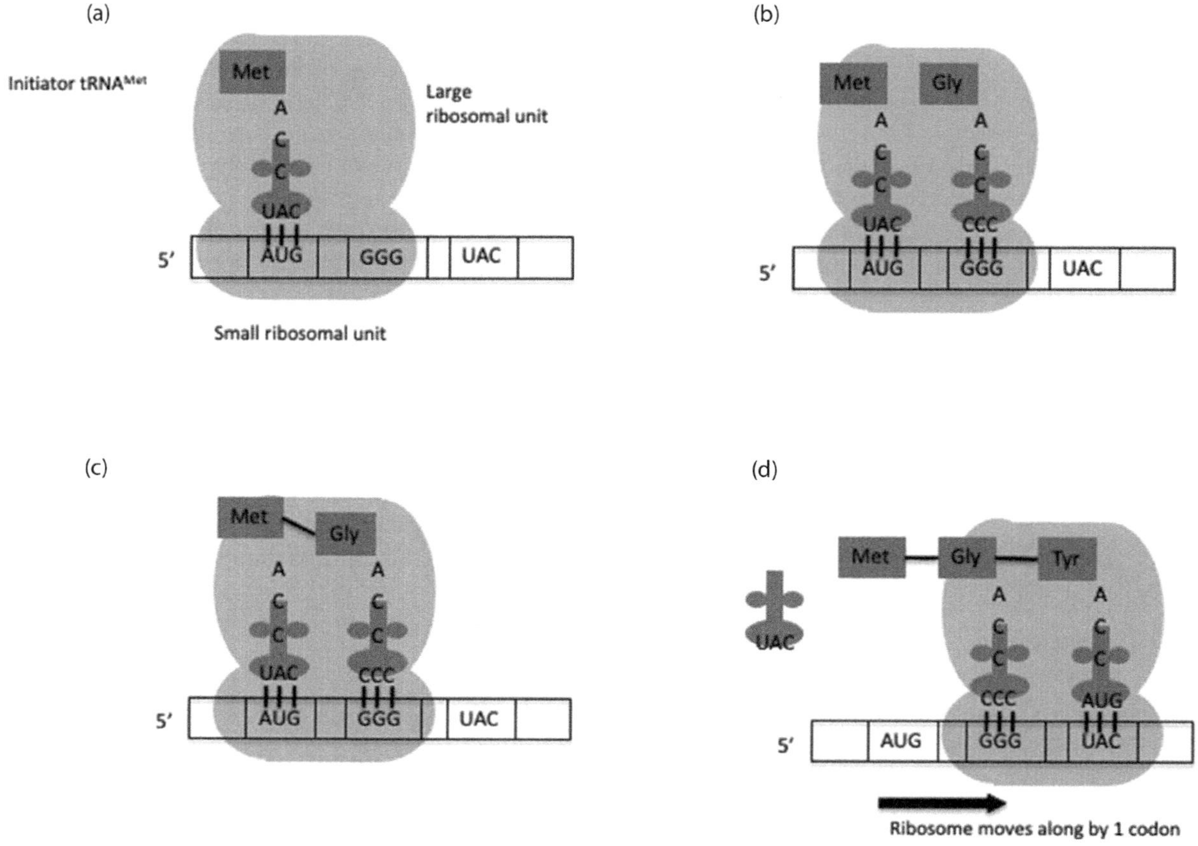

Figure 1.5 Translation of mRNA into protein (a–d). For explanation, see text.

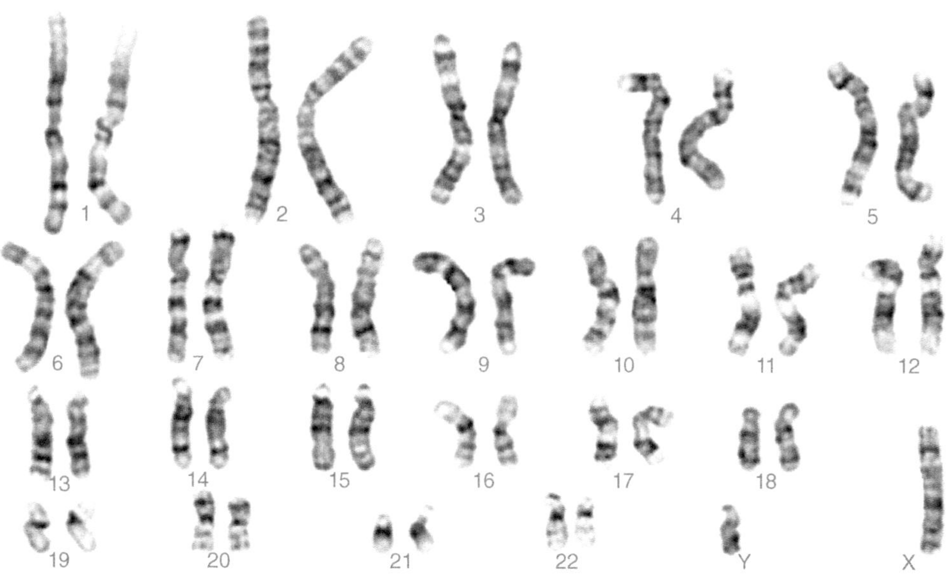

Figure 1.6 Normal human male karyotype (46,XY). Source: QC Science Photo Library.

middle; submetacentric chromosomes, with two arms of clearly unequal length; and acrocentric chromosomes with centromeres at or near an end. Each chromosome has a long arm called the q-arm and a short arm called the p-arm. The banding patterns visible after G-banding characterize each individual chromosome, and allow cytogeneticists to classify and number the chromosomes, and to identify possible rearrangements [5].

1.6.1.2 Fluorescent In Situ Hybridization: Count the Dots

Although fluorescent in situ hybridization (FISH) was the first step from classic cytogenetics to molecular cytogenetics, the method gives only limited information compared with currently developed molecular cytogenetics methods. Fluorescent in situ hybridization is a relatively easy and quick method that has allowed fast screening of, for example, prenatal samples for the most common aneuploidies such as trisomy 13, 18, and 21. The sampled cells are spread on a glass slide, fixed, and the DNA is denatured so that the single-stranded DNA is accessible for fluorescently labeled DNA probes. These probes are chosen so that they are complementary to the DNA region of interest, are allowed to form double-stranded DNA with the DNA in the sample ("to hybridize"), after which the location in the nucleus where the probe has bound can be seen as a fluorescent spot. The FISH probes are carefully chosen to serve a particular purpose. For a quick chromosome count, such as for aneuploidy detection in prenatal samples, centromere probes are usually chosen, because they give large, easy-to-read signals. Moreover, they allow a count of the chromosomes in metaphase as well as interphase nuclei. For the detection of more specific chromosome regions, such as in translocations involving small fragments or microdeletion syndromes such as DiGeorge syndrome, more care has to be taken in the design of the FISH probes. If the probes are mixtures covering a whole chromosome, they are called chromosome paints and are very useful for visualizing a chromosomal translocation.

More on FISH and its application in preimplantation genetic testing can be found in Chapters 4 and 13.

1.6.2 The Detailed View: DNA and RNA Analysis

1.6.2.1 Polymerase Chain Reaction

Before the advent of the polymerase chain reaction (PCR), the only way to amplify a DNA fragment of interest was to clone it into a vector (say a plasmid), introduce the vector in a host (e.g. *Escherichia coli* bacteria), culture the bacteria, and then isolate the expanded plasmid with the DNA of interest. This was very time-consuming, necessitating a large amount of DNA and special equipment for bacterial culture. PCR very quickly took its place in many applications in molecular biology: PCR products are used as templates for restriction enzymes or sequencing reactions, and probes used in other applications (FISH, hybridizations of all kinds) are synthesized using PCR; even making transgenic animals has become much simpler thanks to PCR.

PCR is in essence a DNA copier machine [1,2]. In a first step, the native DNA strands are denatured (i.e. separated) by heating, typically to 95°C (Figure 1.7). In the second step, at a lower temperature, two single-strand short DNA fragments bind to the complementary DNA(cDNA). These short DNA fragments are called the primers and are chosen so that they delineate a fragment of the genomic DNA that has to be amplified, for instance because a disease-causing variant resides in that fragment. The primers then serve as an anchor for the polymerase, which synthesizes the cDNA strand resulting in double-stranded DNA at a higher temperature, typically 72°C, during the third part of the reaction.

Ordinary DNA polymerases are not heat resistant and will thus be destroyed at each PCR cycle. Therefore, the discovery of thermophilic DNA polymerases was a major breakthrough in PCR technology. *Taq* DNA polymerase was the first and most widely used thermophilic DNA polymerase and was first described in *Thermus aquaticus*, a thermophilic bacterium isolated from the hot springs of Yellowstone National Park in the USA.

Like any ordinary chemical reaction, PCR reactions have typical kinetics. First, during the exponential phase, the number of PCR fragments generated increases in an exponential fashion. At a certain point, the reaction components are depleted and PCR reaches a plateau phase so that the number of PCR molecules does not increase further. When PCR fragments are analyzed on agarose gels, or other means of fragment analysis, the reaction usually has reached this plateau phase.

1.6.2.2 Sequencing

The most widely used chemistry tool still in use for DNA sequencing is called Sanger sequencing. The

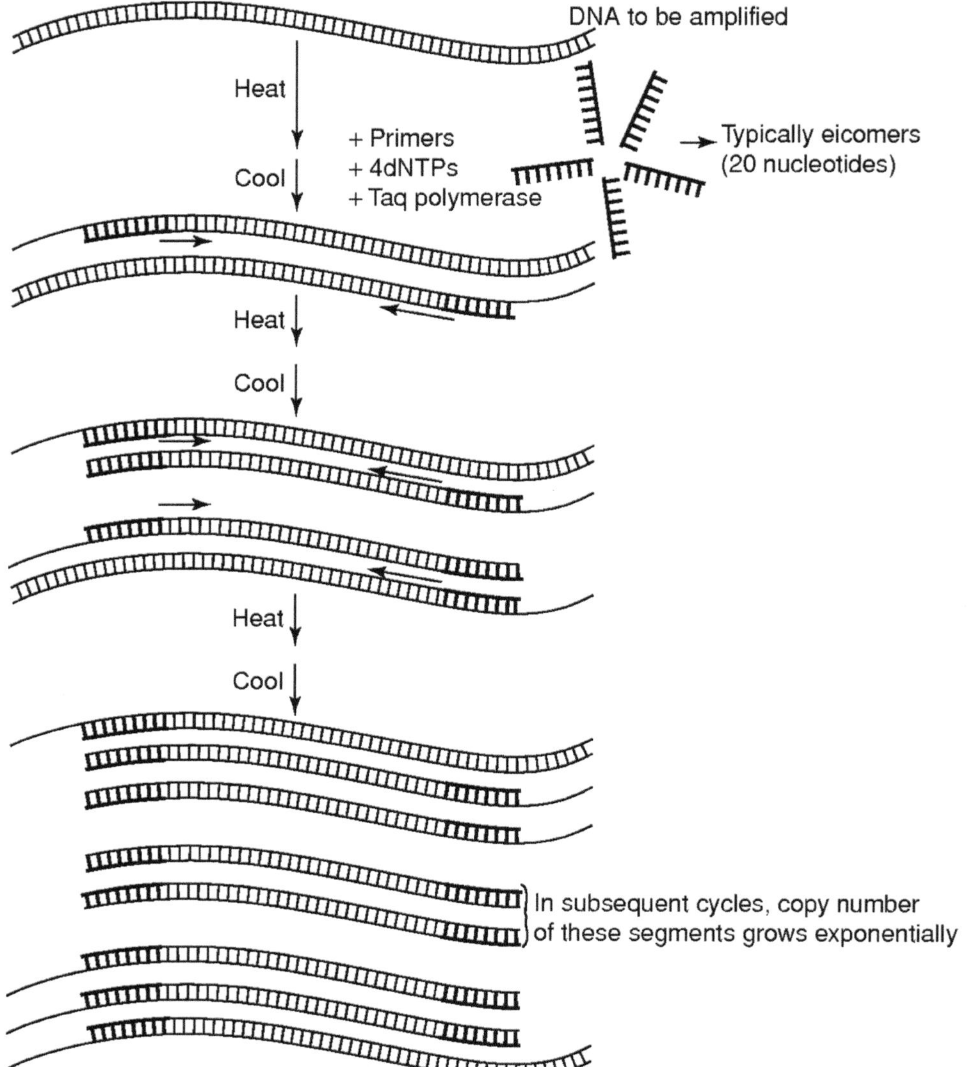

Figure 1.7 Polymerase chain reaction scheme. Source: Fig. 27.7 from Ringo [3].

chemical principle has remained the same, whether using radioactively labeled nucleotides for detection, or the currently ubiquitous fluorescent sequencing. The starting material nowadays is usually a PCR product, but before the advent of PCR, DNA fragments cloned into vectors and flanked by known primers were used. The principle is shown in Figure 1.8: the DNA of interest is denatured, and primers are allowed to bind to the complementary sequences. Then, a polymerase reaction is initiated. In the reaction mix, four nucleotides with a different chemistry (the dideoxynucleotides) are added alongside the usual deoxynucleotides. The four dideoxynucleotides are dideoxyA, dideoxyC, dideoxyG, and dideoxyT. Each dideoxynucleotide is labeled with a different fluorescent dye, represented by shades of gray in Figure 1.8. In the example, dideoxyG is labeled black, dideoxyT is white, dideoxyA is light gray, and dideoxyC is dark gray.

When the polymerase synthesizes the complementary strand, each time a normal nucleotide is incorporated, the synthesis continues. If a dideoxynucleotide is incorporated, the synthesis stops. This reaction creates a mixture of fragments of different lengths, each differing by only one nucleotide in size. The last nucleotide of the fragment is always a fluorescent dideoxynucleotide and is complementary to the last

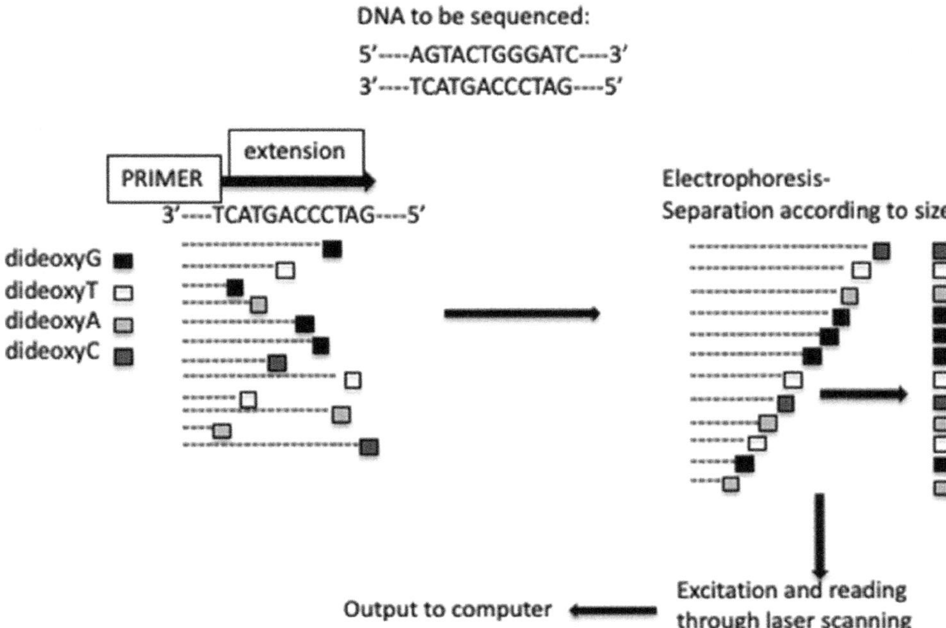

Figure 1.8 Sanger sequencing: dideoxy termination method. For explanation, see text.

nucleotide in the analyzed fragment. By separating the different fragments according to their length, and analyzing which fluorescence they emit, the sequence can be read.

1.6.2.3 Quantitative Real-Time Reverse Transcription PCR

To determine if a specific gene is transcribed in a particular cell type, and to what level, quantitative real-time PCR is often used. After RNA extraction from the cells, the mRNA is reverse transcribed into cDNA. This cDNA can then be used as a template for real-time PCR. The difference between regular PCR and real-time PCR is that the number of fragments generated is read during the PCR reaction, through the exponential phase and to the plateau phase, and not at the end after the PCR has reached this plateau phase. For this, special PCR machines are necessary that can read fluorescence incorporated in the PCR fragments after each PCR cycle. The more cDNA of the gene of interest that is present in the original sample, the earlier (i.e. the fewer number of PCR cycles) the exponential phase of the PCR will start. An often-used measure of the abundance of the cDNA in the original sample is the C_t or threshold cycle. The C_t is defined as the cycle where the number of detected PCR fragments reaches a previously set threshold, usually when the exponential phase has definitely started. It is a measure of the number of fragments in the original sample. The abundance of the mRNA of interest is usually normalized to the abundance of mRNA from household genes, supposed to be expressed in all cells, such as *HPRT* or *GAPDH*. This allows for comparison of the expression of a gene of interest between different samples (e.g. *OCT4* in human embryonic stem cells), or changes can be measured over time.

1.6.3 The Holistic View: Modern Tools in Genetic Analysis

1.6.3.1 Microarrays

Microarrays were developed in the 1990s when it first became desirable to analyze the whole genome or transcriptome of cells. They are always built following the same principles: DNA fragments are spotted onto a suitable support, mostly glass, according to a very precise location, hence "array." The sample to be analyzed can be DNA or cDNA in solution and fluorescently labeled. After hybridization to the support, the amount of fluorescence for every spot in the array can be read in a high-resolution laser scanner. The abundance of a given DNA fragment or cDNA in a sample will determine the amount of fluorescence.

Hence, the amount of fluorescence is a measure for the abundance of the DNA fragment or cDNA.

Originally, DNA fragments from large BAC (bacterial artificial chromosomes) libraries were spotted onto the glass slides. The more versatile and amenable oligonucleotide arrays (i.e. very short DNA fragments of only about 20 bp) have now largely taken over.

Single nucleotide polymorphism (SNP) arrays interrogate SNPs dispersed over the human genome and can be used for haplotyping (i.e. identification of the maternal or paternal origin of the chromosome) and copy number analysis.

Chapters 2, 4, and 13 will explore this topic more deeply.

1.6.3.2 Whole-Genome Sequencing

Sanger sequencing allows for the determination of the sequence of only a short stretch of DNA. In the wake of the Human Genome Project, it was considered necessary to develop methods to sequence the whole genome of one individual in one go. Several large biotech companies took up the challenge and developed methods based on different chemistries and using different platforms [6]. This new technology was termed "next generation sequencing," a term that a couple of years after its introduction is already obsolete and replaced by "third generation sequencing." As the fourth and fifth generation are certainly already in the pipeline, we will refer to these technologies as "whole-genome sequencing" or "high-throughput sequencing."

Whole-genome sequencing includes a number of methods that are grouped broadly as template preparation, sequencing and imaging, and data analysis. According to what methodology each provider applies in each step, different types of data are produced from each platform. Factors to take into consideration when choosing a platform include depth of the sequencing (i.e. the efficiency with which rare sequences are detected such as is required for RNA sequencing), the length of the sequences (which vary from 26 to more than 1000 bp), whether the platform amplifies the DNA fragments first before sequencing, whether the sequencing is done on single molecules, and of course the cost of the platform and of running the samples. A number of reports provide more detail on the chemistry and the possibilities of the different platforms [6–9]. The authors of these studies also point out the difficulties of whole-genome sequencing, which are mainly a result of the large amount of data produced and the complex bioinformatics necessary to isolate relevant information from these data.

The applications of whole-genome sequencing are innumerable. The most prominent application has been the investigation of the genetics of rare human monogenic diseases. Because these diseases are so rare that only a handful of individuals worldwide are affected – sometimes only one family – they were not amenable to gene identification using classical genetics. By sequencing the whole genome in family trios (parents and one affected child) and comparing the whole sequence obtained using complex bioinformatics, genes for several of these rare diseases have now been identified. As most mutations in human genetic disease can be found in the exons, targeted sequencing of exons alone (called exome sequencing) has now become the method of choice as it is much more cost-effective than to sequence the whole genome [10]. If exome sequencing does not yield results, whole-genome sequencing can be considered.

High-throughput sequencing on RNA is now the method of choice for transcriptome analysis and is referred to as RNA sequencing, which can even be applied at the single cell level. Methylome analysis is another important application, as it can establish which parts of the genome are methylated and which are not. This is important since methylation is associated with gene silencing, and variations in methylation can be related to development, cancer, and tissue identity.

1.7 Causes and Effects in Human Genetic Disease

1.7.1 Chromosomal Abnormalities

Chromosome abnormalities have traditionally been defined as alterations that produce a visible change in the chromosomes. However, as we have seen previously, this very much depends on the technique used. If classic G-banding is considered, the minimal visible change is about 4 Mb. A more functional definition states that any abnormality arises from specific chromosomal mechanisms, such as incorrect segregation of chromosomes during mitosis or meiosis, misrepair of broken chromosomes, or improper recombination events. Chromosomal abnormalities can be constitutional (i.e. they are present in every cell of the individual) or the individual can be mosaic (i.e. two or more cell lines are present originating from the same zygote). In postnatal cytogenetics, usually only two cell lines will be detected but, as we will see in Chapter 13, preimplantation embryos can carry many different cell lines and are then called chaotic mosaics.

Two categories are distinguished: chromosomal abnormalities with altered copy number (i.e. numerical abnormalities) or the structural abnormalities where chromosomes show abnormal structure.

1.7.1.1 Numerical Abnormalities

Triploidy and tetraploidy are forms of polyploidy – all cells in the individual carry respectively three or four haploid genomes. Triploidy is usually caused by the fertilization of the oocyte with two spermatozoa, while tetraploidy is due to an incomplete first embryonic division: the DNA is replicated to 4N but the cell does not divide. Both triploidy and tetraploidy are lethal.

Aneuploidy is where one or more individual chromosomes are either missing or present with an extra copy. The best-known trisomy is Down syndrome, where one extra chromosome 21 is present (47,XY,+21). Monosomies are lethal in the embryonic period except for monosomy of the X chromosome, as in Turner syndrome (45,X). One mechanism through which aneuploid cells arise is nondisjunction, which can arise during meiosis leading to abnormal gametes, or during mitosis leading to mosaicism. During meiosis, paired chromosomes fail to separate or sister chromatids fail to disjoin during mitosis. Another mechanism is anaphase lag, in which a chromosome or a chromatid lags behind the other chromosomes during movement in anaphase, and is lost.

The consequences of numerical abnormalities are dire: nullisomies and monosomies (except Turner syndrome) are never viable, while only individuals with trisomies of 13, 18, and 21 may survive to term. Of these, only those with trisomy 21 can reach adult life. Sex chromosomes again form an exception: 47, XXX, 47,XXY and 47,XYY are all known to cause very few clinical problems and these individuals have normal lifespans. The effects of aneuploidies on fertility are discussed in Chapter 7.

1.7.1.2 Chromosomal Rearrangements

Chromosome breaks can be caused by DNA damage or faulty recombination during meiosis. Cellular mechanisms will try to repair these breaks by rejoining the ends, or by adding telomeres to the ends. If such repair mechanisms lead to a chromosome with no centromere (i.e. an acentric chromosome) or a dicentric chromosome, these chromosomes will be lost during segregation in the next mitosis.

When two breaks occur in the same chromosome, a fragment can be lost (deletion) or inserted after it is inverted (inversion) or included in a circular chromosome (a ring chromosome).

When two chromosomes each suffer one break, the resulting fragments can be exchanged. This is called a translocation. If acentric fragments are exchanged, this leads to a stable centric chromosome that can be passed on in mitosis, and this is termed a reciprocal translocation. Any acentric or dicentric product arising from exchange of a centric and an acentric fragment are lost during mitosis. Reciprocal translocations can lead to the birth of offspring with severe deficiencies; however, more often they will lead to repeated miscarriages or even subfertility. This is further discussed in Chapter 7.

Robertsonian translocations are a peculiar type of translocation involving chromosomes 13, 14, 15, 21, or 22. These chromosomes have very small short arms that are very similar in DNA sequence. When the short arms of two of these chromosomes break, acentric and centric fragments are exchanged resulting in acentric and dicentric fragments. The acentric fragment is lost, while in the dicentric fragment the centromeres are so close to each other that they fuse and form one large centromere (centric fusion). This chromosome is stable during mitosis.

Individuals who carry a balanced rearrangement have a normal phenotype, although the term "balanced" has come under pressure since the advent of higher resolution chromosomal analysis. However, carriers' gametes run into problems when entering meiosis. As explained earlier, in order to proceed through meiosis, the chromosomes have to pair in order to allow recombination (see Chapter 3). The cell that carries two abnormal chromosomes has to form special structures to allow for the pairing of the abnormal chromosome with the normal chromosomes. In a carrier of a Robertsonian translocation, no less than six different haploid cells can be formed. Only one of these gametes will lead to a normal individual, while a second one will lead to a balanced (and thus phenotypically normal) individual; the other four will lead to either monosomy or trisomy.

If chromosomes 13 or 21 are involved, Robertsonian translocation carriers are at risk of offspring with trisomy 13 or 21.

1.7.2 Monogenic Diseases

1.7.2.1 Modes of Transmission

How a monogenic disease is transmitted mainly depends on two factors: firstly, whether the disease is dominant or recessive and, secondly, whether the gene is located on an autosome or on a sex chromosome (mostly the X-chromosome). In autosomal recessive disorders the carriers have a normal phenotype, while individuals with a mutation in both genes are affected. This can mostly be explained by the fact that recessive mutations are usually loss-of-function mutations. If one copy of a gene coding for an enzyme is therefore eliminated, the second copy ensures that sufficient enzyme activity remains. However, if the two copies are disabled, then the individual carrying this disease is affected. This explains why in recessive diseases, both parents of an affected child are phenotypically normal carriers. The most frequent autosomal recessive diseases include hemochromatosis and cystic fibrosis in Caucasian populations, sickle cell anemia in African populations, and β-globinopathies in Mediterranean countries.

In autosomal dominant inheritance, individuals who carry only one copy of the defective gene are affected. The genes responsible for the diseases often carry a gain-of-function mutation. The protein that is synthesized from the gene causes problems, for example in Huntington's disease, because the abnormal protein forms toxic aggregates. Loss of function can also lead to dominant transmission if there is so-called haploinsufficiency, i.e. one copy of the gene does not produce sufficient protein for normal functioning. Other well-known examples of autosomal dominant diseases include neurofibromatosis (von Recklinghausen disease) and myotonic dystrophy or Steinert disease.

Monogenic diseases that lie on the X-chromosome have a particular mode of transmission, because women carry two X-chromosomes and men only one. Thus, in male patients, mutations on their one X-chromosome is not balanced by the second X-chromosome as it is in women. Here too, both recessive and dominant modes of inheritance exist. Examples of well-known recessive X-linked disorders are color blindness (very common but quite harmless) and the hemophilias. Dominant X-linked disorders are quite rare: OMIM (Online Mendelian Inheritance in Man) lists 821 of them. One striking example is incontinentia pigmenti: in affected males, the disease is usually lethal in the prenatal period. Affected girls show abnormalities in skin, hair, and central nervous system and a high rate of X-chromosome skewing. This is because the cells with an active mutation-carrying chromosome are lost around birth. Fragile X syndrome is also classified as dominant, but we will see later that this is related to the particular type of mutation.

1.7.2.2 Mutation Types

Changes in one or a few base pairs are the most common mutations. A missense mutation means that one base pair in a sequence is changed and causes one amino acid in the protein to be changed. This can lead to both loss-of-function mutations, such as when the active site of an enzyme is changed, and gain-of-function mutations because the three-dimensional structure of the protein is changed. Nonsense mutations change one codon to a stop codon, leading to a shortened mRNA, which is usually degraded and not translated into protein. This type of mutation thus mostly leads to loss of function. However, if the stop codon is at the end of the mRNA, protein can still be translated. This truncated protein can cause many problems. In frameshift mutations, deletions or insertions of a few base pairs cause a frameshift in the downstream translational reading frame. A well-known example is the most common mutation in Tay–Sachs disease: a 4-bp insertion in the α-chain of hexosaminidase A. Other examples include splice site mutations, which cause whole exons to be missing or introns to be included in the mature mRNA, and mutations in the promoter regions, which cause insufficient or excessive mRNA to be transcribed.

Dynamic mutations are a special class of mutation, and are mentioned here separately because a significant number of patients who are assessed for monogenic disease by preimplantation genetic testing (see Chapter 13) undergo treatment for these types of disease. Small repeats, usually of three nucleotides, are present in many genes and are usually transmitted in a stable, Mendelian way [11]. However, if their size increases over a certain threshold, they become unstable and tend to enlarge during replication and/or meiosis. In some of these diseases, the triplet repeat codes for a stretch of amino acids, which in Huntington's disease comprises glycines; if the glycine repeat becomes too large, it will cause the protein to misfold and form toxic aggregates. In other diseases, the repeats are located outside the coding sequence in the 5′ or 3′ untranslated region, or in an intron. This very long repeat then usually prevents expression of the gene, as is the case in fragile X syndrome. However, myotonic dystrophy type 1 (DM1) has a different molecular pathology: the mRNA transcribed from the *DMPK* gene that carries an expanded repeat in the 3′ region forms aggregates and sequesters proteins responsible for splicing mRNA, thus causing dysfunctions in the genes that are no longer properly spliced. Moreover, the enlarged repeat has an effect on the transcription of genes close to the *DMPK* gene. This partly explains the pleiomorphic character of DM1.

With the advent of technologies allowing a more in-depth analysis of our genomes, many genes involved in human infertility have been identified. These are discussed in Chapter 8 for the genes involved in male infertility, and in Chapter 9 for those involved in female infertility.

References

1. Strachan T, Goodship J, Chinnery, P. *Genetics and Genomics in Medicine.* New York: Garland Science, 2015.
2. Nussbaum R, McInnes R, Willard HF, Hamosh A. *Thompson and Thompson Genetics in Medicine*, 8th ed. Saunders Elsevier, 2016.
3. Ringo J. *Fundamental Genetics.* Cambridge: Cambridge University Press, 2004.
4. www.ensembl.org. The Ensembl project produces genome databases for vertebrates and other eukaryotic species.
5. Shaffer LG, McGowan-Jordan J, Schmid M. *ISCN 2013: An International System for Human Cytogenetic Nomenclature.* Basel: Karger, 2013.
6. Metzker ML. Sequencing technologies: the next generation. *Nat Rev Genet* 2010;11:31–46.
7. Schadt EE, Turner S, Kasarskis A. A window into third-generation sequencing. *Hum Mol Genet* 2010;19(R2):R227–R240. https://doi.org/10.1093/hmg/ddq416
8. van Dijk EL, Auger H, Jaszczyszyn Y, Thermes C. Ten years of next-generation sequencing technology. *Trends Genet* 2014;30(9):418–26. https://doi.org/10.1016/j.tig.2014.07.001
9. an Dijk EL, Jaszczyszyn Y, Naquin D, Thermes C. The third revolution in sequencing technology. *Trends Genet* 2018;34(9):666–81. https://doi.org/10.1016/j.tig.2018.05.008
10. Gilissen C, Hoischen A, Brunner H, et al. Disease gene identification strategies for exome sequencing. *Eur J Hum Genet* 2012;20:490–7. https://doi.org/10.1038/ejhg.2011.258
11. Paulson H. Repeat expansion diseases. *Handb Clin Neurol* 2018;147:105–23.

Application of Whole-Genome Technologies to Assisted Reproductive Treatment

Elia Fernandez Gallardo, Thomas Lefevre, Koen Theunis, and Thierry Voet

2.1 Introduction

Preimplantation genetic testing (PGT) allows the detection of genetic abnormalities in biopsies that comprise 1–10 cells from preimplantation embryos and is performed to avoid the transmission of inherited and de novo genetic abnormalities to the offspring (Figure 2.1) (see Chapter 13). The minute amount of genomic DNA in a single cell represented a challenge for whole-genome profiling of embryo biopsies on development of PGT in the 1990s because whole-genome analysis technologies required micrograms of input DNA. Before the adaptation of these technologies to single-cell input by whole-genome amplification (WGA) methods, PGT was performed using targeted approaches according to the couple's indication [1] (Table 2.1). For instance, fluorescence in situ hybridization (FISH) was used to detect unbalanced karyotypes in the embryos from balanced translocation carriers or from couples with recurrent miscarriage or implantation failure. In case of Mendelian disorders, embryo biopsies were subjected to multiplex polymerase chain reaction (PCR) of the risk allele(s) together with several cosegregating polymorphic markers. These targeted approaches were developed for each family specifically, rendering them labor-intensive, costly, and time-consuming. Moreover, some mutations (e.g. a priori unknown small deletions and duplications or complex chromosomal rearrangements) were practically impossible to diagnose using these strategies. The development of WGA technologies in the early 2000s, their application in genomic array technologies thereafter, and the decrease in cost of next generation sequencing (NGS) helped to overcome these limitations and enabled whole-genome profiling of single cells. Furthermore, the improvements in embryo culture made trophectoderm (TE) biopsy possible at the blastocyst stage, enabling 5–10 cells to be biopsied and tested. Besides increasing the diagnostic accuracy, this allowed for the detection of the mosaic status of genetic variants genome-wide [1]. In parallel, the advancement of embryo cryopreservation techniques expanded the time frame required for embryo diagnosis and therefore also contributed to the development and application of new PGT technologies and data analysis.

The application of whole-genome analysis technologies to ART has reduced the time and cost of PGT and permitted the detection of more than one class of genetic variants with a single approach. Currently, the nomenclature for each possible PGT is as follows: PGT-A denotes preimplantation genetic testing for aneuploidies genome-wide, PGT-M for monogenic disorders, PGT-SR for structural rearrangements [2], and comprehensive PGT for monogenic disorders and aneuploidies genome-wide simultaneously. Different technologies may be applied depending on the needs of the prospective parents. In this chapter we present the current single-cell whole-genome technologies used for PGT including their analysis workflow, their corresponding application possibilities (Table 2.1), as well as their limitations.

2.2 Whole-Genome Amplification

As a single diploid cell contains approximately 7 pg of genomic DNA, standard workflows for single-cell genome array or sequence analyses require first the amplification of the DNA present in a cell's lysate by WGA to obtain enough DNA for the assay. Ideally, WGA (1) ensures a high coverage breadth of the human genome, (2) allows all regions of the genome to be evenly amplified, and (3) maintains the same DNA sequence composition, thus requiring high-fidelity procedures and DNA polymerases.

While over the last decades, different WGA technologies have been developed, none of them is faultless and all produce different types of artifacts, primarily unevenness in genomic amplification with

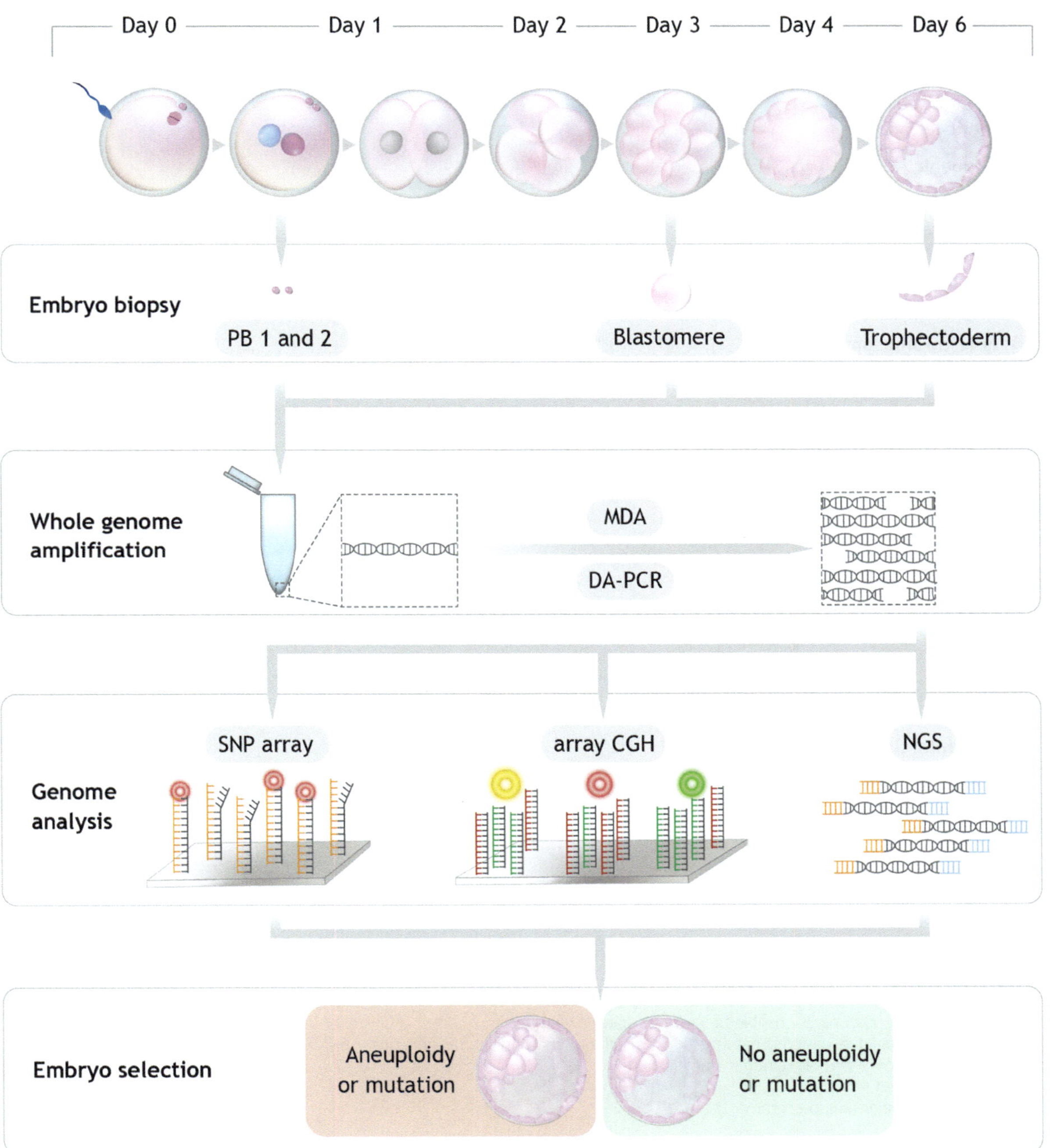

Figure 2.1 Overview of the preimplantation genetic testing workflow. The cell(s) biopsied from the embryo are first whole-genome amplified and then subjected to the genome analysis technology of choice. Based on the results, the genetically normal embryo is chosen for transfer. Generally, whole-genome amplified DNA by multiple displacement amplification (MDA) is preferred for single nucleotide polymorphism (SNP) genotyping, while displacement amplification and PCR (DA-PCR) whole-genome amplified DNA is favored for whole-genome aneuploidy detection either by array comparative genomic hybridization (array CGH) or by next generation sequencing (NGS). The biopsy strategy has implications for the abnormalities that can be detected: a polar body (PB) biopsy allows the detection of only maternally transmitted chromosomal aberrations; a blastomere biopsy might not be representative of the whole embryo if chromosomal abnormalities occur postzygotically, but it can be used for inherited mutation detection; a TE biopsy contains 5–10 cells and may thus permit determination of the mosaic status of genetic abnormalities.

Table 2.1. Preimplantation genetic diagnosis and screening tests

	Microsequencing	QF-PCR	FISH	qPCR	aCGH	Genome-wide haplotyping	Low-coverage sequencing
Genetic lesion							
Monogenic disorders	+	+	–	–	–	+	–
Combination of monogenic and chromosomal disorders	–	–	–	–	–	+	–
Whole-chromosome aneuploidy	–	±[a]	±[b]	+	+	+	+
Balanced chromosomal rearrangements	–	–	–	–	–	+	–
Unbalanced translocations	–	±[a]	+	±	±[c]	+	+
Complex rearrangements	–	–	±[b]	±	±[c]	+	±[c]
Submicroscopic deletions	–	±[a]	+	–	–	+	–
Submicroscopic duplications	–	±[a]	–	–	–	+	–
Uniparental disomy	–	±[a]	–	–	–	+	–
Mechanistic origin of trisomies (mitotic vs. meiotic)	–	–	–	–	–	+	–
Familially inherited	+	+	+	+	+	+	+
De novo mutations	+	–	+	+	+	–	+
Methodology							
WGA required	–	–	–	–	+	+	+

+, Possible; –, not possible; ±, possible but requires specific work-up.
aCGH, array comparative genomic hybridization; FISH, fluorescence in situ hybridization; QF-PCR, quantitative fluorescence PCR; qPCR, quantitative PCR; WGA, whole-genome amplification.
[a] Limited by primer set.
[b] Limited by number of fluorochromes.
[c] Limited by size.
Source: adapted from Vermeesch et al. [1].

the occurrence of locus and allele dropouts, chimeric DNA molecules, and nucleotide misincorporations. Following genome analysis of the cell's WGA product, these artifacts may be falsely interpreted as DNA copy number alterations (CNAs), structural variants, or single nucleotide variations (SNVs), respectively. Principles for WGA are based on PCR, multiple displacement amplification (MDA), or a combination of displacement preamplification and PCR (DA-PCR) (Figure 2.2) [3]. The WGA method of choice strongly depends on the desired genetic variant to be detected.

The first WGA methods were based on PCR amplification of the genome using different designs of random primers. Degenerate oligonucleotide primed

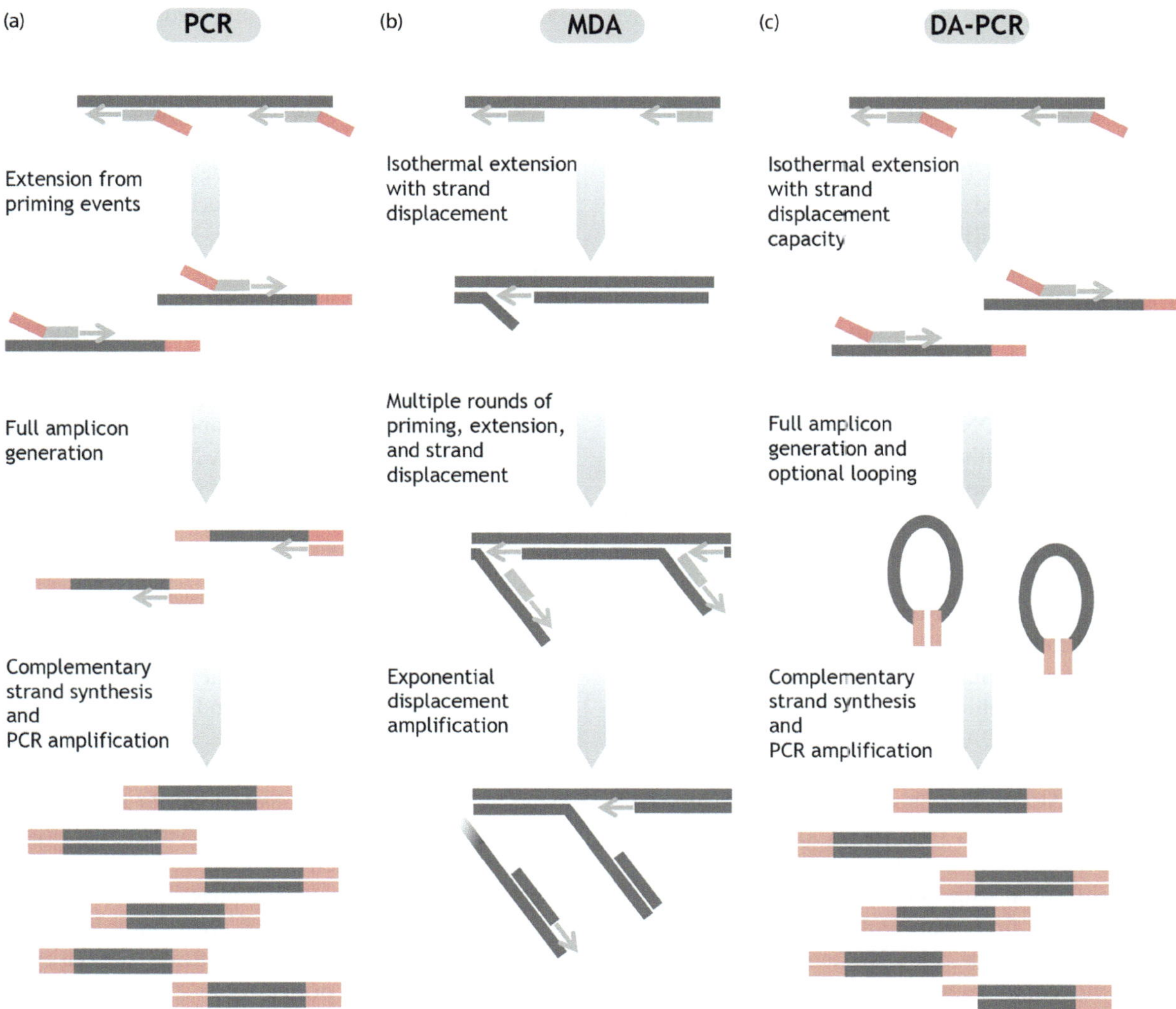

Figure 2.2 Whole-genome amplification technologies. (a) PCR-based methods use either random or degenerate priming. This primer (gray) can be flanked with a defined sequence (pink) that can be used, after generation of the full amplicon, to specifically amplify these short amplicons using PCR. (b) Multiple displacement amplification (MDA) is an isothermal method that uses random priming and a polymerase with strand displacement activity and high processivity. This enables high genome coverage but results in a nonuniform representation of the genome. (c) Methods combining displacement (pre-)amplification with PCR use either random or quasi-random primers and start with rounds of displacement preamplification. The primers (gray) also contain a common sequence (pink) that enables looping of the full amplicons. These looped structures protect the full amplicons from further preamplification.

(DOP) PCR uses hybrid primers containing a 6-base degenerate sequence flanked with a 13-base defined sequence in order to amplify small segments of the genome (Figure 2.2a). A permissive PCR using low annealing temperature followed by high annealing temperature PCR results in short amplicons spanning the genome. Other examples of PCR-based methods are primer extension preamplification (PEP) PCR, and tagged random primer PCR (T-PCR) [3]. Over the years, these PCR-based methods have been improved by replacing the original polymerase with the higher fidelity DNA polymerase *Pwo*. A method that uses a mix of the above concepts is the GenomePlex technology (Sigma-Aldrich), which has been used to amplify human blastomeres and analyze the amplification products using aCGH (see next section), or to enable single-nucleus sequencing to study tumor evolution [3]. Another approach for PCR-mediated WGA, which does not depend on random priming, is ligation-mediated PCR

(LM-PCR) [3]. For instance, single-cell comparative genomic hybridization or SCOMP applies DNA fragmentation using the restriction endonuclease *MseI*, thereby generating short DNA fragments, followed by adapter ligation and PCR, a method later commercialized as Ampli1 (Silicon Biosystems).

In MDA-based single-cell WGA, commercially available as the REPLI-g Single Cell Kit (Qiagen) or the GenomiPhi V2 DNA Amplification Kit (GE Healthcare), cell lysis and DNA denaturation are generally performed in an alkaline environment at elevated temperatures (Figure 2.2b). After neutralization of the cell lysate, random hexamer primers are annealed to the DNA template, and DNA synthesis is carried out isothermally using the bacteriophage Phi 29 DNA polymerase at 30°C. Due to the strong strand displacement activity of Phi 29, it is able, while synthesizing new strands, to displace the 5′ end of neighboring primed strands. In turn, new primers can anneal to the displaced DNA strands and proceed with additional amplification of the DNA, resulting in exponential gain.

Methods combining displacement preamplification with PCR (DA-PCR) use primers with defined and random or quasi-random sequence, as is the case for MALBAC or PicoPlex (originally developed by Rubicon Genomics), respectively (Figure 2.2c). DA-PCR starts with several rounds of strand displacement preamplification with intervening DNA denaturation steps, creating first semi- and then full-amplicons of the denatured single-cell DNA template. Full amplicons with complementary sequences at their DNA ends can be produced from the second preamplification round onward, leading to the formation of a hairpin loop by annealing of the complementary sequences at the ends of these DNA-amplicon fragments in a denaturation–renaturation process, thus presumably protecting them from further preamplification. Next, these full amplicons are PCR amplified.

Unfortunately, all WGA methods suffer from biased amplification of the genome of a single cell. Moreover, different WGA methods will show differences in the breadth of genomic coverage, amplification biases due to local richness in guanine and cytosine bases (GC-amplification bias) or other local base compositions, preferential allelic amplification, allelic dropout (ADO), chimera formation, and the amount of nucleotide copy errors. These biases have to be taken into account when analyzing single-cell genome data in order to distinguish them from true DNA copy number alterations, structural variants, and nucleotide changes. Therefore, there is no one-size-fits-all method that would enable the detection of all genetic variants sensitively and specifically. Thus, the WGA method should be chosen based on the genetic variant that needs to be detected.

Because of the 3′ to 5′ exonuclease proofreading activity of the Phi 29 polymerase, MDA has an error rate per amplification cycle in the range of 10^{-6}, at least an order of magnitude lower than the polymerases of other WGA methods [4]. Additionally, it allows the interrogation of up to 90% of the cell's genome. These qualities make MDA the preferred method for both genotyping and de novo base-mutation detection [4]. However, MDA is less suited for detecting DNA copy number alterations because of its pronounced unevenness in genomic amplification, while the reverse is true for methods based on PCR and DA-PCR [3,4].

Apart from artifacts introduced by WGA, the cell-cycle state can also influence the detection of genetic variation in a single cell. Given the DNA is replicated semi-randomly from early and late replicating domains typical for the cell type, different loci of a diploid cell in S-phase can have DNA copy number states alternating between two, three, or four copies. These temporary copy number changes are easily misinterpreted as genetic variants [5]. On the other hand, the use of cells in G2 phase can be used to reduce ADO and false positive error rates. However, in most research and diagnostic settings, the cell cycle phase is not known a priori.

Over the years, many attempts have been undertaken to improve WGA for single cells. Although MDA has the lowest base-misincorporation rate, sequencing data from MDA samples still contain many C to T transition artifacts due to cytosine deamination obtained through single-cell lysis at elevated temperatures. Therefore, Dong et al. developed single-cell multiple displacement amplification (SCMDA) that performs cell lysis and DNA denaturation on ice to minimize the cytosine-deamination artifacts [6]. Additionally, they developed a computational method called SCcaller to improve SNV calling by modeling and correcting local allelic amplification bias. Other groups have focused on improving the unevenness in amplification of MDA by limiting the reaction volume, limiting the amplification gain, or emulsification of the DNA followed by saturated amplification in each droplet [3,4]. It has now been shown by various groups that performing MDA in nanovolumes in microfluidic devices has a positive effect on the uniformity of amplification as well as the level of contamination. The use of nanovolumes or limiting reaction time or reagents results in more even coverage; however, a lower amplification gain comes at the cost of lower genome coverage [4]. To obtain coverage uniformity as well as a high coverage breadth, emulsion and digital droplet MDA has been developed. In these methods single-cell genomic DNA fragments are divided in a large number of picoliter droplets and amplified to saturation using MDA [3].

Another approach to acquire uniform coverage along genomes of single cells is by performing linear amplification, using T7 promoters and in vitro transcription, in contrast to exponential amplification used in the conventional WGA methods. Apart from more uniform coverage, linear amplification is also beneficial in the context of polymerase errors as these errors are not propagated to the next amplification cycle. The first method where linear amplification was implemented was LIANTI [7]. SCI-L3 also makes use of linear amplification, and although both methods use transposition to fragment the genome, in SCI-L3 the T7 promoter is introduced by ligation instead of transposition as is the case with LIANTI [8]. To tackle the low throughput of LIANTI, SCI-L3 uses a combinatorial indexing approach that enables processing of thousands of cells at once.

Other groups minimized amplification bias by omitting WGA and performing sequencing library preparation directly onto the genomes of single cells without any preamplification. The first method implementing this was Strand-seq, a method that uses bromodeoxyuridine incorporation during DNA replication of the cell, enabling the construction of directional sequencing libraries to independently sequence Watson and Crick strands [9]. Later, Strand-seq was modified to detect aberrant chromosome copy numbers [10]. Because genome coverage was very low, this resulted in a lower resolution for detecting DNA copy number alterations. While these methods rely on enzymatic fragmentation followed by end repair, A-tailing, and adapter ligation, other groups have used tagmentation to fragment the genomes of single cells by inserting adapters [11]. In both ligation-based and tagmentation-based methods, the sequencing adapter and barcode to multiplex multiple samples are introduced by PCR amplification of the fragmented genomes. An optimized and miniaturized version of tagmentation-based direct library preparation is called DLP+ and has been recently used to generate a resource of 51 926 single-cell genomes and matched cell images from various cell types [12]. The downside of these direct library preparation methods is that although coverage along the genome is even, the total genome coverage is rather low and therefore not suited to study SNVs in rare samples like blastomeres.

2.3 Single Nucleotide Polymorphism Array

DNA microarrays are glass slides coated with millions of single-stranded DNA fragments of a known genomic sequence to which a denatured and labeled DNA sample can be hybridized for genetic analysis. Two main microarray technologies are relevant in the field of ART, in particular for PGT: aCGH (array for comparative genomic hybridization) and SNP (single nucleotide polymorphism) array [1] (see Figure 2.1). While both technologies can determine genome-wide (segmental) aneuploidy from an embryo biopsy, SNP arrays also allow concurrent haplotyping. These technologies require about 1 μg of input DNA, and thus the embryo biopsy should be first subjected to WGA for which the method of choice will depend on the genetic abnormalities that need to be detected. For CNA detection, (DA)PCR-based methods are used to obtain a more even amplification of all regions of the genome. In contrast, for the detection of SNVs genome-wide, MDA-based approaches are preferred.

Array CGH was the first technique allowing comprehensive chromosome screening (CCS), i.e. analysis of the whole embryo's genome, from a single-cell biopsy. The DNA probes in each spot on the slide correspond to widespread but unique positions in the genome on which a test DNA sample labeled with one fluorophore (e.g. green) is cohybridized with a reference DNA sample labeled with a different fluorophore (e.g. red). Based on molecular competition during hybridization, a color ratio is obtained for each spot on the array using a fluorescence scanner and this is representative of the copy number of each tested genomic position in the test DNA sample. For instance, a greenish fluorescent spot would be indicative of a DNA gain in the test versus the reference DNA sample, while reddish and yellowish spots would be suggestive of a DNA loss and a balanced DNA region, respectively (see Figure 2.1). Even

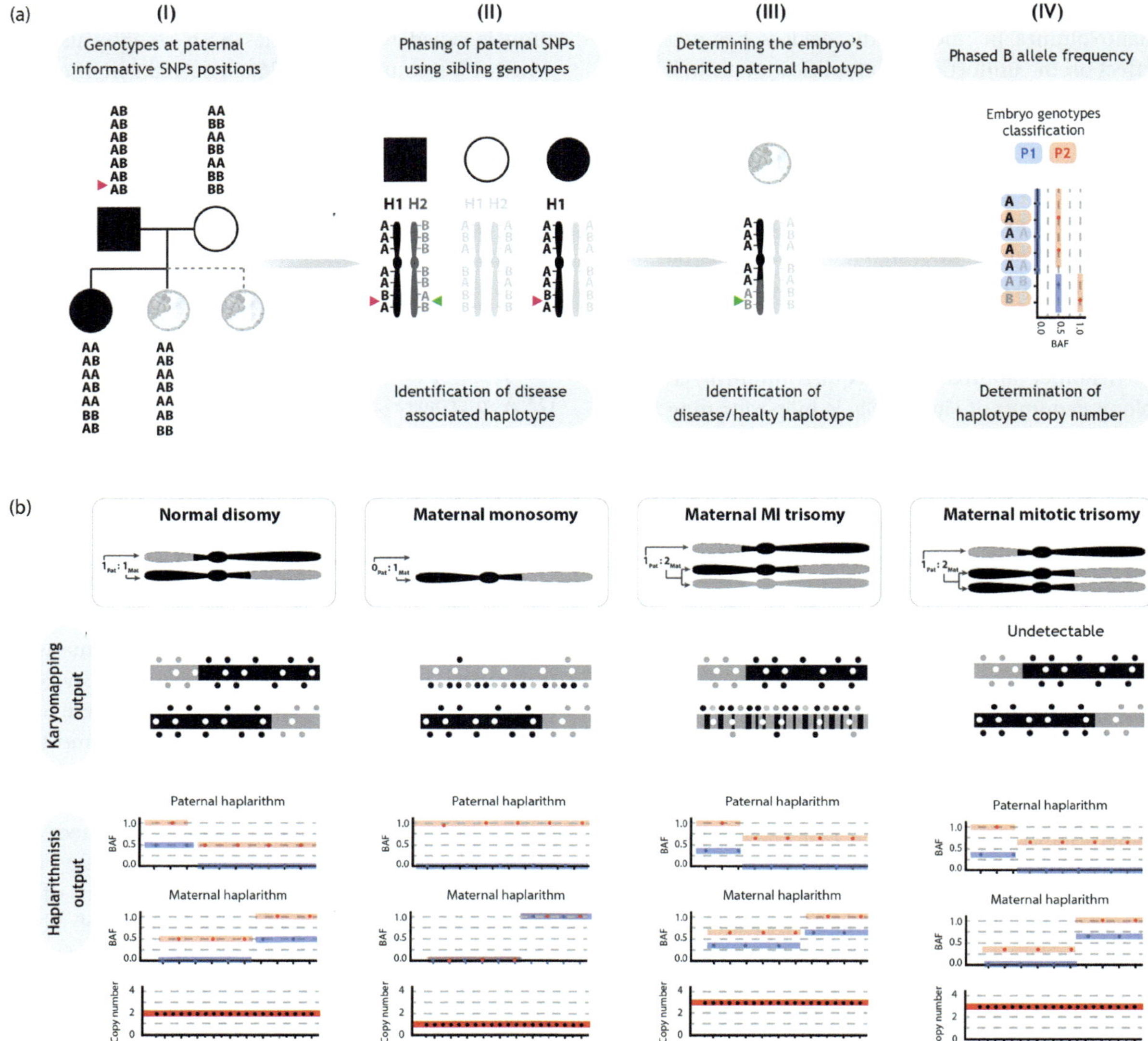

Figure 2.3 Genome-wide haplotyping and copy number profiling for PGT. (a) Analysis workflow. The two main principles for haplotyping used in PGT [1] include discrete genotype calls (AA, BB, AB), i.e. karyomapping (steps I–III); and B-allele frequency (BAF) analysis using haplarithmisis (steps I–IV). A family where father and sibling are affected by an autosomal dominant disorder is presented in the example. For simplicity, the analysis of only one paternal chromosome is depicted. (I) First, genome-wide genotyping data of father, mother, sibling, and embryo(s) are obtained and informative SNPs are selected (i.e. SNPs for which the father is heterozygous and the mother homozygous). (II) Then, paternal haplotypes for homolog 1 and 2 (H1 and H2) are reconstructed (i.e. phased) by identifying which of the sibling allele variants (A or B) are inherited from the father at each SNP position. These variants are allocated to H1 of the father and H1 of the sibling, and are identified as the disease haplotype (red arrow indicates the disease locus). (III) Using the parental phased SNP haplotypes, the inherited haplotype of the embryo can be deduced, by determining which allele variants (located in H1 or H2 of the father) the embryo has inherited. In this example, the embryo has inherited the healthy haplotype (green arrow). (IV) In haplarithmisis, SNP B-allele frequencies in the embryo are analyzed by classifying the SNPs into two categories: P1 (the embryo is homozygous if it inherited H1 from the father and heterozygous if it inherited H2) or P2 (the embryo is homozygous if it inherited H2 from the father and heterozygous if it inherited H1). The BAF is plotted and segmented along SNPs of the same category (red line for P1 and blue line for P2). (b) Result interpretation. Schematic examples of karyomapping (top row) and haplarithmisis (bottom row) together with copy number for four possible situations: normal disomy, maternal monosomy, and maternal meiotic and mitotic trisomy. For karyomapping, different colors (black, gray) along a single chromosome plot indicate meiotic homologous recombination sites. As a measure of reliability for the haplotype calls, three types of SNP are pictured together with the haploblocks (dots): above, the SNPs where allele dropout (ADO) would not affect their phasing (key SNPs) and thus the haplotype call is reliable; below, the SNPs where ADO would affect the phasing (i.e. nonkey SNPs); middle, in white, are the SNPs where no genotype could be called. For haplarithmisis the distance between the red and blue lines can confirm DNA copy number aberrations of haplotypes. In a normal

though aCGH significantly improved the resolution of its predecessor technique for PGT-A (i.e. ranging from FISH with probes for a limited number of chromosomes, to CCS with 2–6 Mb resolution genome-wide), its labor, time, cost, and sensitivity has now been outperformed by CNA detection using low coverage NGS [1].

SNP array technology takes advantage of the presence of polymorphic SNVs in our genome. SNVs are single-nucleotide substitutions (e.g. A/T for C/G) in comparison with the reference genome. If this SNV reaches a frequency above 1% in the population, it is termed an SNP. In a diploid situation, a cell is either homozygous (AA or BB) or heterozygous (AB) for a particular SNP, where A and B refer to the two possible variant alleles at that base position. Currently, more than 84 million SNPs are known in the human genome, but typically SNP arrays interrogate about 300 000 to 1 million known positions.

Genome-wide genotyping of SNPs permits the reconstruction of haplotypes, as well as detection of copy number alterations (Figure 2.3). In a diploid cell, each SNP allele belongs to either of the two parental haplotypes, which represent the string of genetic variants along a single chromosome inherited at fertilization. In embryo biopsies, SNP haplotyping can be used to infer the inheritance of the genetic risk allele of monogenic disorders through linkage with nearby SNP alleles on the same DNA molecule for nearly all positions in the genome, circumventing the laborious family-specific optimization needed for PCR-based PGT-M. For this reason, SNP array technology is applied for comprehensive PGT, including PGT for both monogenic disorders and genome-wide aneuploidies. To determine on which haplotype the genetic risk allele resides, maternally and paternally inherited haplotypes have to be phased, i.e. the alleles coinherited on the same chromosome have to be identified (Figure 2.3aII). For that, DNA of the mother, father, and first-degree relative(s) of the parents with known disease status is required [1] (Figure 2.3aI). These first-degree relative(s) can comprise an affected or unaffected sibling, both grandparents, or a single grandparent related to the affected/carrier parent.

Despite the need for the relatives' DNA, SNP arrays offer advantages compared to previous technologies. First, WGA artifacts like ADO and preferential amplification of certain alleles can be discriminated from real deletions or duplications since the genotype calls are available for many consecutive SNPs. Only when loss of heterozygosity (LOH) is detected along several consecutive SNPs of a chromosome (i.e. SNPs are hemizygous along that DNA segment), a true deletion can be confirmed. In addition to the genotype, a B-allele frequency (BAF) value can be extracted for each SNP from the array readout. BAF values range from 0 to 1, with BAF values of 0, 0.5, and 1 denoting homozygous AA, heterozygous AB, and homozygous BB genotypes, respectively. Furthermore, distorted BAF values of 0.33 or 0.67 indicate a trisomy with AAB or BBA genotypes, respectively.

Two different genotyping chemistries have been developed to interrogate SNPs on arrays. One is the Affymetrix chemistry, which uses DNA probes encompassing the SNP positions that are specific for a particular variant allele of the SNP. Fragmented, labeled, and denatured sample DNA is hybridized, generating a specific fluorescence intensity for each probe corresponding to each of the SNP's variants present in the DNA sample. Alternatively, Illumina chemistry uses probes which contain the sequence upstream of an SNP, to which the DNA sample hybridizes and which is then subjected to a one-nucleotide extension. Each nucleotide pair (AT and CG) is labeled with a different fluorochrome, which is then detected by the array scanning platform. In both cases, the fluorescent signals are interpreted algorithmically for the SNP genotype, copy number, and LOH.

Comprehensive analyses of SNP array data for PGT can currently be performed by two main bioinformatic pipelines: karyomapping and haplarithmisis [1]. Both methods have been applied clinically and are based on the following general principles. First, the

Figure 2.3 (*cont.*) diploid biopsy, both maternal and paternal haplarithms show a distance of 0.5 between the red and blue lines, which are located at 0 and 0.5 or, in case of the alternative haplotype, at 0.5 and 1. In a maternal monosomy, the maternal haplarithm shows a distance between the red and blue lines of 0, which locate at 0 or 1 while the paternal haplarithm shows a distance of 1. In a maternal trisomy, 0.33 and 0.67 are the distances between the lines in the maternal and paternal haplarithms, respectively, indicating the presence of a third allele. Segmented haplarithm BAF values are also used to deduce the origin of the haplotype anomaly.

SNP genotypes of the parents, relative(s) of known disease status, and embryo biopsies are determined (Figure 2.3aI). Then, the informative SNPs (where one parent is homozygous and the other heterozygous, which constitute 20–40% of all the SNPs) (Figure 2.3aI) are selected for phasing, whereby the relative's SNP genotypes are used to assign parental SNP alleles to one of the two homologs and the disease-carrying haplotype is identified (Figure 2.3aII). In turn, by comparing the parental phased genotypes with the embryo's, the inherited haplotypes (or haploblocks) by the embryo can be determined (Figure 2.3aIII). From this, karyomapping produces the output plots where each of the embryo's chromosomes is colored according to the inherited parental haplotypes (Figure 2.3b).

An advantageous feature of haplarithmisis over karyomapping is that it uses BAFs instead of genotype calls of informative SNPs (Figure 2.3aIV). This allows, in addition to haplotype reconstruction, confirmation of copy number anomalies of haplotypes, as well as determination of the parental and segregational origin of haplotype anomalies. Haplarithmisis outputs two plots per chromosome: the maternal and paternal BAF haplarithms, from which haploblocks can be deduced (Figure 2.3b). The DNA copy number is calculated and plotted separately. While both pipelines allow detection of the parental origin of aneuploidies, karyomapping is limited in detecting trisomies of mitotic origin (Figure 2.3b).

In summary, SNP array technology and its subsequent bioinformatic analysis can detect the inheritance of variants causing monogenic disorders as well as copy number variations genome-wide in embryo biopsies, provided parental and first-degree relative DNA and disease status are available.

2.4 Next-Generation Sequencing

The most recent approaches for PGT-A make use of NGS [1], which can provide information for each base in the genome, in principle allowing the detection of SNVs, imbalanced structural variations, and aneuploidies simultaneously [13] from the same embryo biopsy. In contrast to arrays, where the resolution of the assay is largely limited by the number of unique probes targeting the chromosomes, NGS can, in theory, achieve base resolution, although sensitivity and specificity for genetic variant detection largely depends on the WGA method and the depth of sequencing applied. As the requirement for sequencing depth increases, the cost of the assay increases accordingly. Therefore, the cost-effective analyses for PGT at the moment include (1) low-coverage sequencing to determine the genome-wide DNA copy number landscape of the embryo biopsy, which has become cheaper than arrays as more samples can be assayed in parallel, and (2) selective genotyping by sequencing approaches [14]. These, together with reduced running time, are the main advantages that NGS provides to PGT.

In general, in order to sequence DNA from an embryo biopsy, the DNA has to be released from the cell, amplified by WGA, and converted into a sequencing library (Figure 2.4). For library preparation, the WGA DNA is fragmented to 300 bp on average, and equipped at both ends with sequencing adaptors, which contain a sequence complementary to the oligonucleotides on the flow cell together with unique sequence barcodes. These indexes allow the assignment of each sequencing read to the correct sample. The NGS platforms often used for PGT are short-read sequencers for which there are various commercial sources. The read output attainable by a sequencing platform and the genome coverage desired per sample will determine the number of input samples per sequencing run. For CNA analysis of embryo biopsies, genome coverage is typically less than half ($<0.5\times$), with approximately 1 million reads per sample [1]. The most commonly used platform for PGT-A is the Illumina MiSeq, which can sequence up to 25 samples in about 4 h.

Illumina sequencing chemistry is based on the sequencing-by-synthesis principle. First, denatured DNA library fragments bind to the oligonucleotides immobilized in the flow cell. Then, fragments are subjected to bridge amplification to create clusters of identical copies of each DNA molecule. The sequencing occurs by adding simultaneously the four kinds of nucleotides that contain a reversible terminator and a specific dye in iterative DNA-synthesis cycles to the flow cell that incorporate into the growing DNA fragments only one nucleotide at a time. After each sequencing cycle, the fluorescent color of each cluster is imaged corresponding to one of the four nucleotides. Subsequently, the reversible terminator and dye are removed and the next cycle starts. The number of cycles determines the read length in number of bases as well as the cost of the sequencing run. For the detection of SNVs or structural variants

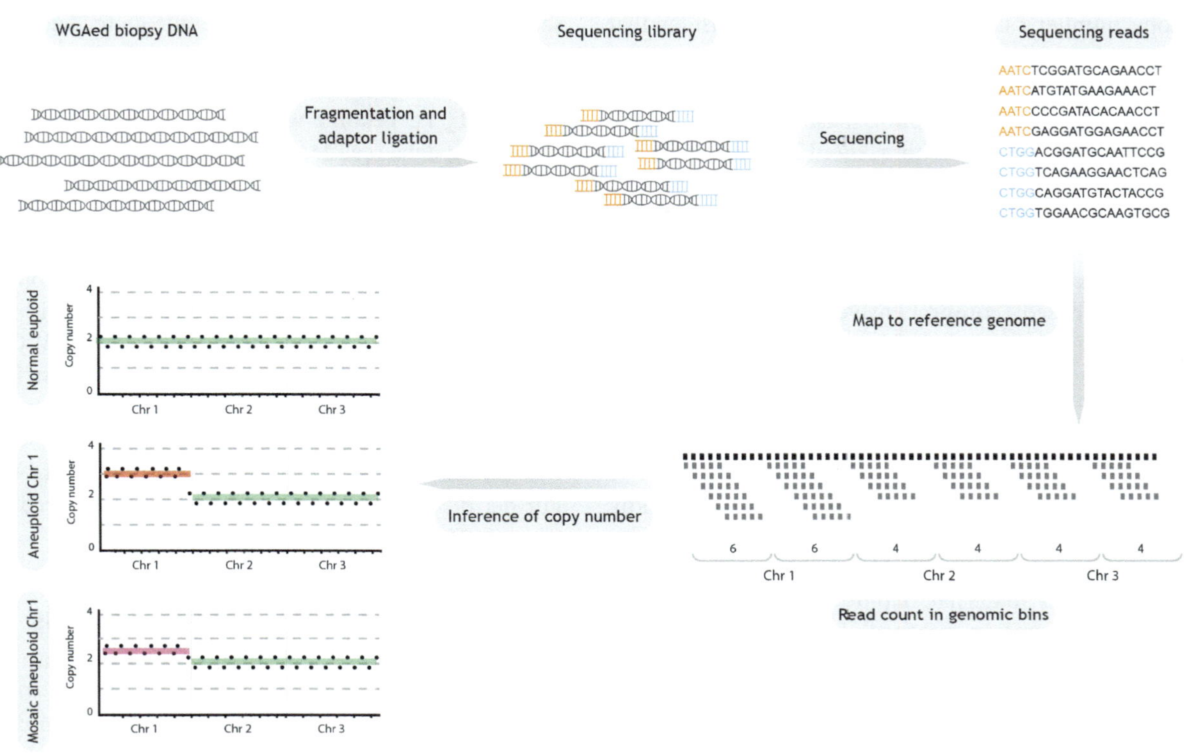

Figure 2.4 Genome-wide copy number profiling by low coverage sequencing for PGT. Schematic workflow of whole-genome sequencing of embryo biopsies to detect copy number aberrations along the genome, starting from whole-genome amplified DNA of the biopsy toward data interpretation. Whereas single-cell biopsies can only result in discrete copy number calls (e.g. 1, 2, and 3 equivalent to $\log_2$ relative copy number values of −1, 0, and 0.58, respectively), when a trophectoderm biopsy is analyzed (5–10 cells), intermediate $\log_2$ values are indicative of mosaic status of the aberrations (left bottom plot).

at high resolution, shallow sequencing does not suffice. To achieve the desired sequencing depth, more sequencing cycles may be required – resulting in longer read lengths – in combination with fewer samples loaded per run, yielding increased breadth and depth of coverage, depending on the characteristics of the sample's DNA library. Furthermore, for the detection of structural variation with high resolution, paired-end instead of single-end sequencing can be applied, whereby both ends of DNA fragments are sequenced, followed by the detection of discordantly mapping read pairs [13].

The detection of CNAs in embryo biopsies using NGS is mainly based on focal read depth analysis [13] (Figure 2.4). Once the sequencing reads are obtained, the first step is demultiplexing, i.e. computationally assigning each read to a specific sample according to the index added during library preparation. All sample-specific reads are then mapped to the reference genome, which is subsequently computationally divided into regions of equally sized bins. For each bin, the number of quality-passed reads mapping to that bin is counted, normalized, $\log_2$ transformed, corrected for per cent GC-amplification bias, and finally transformed to DNA copy number following segmentation of values obtained for consecutive bins along the genome. Ultimately, the copy number is usually represented by a line or colored bar along each chromosome, with line breakpoints or color differences indicating copy number alterations, respectively (Figure 2.4).

Because of the high rate of postzygotic CNAs present in embryos [15], the TE biopsies analyzed for comprehensive PGT might contain only a fraction of cells with a CNA. This phenomenon is called mosaicism and, despite remaining a controversial topic for the IVF community [15], there is evidence that the fraction of normal/abnormal cells in the embryo might influence the embryo's implantation potential [16]. Despite its doubtful diagnostic meaning, NGS enables the detection of low-grade (down to 20%) mosaicism in TE biopsies. For that, the DNA copy number calling algorithm should consider that $\log_2$ relative copy number values can lie between theoretical values for deletions and duplications, indicative of a certain level of mosaicism [17] (Figure 2.4). Importantly, this algorithm should be validated for each WGA and sequencing workflow by cell line mixture experiments using different proportions of known aneuploid/normal cells before using it for diagnostic purposes [17].

NGS-based comprehensive PGT technologies have also been developed as an alternative to SNP arrays, overcoming the need for different platforms for different types of PGT [14]. Haploseek is an adapted bioinformatic pipeline that can be used to infer the haplotype from low-coverage (0.3–1.4×) genome sequencing data using a Hidden Markov model. However, this methodology still relies on SNP microarray data from parental DNA to infer paternal and maternal haplotypes. Alternatively, a library preparation protocol has been adapted to avoid the costly approach of calling all SNPs in a genome. The protocol is based on the genotyping-by-sequencing (GBS) principle whereby only a selection of the SNPs genome-wide are sequenced. Such protocols become considerably cheaper than whole-genome SNV detection because less reads per sample are needed to reach the adequate depth of coverage.

With the advent of noninvasive prenatal testing (NIPT) [1], i.e. genome-wide copy number analysis of circulating fetal cell-free DNA in maternal blood, coupled with the successful isolation of embryonic cell-free DNA in spent culture media as well as in blastocoel fluid, noninvasive PGT (niPGT) was proposed as a timely and desired alternative to biopsy. Embryonic cell-free DNA from blastocoel fluid or from spent culture media can be subjected to WGA and, subsequently, to aCGH or low-pass sequencing for whole-genome analysis [18]. Current publications show heterogeneous results, likely due to the low abundance and poor integrity of the cell-free DNA. Further discussion of this methodology is covered in Chapter 13.

2.5 Beyond Whole-Genome Analysis

Whole-genome analysis technologies in ART are in constant evolution not only to overcome the existing drawbacks of current methodologies, but also to discover new molecular biomarkers for embryo competency. A schematic summary of the technologies introduced in this section is provided in Figure 2.5.

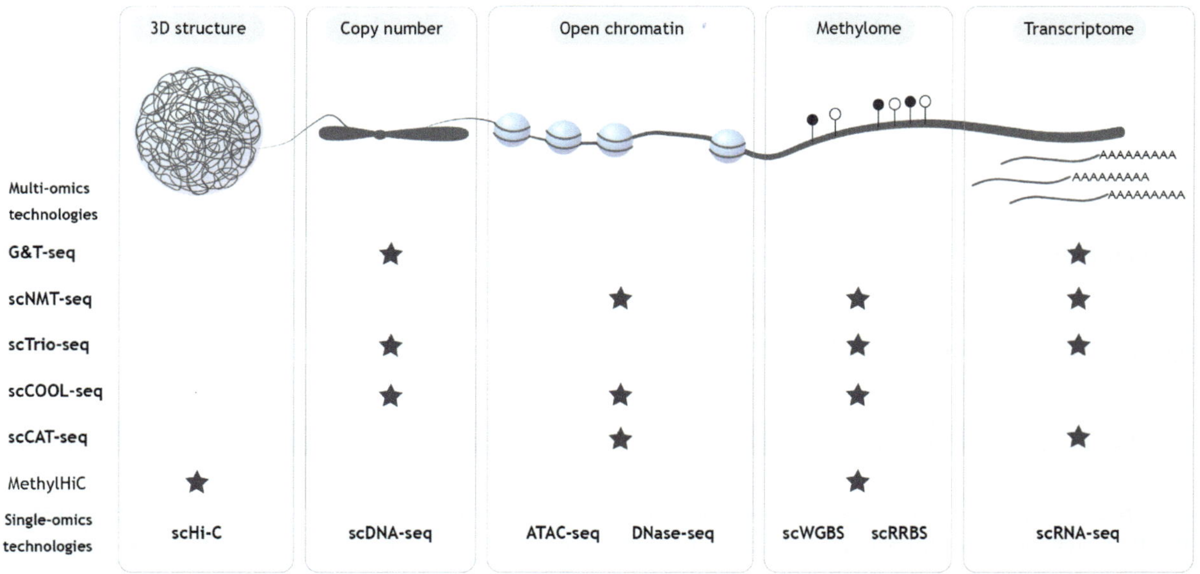

Figure 2.5 Single-omics and multi-omics technologies relevant for research on embryo development. Besides chromosomal copy number, other molecular layers interact to regulate biological processes in the cell. For example, the three-dimensional conformation of the chromatin in the nucleus, the nucleosome positioning along the DNA (open chromatin), as well as DNA methylation on cytosines interact to regulate gene transcription. Technologies are listed according to the molecular layer(s) that they can investigate. Relevant literature for each technique and its application in human embryos is provided in the text.

2.5.1 Transcriptomics as a Tool for Assessing Embryo Competence

Being the first product of gene expression, RNA is a valuable and quantifiable indicator of phenotype at preimplantation stages. For this reason, considerable efforts have been made into finding associations between the genetic constitution of the embryo, its developmental potential, and its transcriptome.

In the era of next generation sequencing, RNA-seq has become the gold standard for whole-transcriptome profiling of tissues and single cells. Due to the limited material that can be sampled from an embryo, single-cell RNA-seq protocols [19] have been used to study preimplantation development. Before library preparation for sequencing, RNA is released from the cell, reverse transcribed, and amplified. A widely used method is Smart-seq2, whereby polythymine oligonucleotides, containing a 5′ PCR handle sequence, anneal to polyadenylated mRNA. Following first-strand and full-length cDNA synthesis using a reverse transcriptase, a tricytosine overhang is added to the first-strand cDNA at the 5′-CAP site of the transcript. This enables a template switching reaction through the addition of a template-switching oligonucleotide (TSO) that hybridizes with the non-templated tricytosine overhangs, allowing the extension of this TSO sequence onto the 3′ end of the cDNA strand. This results in RNA and cDNA hybrids that contain PCR handle sequences at both ends and thus full-length cDNA molecules can subsequently be amplified by PCR. For gene expression analysis, the number of sequencing reads mapping to each gene are first counted and then normalized for transcript length and sequencing depth [19]. Its application to human preimplantation embryos has tremendously advanced our knowledge of gene regulation during development, including better understanding of processes such as maternal RNA degradation, embryonic genome activation (EGA), cell lineage differentiation, and activated gene expression modules [14].

Different techniques and experimental approaches have been used to find clinically relevant marker genes of embryo genomic composition [14]. For instance, reverse transcription quantitative PCR (RT-qPCR) or bulk RNA-seq has been applied to previously genetically tested embryos to extract transcriptomic information. In addition, methods for simultaneous DNA and RNA examination of the same trophectoderm biopsy have been developed and applied. Other authors have chosen a single-cell approach, processing half of the cells of the embryo with single-cell RT-qPCR and the other half with single-cell aCGH. Alternatively, given that CNAs cause changes in gene expression, the genomic composition of the cell was inferred from single-cell RNA-seq data. In these studies, differential gene expression analysis between genetically normal and aneuploid samples led to heterogeneous results, probably due to the substantial variation in terms of sample type and size, embryonic stages analyzed, genomic composition of embryos, as well as the technologies used for gene expression analysis [14]. Nevertheless, individual studies have pointed out relevant observations. For example, aneuploid cells from day 4–7 embryos show downregulation of genes involved in proliferation, metabolism and protein processing in addition to upregulation of immune response genes, coinciding with signatures of aneuploidy previously described for other cell types [20]. On the other hand, a different gene signature for aneuploid embryos was described before day 3, leading to the hypothesis that inherited transcripts from the oocyte might play a role in embryo genomic instability. At the moment, none of the described genes are confirmed to be dysregulated in the same direction under the same conditions in different studies, highlighting the need for further research [14].

2.5.2 Single-Cell Epigenomic Technologies to Uncover Embryonic Genome Regulation

Owing to the characterization of the embryonic transcriptome from the zygotic to blastocyst stages and to postimplantation stages by single-cell RNA-seq technologies [14], interest arose in uncovering the epigenetic regulation of gene expression characteristics such as EGA and cell lineage differentiation. Even though most epigenomic data originates from mouse, several studies have been conducted in human embryos [21].

To restore its totipotency, the zygote undergoes drastic epigenetic remodeling [21], namely erasure and reestablishment of DNA methylation, posttranslational modifications of histones, chromatin remodeling, and higher-order genome reorganization. These epigenetic regulators modify the structural and molecular conformation of the DNA in different ways such that specific areas of the genome become accessible or inaccessible for gene transcription. During preimplantation development, global DNA

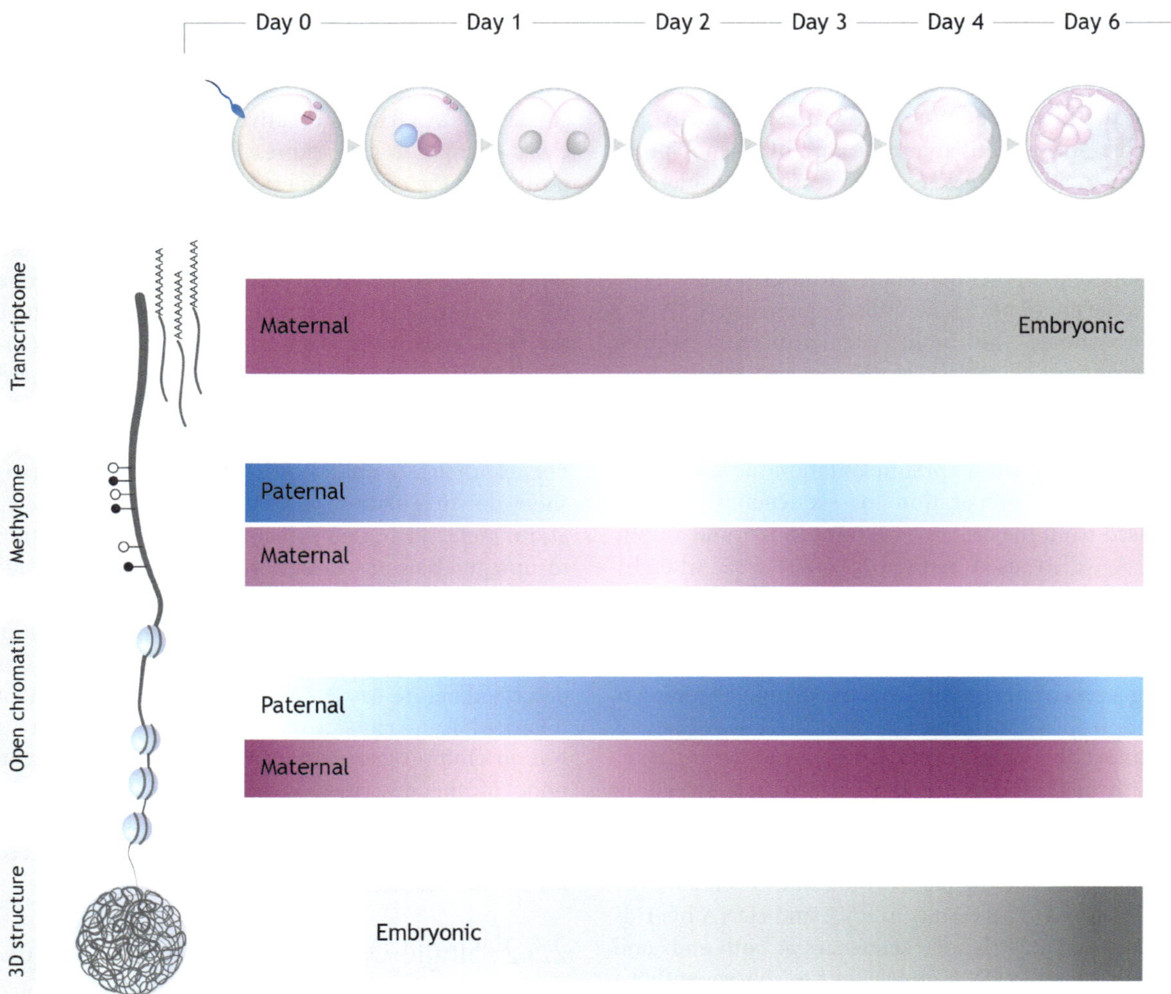

Figure 2.6 Overview of the epigenomic and transcriptomic landscape along human embryo preimplantation development. Color variation along the bars denotes transcript origin and proportional methylation, chromatin accessibility, and chromatin three-dimensional structure during development, respectively [21]. DNA becomes more accessible for gene transcription when methylation is low (light color), chromatin is open (dark color), and three-dimensional structure is not established (light color).

accessibility transitions from repressive in the mature gametes to extremely permissive during EGA, and gradually decreases when cells become committed to a cell lineage (Figure 2.6).

Techniques for the measurement of DNA methylation in CpG dinucleotide sequence context are commonly based on bisulfite treatment, whereby unmethylated cytosines are deaminated to uracils, allowing the detection of methylation at single-base resolution following NGS. Specifically, whole-genome bisulfite sequencing (scWGBS) and reduced representation bisulfite sequencing (scRRBS) have been optimized for low-input or single-cell samples through integration of post bisulfite adapter tagging (PBAT) and the utilization of a single-tube workflow, respectively [22]. Single-cell WGBS libraries are generated after bisulfite treatment through random priming and extension, followed by PCR amplification. In scRRBS, the DNA is digested using *Msp*I, followed by adapter ligation, bisulfite treatment, and amplification. Application of these techniques in human embryos uncovered demethylation and remethylation dynamics of maternal and paternal DNA, of gene promoters as well as of noncoding elements [23].

In concordance with DNA methylation dynamics, chromatin conformation and three-dimensional structure of the nucleus indicates promiscuous transcription of the DNA during the establishment of EGA, including expression of retrotransposable elements, which are usually repressed in somatic cells. To reveal the transcriptionally active regions of the genome, human embryos have been subjected to DNaseI hypersensitivity mapping (DHS) or to the assay for transposase accessible chromatin-sequencing (ATAC-seq) [24], whereby nucleosome-depleted or accessible DNA is fragmented with specific enzymes (i.e. DNaseI or Tn5 transposase, respectively), ligated to adapters, and then amplified and sequenced. To study the three-dimensional conformation of the nucleus during human preimplantation development [25], Hi-C was used. First, cells are fixated to achieve cross-linking of interacting loci, followed by library preparation and sequencing, resulting in a readout of interacting fragments in a contact matrix. Despite the low input achieved with these technologies (20–100 cells in the same reaction) down to single-cell resolution, no study of chromatin conformation or three-dimensional nuclear organization at single-cell resolution during human embryonic development is yet available.

2.5.3 Single-Cell Multiomics

Newly developed technologies allow the study of multiple molecular layers within the same single cell, providing insight into the interaction between the different players in gene expression regulation (Figure 2.5). Available single-cell multiomics techniques [26] permit the simultaneous investigation of the following molecular layers: genome and transcriptome (G&T-seq); genome, chromatin accessibility, methylome and ploidy (scCOOL-seq); chromatin accessibility and transcriptome (scCAT); chromatin accessibility, methylome and transcriptome (scNMT-seq); and the genome, methylome and transcriptome (scTrio-seq2) (Figure 2.5). These techniques combine previously existing single-cell omics technologies but adapt cell manipulation and, in some cases, add an initial step for physical separation of DNA and RNA, using polyadenylated magnetic beads or separation of the nucleus and the cytoplasm by centrifugation [26].

Human embryos have been subjected to scCOOL-seq as well as scTrio-seq2 along preimplantation and periimplantation stages, respectively. These studies provide a detailed map of epigenetic regulation from conception to implantation, confirming the correlation between DNA methylation and chromatin accessibility described previously, and highlighting the different dynamics in maternal and paternal chromatin as well as in different cell lineages at differentiation and during implantation [14].

For PGT-A in particular, single-cell multiomics technologies that permit determination of the genomic landscape of the cell together with other molecular changes such as RNA, epigenomic and/or chromatin structure have great potential to shed light on the fate and impact of aneuploid cells during preimplantation development and beyond [14]. The high cost, together with the challenging data analysis and integration, confine multiomics technologies to a small number of research groups. Nevertheless, they represent an exceptional tool for studying the dysregulation caused by postzygotic abnormalities as well as for uncovering their possible mechanistic origins. Even though these technologies might not reach daily practice in the IVF clinic, they may allow the identification of relevant biomarkers for the quality assessment of embryos.

Acknowledgments

E.F.G. and K.T. are supported by PhD fellowships from the Research Foundation Flanders (FWO: 11A9419N and 1126018N). T.L. and T.V. are supported by FWO G081318N, G0C6120N and KU Leuven C14/18/092.

T.V. is coinventor on the following patents: Methods for haplotyping single cells (WO/2011/157846); High-throughput genotyping by sequencing low amounts of genetic material (WO/2014/053664); and Haplotyping and copy number typing using polymorphic variant allele frequencies (WO/2015/028576).

References

1. Vermeesch JR, Voet T, Devriendt K. Prenatal and pre-implantation genetic diagnosis. *Nat Rev Genet* 2016;17:643–56.
2. Zegers-Hochschild F, Adamson GD, Dyer S, et al. The international glossary on infertility and fertility care, 2017. *Hum Reprod* 2017;32:1786–801.
3. Gawad C, Koh W, Quake SR. Single-cell genome sequencing: current state of the science. *Nat Rev Genet* 2016;17:175–88.
4. de Bourcy CFA, De Vlaminck I, Kanbar JN, et al. A quantitative comparison of single-cell whole genome amplification methods. *PLoS One* 2014;9:e105585.
5. Van der Aa N, Cheng J, Mateiu L, et al. Genome-wide copy number

profiling of single cells in S-phase reveals DNA-replication domains. *Nucleic Acids Res* 2013;41:e66.

6. Dong X, Zhang L, Milholland B, et al. Accurate identification of single-nucleotide variants in whole-genome-amplified single cells. *Nat Methods* 2017;14:491–3.
7. Chen C, Xing D, Tan L, et al. Single-cell whole-genome analyses by Linear Amplification via Transposon Insertion (LIANTI). *Science* 2017;356:189–94.
8. Yin Y, Jiang Y, Lam K-WG, et al. High-throughput single-cell sequencing with linear amplification. *Mol Cell* 2019;76:676–90.e10.
9. Porubský D, Sanders AD, van Wietmarschen N, et al. Direct chromosome-length haplotyping by single-cell sequencing. *Genome Res* 2016;26:1565–74.
10. van den Bos H, Spierings DCJ, Taudt AS, et al. Single-cell whole genome sequencing reveals no evidence for common aneuploidy in normal and Alzheimer's disease neurons. *Genome Biol* 2016;17:116.
11. Vitak SA, Torkenczy KA, Rosenkrantz JL, et al. Sequencing thousands of single-cell genomes with combinatorial indexing. *Nat Methods* 2017;14:302–8.
12. Laks E, McPherson A, Zahn H, et al. Clonal decomposition and DNA replication states defined by scaled single-cell genome sequencing. *Cell* 2019;179:1207–21.e22.
13. Van Loo P, Voet T. Single cell analysis of cancer genomes. *Curr Opin Genet Dev* 2014;24:82–91.
14. Tšuiko O, Fernandez Gallardo E, Voet T, et al. Preimplantation genetic testing: single-cell technologies at the forefront of PGT and embryo research. *Reproduction* 2020;160: A19–A31.
15. Popovic M, Dhaenens L, Boel A, et al. Chromosomal mosaicism in human blastocysts: the ultimate diagnostic dilemma. *Hum Reprod Update* 2020;26:313–34.
16. Spinella F, Fiorentino F, Biricik A, et al. Extent of chromosomal mosaicism influences the clinical outcome of in vitro fertilization treatments. *Fertil Steril* 2018;109:77–83.
17. Munné S, Wells D. Detection of mosaicism at blastocyst stage with the use of high-resolution next-generation sequencing. *Fertil Steril* 2017;107:1085–91.
18. Leaver M, Wells D. Non-invasive preimplantation genetic testing (niPGT): the next revolution in reproductive genetics? *Hum Reprod Update* 2020;26:16–42.
19. Picelli S. Single-cell RNA-sequencing: the future of genome biology is now. *RNA Biol* 2017;14:637–50.
20. Chunduri NK, Storchová Z. The diverse consequences of aneuploidy. *Nat Cell Biol* 2019;21:54–62.
21. Wang Y, Liu Q, Tang F, et al. Epigenetic regulation and risk factors during the development of human gametes and early embryos. *Annu Rev Genomics Hum Genet* 2019;20:21–40.
22. Clark SJ, Smallwood SA, Lee HJ, et al. Genome-wide base-resolution mapping of DNA methylation in single cells using single-cell bisulfite sequencing (scBS-seq). *Nat Protoc* 2017;12:534–47.
23. Zhu P, Guo H, Ren Y, et al. Single-cell DNA methylome sequencing of human preimplantation embryos. *Nat Genet* 2018;50:12–19.
24. Wu J, Xu J, Liu B, et al. Chromatin analysis in human early development reveals epigenetic transition during ZGA. *Nature* 2018;557:256–60.
25. Chen X, Ke Y, Wu K, et al. Key role for CTCF in establishing chromatin structure in human embryos. *Nature* 2019;576:306–10.
26. Stuart T, Satija R. Integrative single-cell analysis. *Nat Rev Genet* 2019;20:257–72.

Meiosis: How to Get a Good Start in Life

Ursula Eichenlaub-Ritter

3.1 Introduction

Meiosis comprises fundamental processes that permit sexual reproduction and species evolution. In addition to producing haploid gametes, it provides for a stochastic distribution of maternally and paternally inherited chromosomes, which undergo allelic recombination thereby increasing genetic variability in the next generation. Thus it provides for diversity within a population, and is essential for the formation of euploid germ cells that will contribute to a euploid healthy embryo after fertilization. Meiosis is therefore the basis for maintaining genomic integrity, high developmental potential, and health of the embryo and offspring, and normal fertility in males and females [1,2]. Furthermore, it is the basis of changes in the genome that are important for adaptation and evolution of species.

3.2 Principles of Chromosome Segregation at Mitosis and Meiosis

During mitosis of a diploid cell there is typically one round of replication of all chromosomes during S-phase (Figure 3.1a). After alignment (also termed "congression") of chromosomes at the spindle equator in mitotic metaphase (M-phase), the separation of replicated sister chromatids at anaphase of mitotic division results therefore in the formation of two diploid daughter cells, each containing an identical set of originally maternally (black) and paternally (gray) derived chromosomes (2N chromosomes/2C chromatids; N refers to number of chromosomes, C to number of chromatids, respectively) corresponding to two homologous chromosomes in a diploid cell (Figure 3.1a). Mitosis and meiosis are briefly introduced in Chapter 1.

In contrast, to obtain haploid gametes for fertilization (1N/1C) two consecutive divisions after only one S-phase result in formation of four haploid recombinant or nonrecombinant gametes in meiosis (Figure 3.1b) that can form a diploid zygote after fertilization (2N/2C). Meiosis, with these two specialized divisions, takes place during germ cell formation of most sexually reproducing animals not only in order to obtain haploid gametes but also gametes and zygotes with new combinations of alleles from originally maternally and paternally derived chromosomes (Figure 3.1b) [1,2]. So, in terms of DNA content meiosis allows progression from 2N/2C to 2N/4C to 1N/2C and finally to a 1N/1C haploid state (Figure 3.1b). After fertilization a zygote with 2N/2C is generated, containing a haploid complement from father and mother.

Meiosis is characterized by specific events that do not normally occur in mitosis. After replication, at prophase I of meiosis, homologous chromosomes originally obtained from the father (gray in Figure 3.1b) and mother (black in Figure 3.1b) first pair side-by-side and there is usually at least one genetic exchange (crossover) between nonsister chromatids of the two homologs that results in recombination [1,2]. This provides for a physical connection between the homologs by a single chiasma or several chiasmata. By attachment of the replicated sister chromatids in each homolog to each other by cohesion proteins [3] and chiasmata, a bivalent chromosome is generated. After resolution of the nuclear membrane the maternal and paternal homolog of each bivalent disjoin except at sites of crossover with a chiasma (Figure 3.1b and prometaphase I in Figure 3.2). Upon joined attachment to opposite spindle poles of the centromeres of the sister chromatids of each homolog (termed "syntelic attachment"), tension is generated at first meiotic M-phase such that the bivalents become aligned at the spindle equator, with the first chiasma located in the equatorial plane (indicated by stippled line in Figure 3.1b). The homologous chromosomes then separate from each other reductionally (i.e. they proceed from 2N to 1N) at

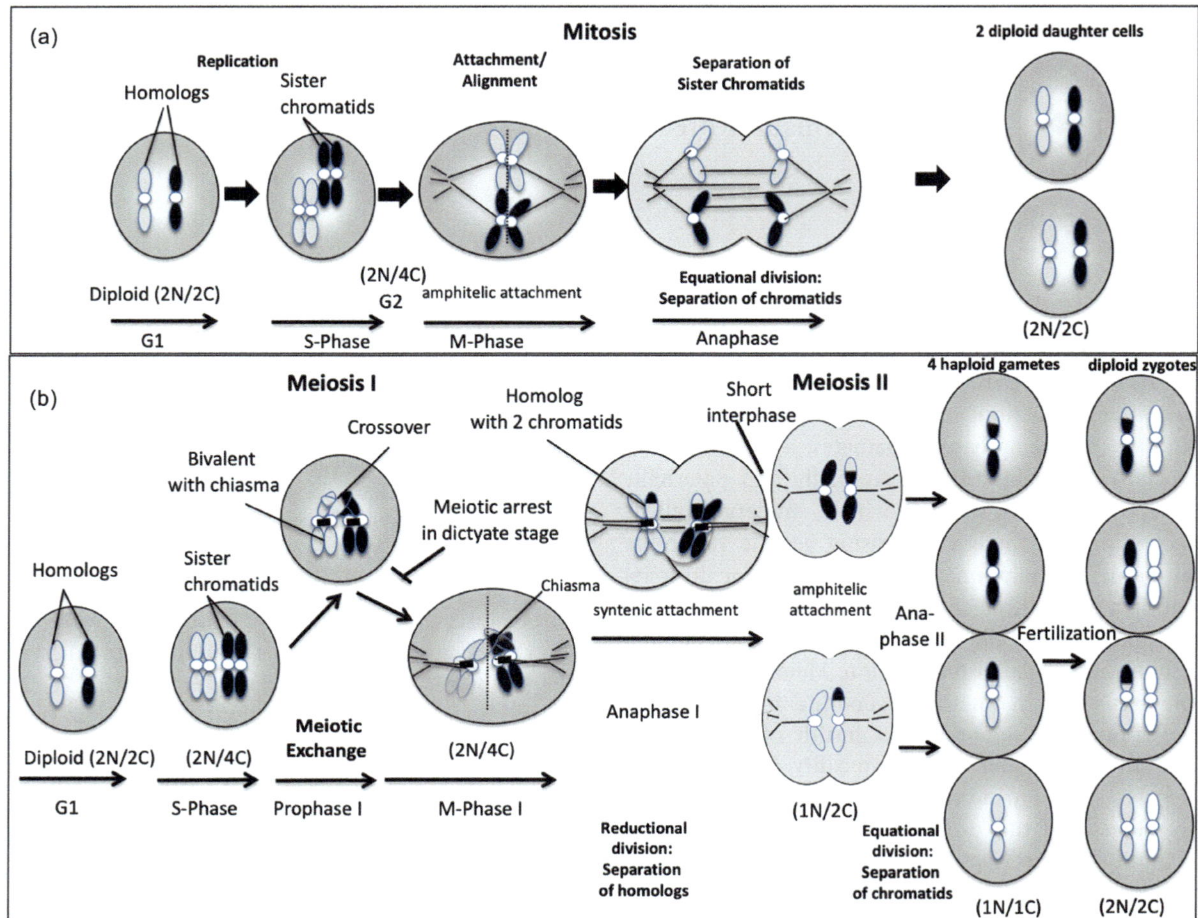

Figure 3.1 Chromosome segregation at mitosis (a) and meiosis (b). Gray and black chromatids symbolize homologs originally derived from father or mother. Note that during mitotic M-phase/anaphase chromosomes disjoin equationally (black and gray) from each other by separation of sister chromatids (a) after amphitelic attachment by microtubules (black lines) to opposite spindle poles. In meiosis I (b) the chromosomes separate reductionally from each other at anaphase I by separation of originally paternally derived and maternally derived homologs with their two sister chromatids attached to the same spindle pole (syntelic attachment). At second meiosis the two sister chromatids of each homolog separate equationally from each other at anaphase II generating haploid gametes that after fertilization can form diploid zygotes. N, number of chromosomes; C, number of chromatids. For further explanation, see text.

anaphase I (Figure 3.1b). For this, the resolution of chiasmata is mediated by loss of cohesion and the release of the physical attachment between the arms of the sister chromatids on each of the homologous chromosomes [2–4]. Sister chromatids usually remain tightly attached to each other at their centromeres until the second meiotic division that proceeds without an intervening S-phase and DNA replication. At the end of meiosis I, two cells with only half the number of chromosomes, each with two sister chromatids (1N/2C), are formed. In spermatogenesis, between first and second meiosis an interphase nucleus can briefly be formed before further meiotic progression to prometaphase II. In oogenesis, there is rapid progression from telophase I to prometaphase II and no interphase with major chromosome decondensation between meiosis I and meiosis II (see below). However, unlike in spermatogenesis with continuous meiotic progression, oogenesis is halted twice. The first arrest is after progression through meiotic prophase I in diplotene when meiotically arrested oocytes with decondensed chromosomes remain blocked in dictyate stage for months to decades within primordial follicles before follicle activation and oocyte growth take place. Only fully grown oocytes resume meiosis I and develop to metaphase II (MII). The second meiotic arrest is at MII when oocytes emit a first polar body and become ovulated with condensed MII chromosomes. Fertilization releases from MII arrest and is followed by completion and second polar body formation in oogenesis (see below).

Unlike after mitotic division, each daughter cell from the first meiotic division usually contains only one or the other copy of the originally maternally or the paternally derived individual chromosomes (1N) in both sexes, each with its two sister chromatids (2C) except for those parts of the chromatid arms that underwent recombination (switch in gray and black on chromatids in Figure 3.1b). The centromeres of the two sister chromatids usually attach to opposite spindle poles (termed "amphitelic attachment," comparable to mitosis) during second meiotic M-phase. The recombinant or nonrecombinant sister chromatids then separate from each other at anaphase II (termed "equational division"). This is mediated by the loss of cohesion between the centromeres of sister chromatids [2–4]. Thus, up to four haploid gametes are generated (Figure 3.1b). The gametes contain only one haploid set of chromatids of each chromosome (1N/1C) with genes (alleles) derived from either father or mother, except for parts where recombination by crossover occurred [1]. The haploid gamete can form a chromosomally normal diploid zygote (2N/2C) after fertilization with another haploid germ cell of the opposite sex (white chromatid in Figure 3.1b) such that a euploid (diploid) one-cell embryo is generated (2N/2C) before first mitotic S-phase and mitotic divisions [1].

In spermatogenesis cytoplasmic meiotic divisions are equal, resulting in the formation of four equal-sized spermatids and sperm at the completion of meiosis. In contrast, only one haploid oocyte is generated during unequal first and second meiotic divisions of oogenesis that generates one large oocyte (1N/1C) and one small first polar body (PB1) after first division (PB1; 1N/2C) and another small second PB (PB2; 1N/1C) after completion of meiosis following fertilization [2,4] (see below).

3.3 Key Events in Meiosis

3.3.1 Stage-Specific Events during Progression through Prophase I of Meiosis

Meiotic entry in both sexes involves progression through a premeiotic S-phase and stages of meiotic prophase I during which chromosome pairing, recombination, and DNA repair occur in leptotene, zygotene, and pachytene stages (Figure 3.2).

After premeiotic S-phase the chromatin of each maternal and paternal homolog with its two sister chromatids (shown in red and blue, respectively, in Figure 3.2) becomes organized on an axis and chromosomes begin to condense progressively at chromomeres that can be recognized microscopically as whorls or bands (blue dots in insets in Figure 3.2) [2]. The chromosome ends – the telomeres – become attached to the inner nuclear membrane and glide during leptotene and zygotene [2]. This eventually results in a clustering at one side in a bouquet-like arrangement with chromosome ends all close to each other at pachytene stage. The assembly of telomeres facilitates homolog pairing but pairing can also occur on homologous regions along the arms of the parental chromosomes [2,5]. Between the homologs a specialized tripartite pairing structure, the synaptonemal complex (SC), is formed that stabilizes pairing and recombination (green line in Figure 3.2). It promotes exchanges by recruiting proteins for DNA breaks, recombination, and DNA repair [2,5,6]. During zygotene, the process of chromosome condensation and pairing continues in a zipper-like fashion until homologs with their two sister chromatids are fully paired (synapsis) at the pachytene stage when the SC extends from one to the other telomere of each chromosome (Figures 3.2 and 3.3a). At sites of recombination, spherical dense nodes can be recognized that contain recombination proteins (termed "recombination nodules"; pink spheres in Figures 3.2 and 3.3a). The numbers and distribution of such nodules correspond to sites of exchange [2,7]. In male meiosis, telomeres of all homologs are attached to the inner nuclear membrane at pachytene stage next to a morphologically distinct site in the nucleus, the sex body. The sex body contains the transcriptionally inactive X and Y chromosomes that are only partially paired in a region containing homologous genes, termed the "pseudoautosomal region" (PAR; see later) [1,2,8]. During diakinesis of meiotic prophase I the chromosomes are maximally condensed and the SC becomes resolved (Figures 3.2 and 3.3a,b) such that the homologs disconnect and remain attached to each other only at sites of recombination that are now recognized as chiasmata. In female meiosis there is an extended meiotic arrest after diakinesis. Chromatin decondenses and oocytes with a nucleus (termed

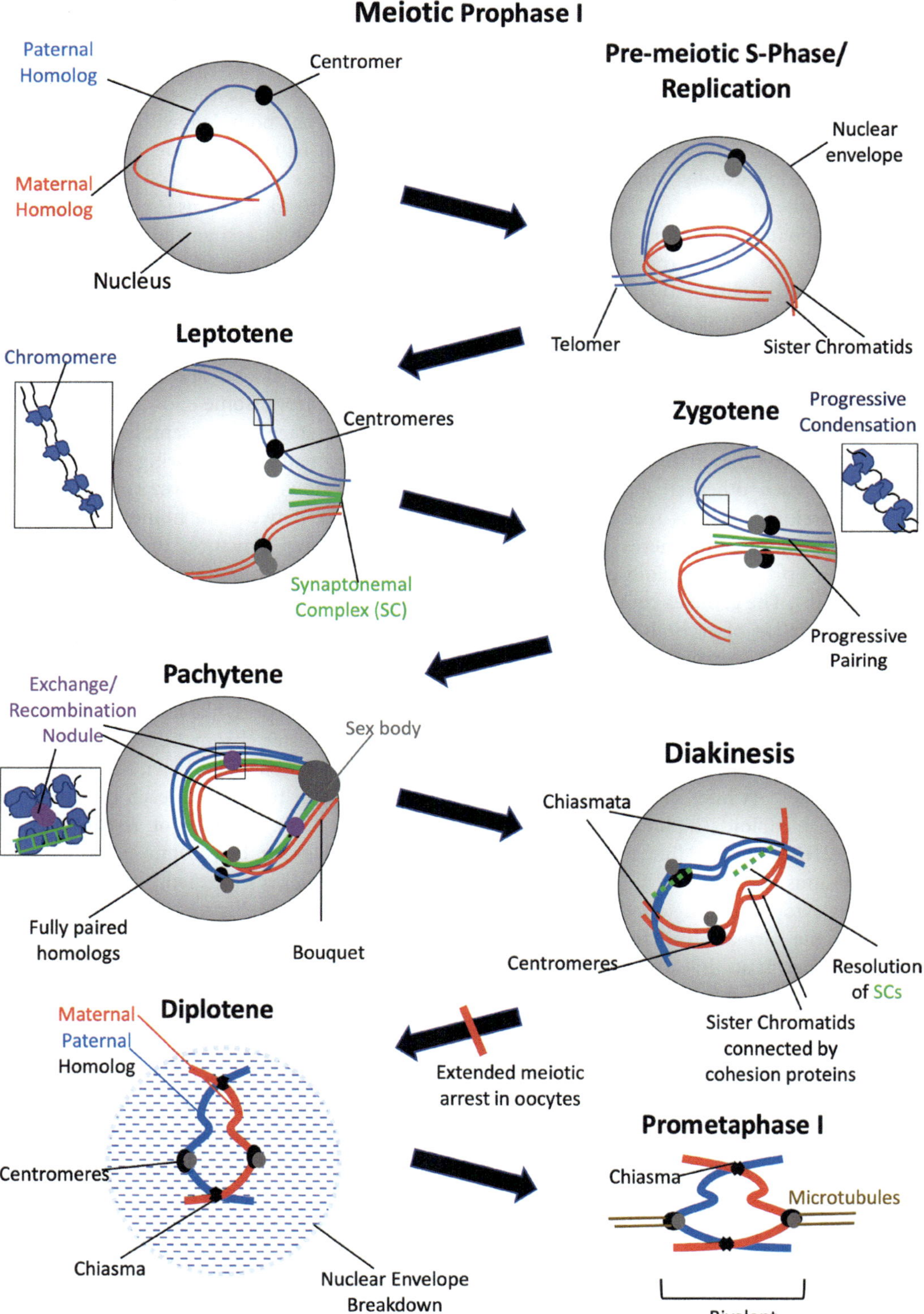

Figure 3.2 Prophase I of meiosis. Initially, homologous chromosomes become replicated during premeiotic S-phase, followed by continuous chromosome condensation at chromomeres (blue spheres in insets), and progressive homologous pairing during leptotene, zygotene, and pachytene.

"germinal vesicle") remain in a meiotically quiescent stage (termed "dictyate stage") within primordial follicles (see below).

Since the SC is solved in diakinesis stage, the homologs lose contacts all along the chromosome arms (Figures 3.2 and 3.3a) while the two sister chromatids remain glued to each other by cohesion complexes upon progression to diplotene stage, when the nuclear membrane breaks down (Figures 3.2 and 3.3b) [2,3]. The chiasmata, together with the cohesion between sister chromatids, hold homologs together within the bivalent throughout prometaphase I and metaphase I until first anaphase of meiosis (Figures 3.1b, 3.2b and 3.3b) [2–4].

3.3.2 The Synaptonemal Complex: Framework for the Recombination Process

To achieve recombination between homologous DNA strands of nonsister chromatids of the DNA of homologous chromosomes during meiotic prophase I, chromatin becomes highly organized within the nucleus. The search and pairing process at prophase I involves conserved mechanisms and structures which permit DNA strands of nonsister chromatids to physically interact with each other at the molecular level involving homologous stretches of DNA.

Thus, the SC is a meiosis-specific large proteinaceous structure that is formed during meiotic prophase I. Together with meiotic cohesin complexes and condensins that affect chromosome compaction and organization [3], the SC plays an important role in the deposition and activity of a structural framework for pairing and recombination during meiotic prophase I of both sexes (Figure 3.3a) [1,2,6]. Structurally, the SC is highly conserved. It consists of three main morphologically recognizable components first identified by electron microscopy: two outer electron-dense lateral or axial elements spaced about 100 nm apart and an inner central element (Figure 3.3a). In the space between the lateral and central element, transversal fibers are usually recognizable that overlap in the center at the central element (Figure 3.3a).

The SC formation proceeds in distinct steps: the lateral protein axes are formed first during leptotene on which loops of chromatin of the paternal and maternal homolog each with two chromatids are attached. Loop sizes appear different and shorter in female compared with male meiosis and therefore the SC is longer in oocytes compared with spermatocytes [2]. In both sexes, the two sister chromatids are physically connected to each other and to the lateral elements by cohesion complexes [3].

From leptotene to zygotene the transversal fibers interact with each other forming a central element associated with further protein complexes. Many components of the SC have by now been identified at defined periods of prophase of meiosis I and at distinct sites of the SC (for detailed information, see Bolcun-Filas and Handel [2] and Dunne and Davies [6]). They appear essential for normal meiotic progression and germ cell survival as mutations severely affect fertility.

At pachytene when homologs are fully paired, ultrastructural analysis by spreading or electron microscopy reveals the presence of electron-dense knobby structures on top of the SC termed "recombination nodules" (RNs) that contain nucleases, recombinase polymerase enzymes, helicases, and DNA mismatch repair proteins (Figures 3.2 and 3.3a) [5,6]. At the end of pachytene, nodules containing a certain family of mismatch enzymes appear spaced at a distance from each other along the chromosome axis consistent with sites where exchanges mature into a crossover and establish a chiasma [7]. At this stage, the relative length of the SC corresponds to the relative length of the different chromosomes in each nucleus. RN numbers maturing to crossovers are influenced by so-called "chiasma interference" that suppresses further crossovers in the vicinity of a once-established exchange. Accordingly, the number of RNs are related to the overall chromosome length, SC length, chromatin conformation status and sex, as are the chiasmata and genetic exchanges by crossovers [2,7]. There is usually no recombination between blocks of heterochromatin. As SCs are longer in female compared with male meiosis, there are more

Figure 3.2 *(cont.)* Telomeres attached to the inner nuclear membrane glide and assemble at one site (in the male at the sex body) in a "bouquet" at pachytene when the synaptonemal complex (green) between nonsister chromatids extends from one to the other end of the homologs. Sites of exchange exhibit recombination nodules (purple) at pachytene that form chiasmata at diakinesis. At diplotene the nuclear envelope breaks down, the synaptonemal complex resolves, and homologs separate except for sites of exchanges. Sister chromatids of each homolog within the bivalent attach to microtubules (gray) of the same spindle pole at prometaphase I and remain attached by chiasmata until anaphase I. For further explanation, see text.

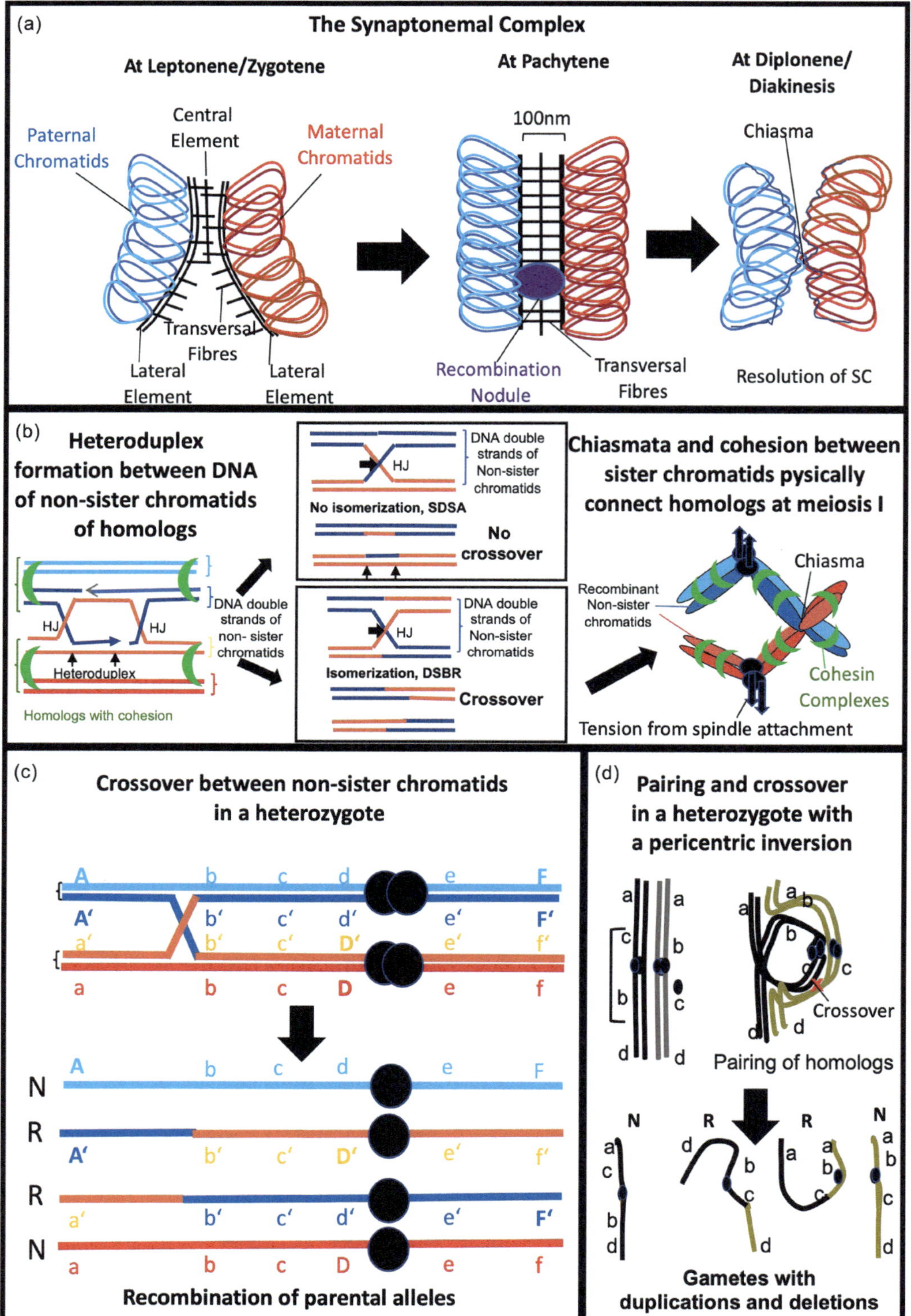

Figure 3.3 Processes at homologous recombination. (a) Synaptonemal complex (SC) formation and meiotic recombination between nonsister chromatids. Replicated sister chromatids (in lighter and darker blue and red color) become attached to a proteinaceous axis, forming

nodules and about 1.6 times more recombination events in female compared with male meiosis in humans, consistent with fine-scale recombination maps obtained by population genetics [7,9–12]. In addition, there tend to be more intermediate recombination events on chromosomes in female meiosis whereas there are more distal exchanges relative to centromeres in male meiosis [8–11].

3.3.3 Molecular Framework for Crossover between Homologs

Studies in model organisms from yeast to mammals have provided information on highly conserved mechanisms and molecules in intrachromosomal meiotic recombination, and a roadmap of events that mediate genetic exchange, homologous recombination, and formation of chromosomally balanced but genetically different germ cells. In short, basically, meiotic recombination is initiated by homology search and by homologous pairing of DNA molecules from the originally maternally and paternally derived homologous chromosomes each containing two chromatids (Figure 3.3b, left side). Introduction of DNA double-strand breaks by conserved endonuclease (SPO11) into one of the sister chromatids of a homolog containing just one DNA double helix (indicated by dark blue lines in Figure 3.3c) involves the activity of meiosis-specific enzymes such as PRDM9, a zinc finger-containing protein binding to DNA [2,13,14]. After pairing and breakage, for recombination one single strand of nucleotides invades the double strand of a nonsister chromatid of the other homolog (as shown for blue line in Figure 3.3b, left side). Binding of the single-stranded DNA and the opening of the homologous double strand of the DNA in the nonsister chromatid of the other homolog (orange in Figure 3.3b, left side) leads to formation of a loop structure (D-loop). In the D-loop there are two crosses by single-stranded DNA molecules, termed "Holliday junctions" (HJ) (Figure 3.3b). Within the loop single strands of DNA from homologous nonsister chromatids become paired (heteroduplex DNA) [2,5,13,14]. If the crossings are resolved by nucleases without further change in the conformation of the DNA by synthesis-dependent strand annealing (SDSA), there is only a very small stretch of DNA that contains heteroduplex DNA from both partners (indicated by small arrows in Figure 3.3b, upper middle panel), and there is no crossover. However, if the DNA in the HJ undergoes a conformational change (termed "isomerization") followed by resolution by nucleases and ligation of DNA strands (by double-strand break repair, DSBR), the double-strand DNA of the maternal and paternal nonsister chromatids left and right of the crossover are exchanged and recombined (Figure 3.3b, middle lower panel), and can mature into a chiasma (Figure 3.3b, right side) that physically connects the homologs together with cohesion between sister chromatids (green in Figure 3.3b, right side). Usually, a large excess of DNA breaks is initiated originally but only some of the breaks, predominantly at sites termed hotspots, mature into a crossover and establish a chiasma (for detailed information, see Bolcun-Filas and Handel [2], Gray and Cohen [5], Baudat et al. [13] and Manhart and Alani [14]).

3.3.4 Reassortment of Parental Homologs at Anaphase I

Homologs usually contain one or several chiasmata at the end of prophase I and are therefore connected

Figure 3.3 (*cont.*) the lateral elements of the SC at leptotene and zygotene stage. Proteins from transversal fibers interact in a zipper-like fashion upon homologous pairing. At contacts in the center of the SC central elements are assembled on which late recombination nodules (violet) appear at pachytene that mark sites of meiotic recombination, and after resolution of the SC at diplotene/diakinesis form a chiasma. (b) After pairing of homologs, each with two chromatids with double strand of DNA (paternal: dark and light blue; maternal: red and orange), double-strand breaks initiate strand invasion of one DNA strand (dark blue) into the double helix of the nonsister chromatid (orange) and formation of a D-loop with heteroduplex DNA. Without isomerization by synthesis-dependent strand annealing (SDSA) there is only a short stretch of heteroduplex (small arrows) and no crossover (upper box). By isomerization and double-strand break repair (DSBR) there is a crossover (lower box) that matures into a chiasma (right side) that together with rings of cohesion complexes (green, left and right side) connects sister chromatids in homologs at meiosis I, thus facilitating bipolar attachment of homologs within the bivalent and tension from spindle attachments leading to alignment at the spindle equator (right side). (c) Crossover between nonsister chromatids of homologs (represented by single line) of a heterozygote with polymorphic markers and different parental alleles (indicated by color and upper and lower letters) leads to formation of nonrecombinant (N) and recombinant (R) chromatids in gametes at completion of meiosis. (d) Pairing in a heterozygote with a pericentric inversion (indicated by bracket on upper left) leads to formation of a ring structure during prophase I (right side). One crossover between nonsister chromatids within the inverted part leads to formation of nonrecombinant (N) and recombinant (R) gametes with duplications and deletions after first and second meiotic division. For further explanation, see text.

within bivalents: all 22 autosomes in the human and the two X chromosomes in female meiosis. In males there are 22 autosomal bivalents and the X and Y sex chromosomes that possess one obligatory exchange and chiasma in the PAR (Figure 3.4a). Since the homologs of different chromosomes can randomly attach to either spindle pole (as shown in Figure 3.4b) at first metaphase, the paternal and maternal homologs can segregate to either spindle pole at the reductional anaphase I, generating a mixture of paternal and maternal homologs in the daughter cells. The haploid gametes therefore usually contain a new assortment of the different originally maternally and paternally derived chromosomes (Figure 3.4c). For instance, by random separation and assortment of three different parental homologs (shown in lighter and darker colors in Figure 3.4c) at anaphase I, two daughters with three maternal or three paternal chromatids can be formed by chance (nonrecombinants, Figure 3.4c, left side) plus six gametes that are recombinant-containing mixtures of originally maternally or paternally derived chromosomes (Figure 3.4c, right side). When their chromatids separate in meiosis II, 2^N different gametes can therefore be generated according to the number of chromosomes in the haploid complement (N referring to haploid chromosome number). In the human the random distribution of homologs from father or mother at the reductional division at meiosis I can therefore result in the formation of 2^{23} combinations of parental chromosomes, so more than 8 million genetically different gametes [1,2,13].

3.3.5 New Assortment of Alleles in Chromatids by Recombination

In addition to the new assortment of parental chromosomes that is based on the random reductional segregation of the maternally and paternally derived homologs at first meiotic anaphase (interchromosomal recombination), intrachromosomal exchanges (crossovers) between homologous regions of the parental chromosomes lead to further recombination of parental alleles on individual chromosomes. While exchanges between sister chromatids are suppressed during meiotic prophase I, those between nonsister chromatids of homologs can result in multiple changes in the presence of different parental alleles on a chromatid of a heterozygote (indicated by different colors and upper or lower case letters in Figure 3.3c). These are eventually inherited through the germline to the next generation (as shown in lower part of Figure 3.3c; each chromatid represented as single line, sister chromatids in lighter or darker colors). One single crossover between two of the nonsister chromatids produces two recombinant and two nonrecombinant chromatids (Figure 3.3c). Two crossovers between two of the four chromatids leads to nonrecombinant distal alleles and only a short stretch of recombinant DNA within the two crossovers (not shown). Since double crossovers involving two, three or all four chromatids are random, they on average induce maximally 50% recombinants as is also the case for the single crossovers involving two chromatids. The closer alleles are located on the chromosomes, the less likely is an exchange between them due to crossover interference. Analysis of genetic traits in offspring or the distribution of polymorphic markers/alleles on chromosomes (see below) of gametes, zygote, blastomeres or offspring can thereby be used to establish a genetic map and deduce the numbers and localization of exchanges by crossovers during the meiotic divisions [9–11]. Thus, by analysis of recombination between polymorphic markers the order and genetic distance of alleles on the chromosomes and hotspots of recombination can be determined and a genetic recombination map of each chromosome be established. The latter is usually shorter compared with the overall physical length of the chromosome since heterochromatic regions do not contain crossovers [5,9,13].

Examining the numbers and localization of RNs at diplotene or of chiasmata at diakinesis and prometaphase I on spread chromosomes of spermatocytes and oocytes using immunofluorescence of proteins in late RNs has been initially used to characterize the events at meiotic prophase I. About 50 crossovers were characteristically observed in human spermatocytes and about 50–70 in human oocytes [7].

3.4 Origin and Impact of Meiotic Disturbances

3.4.1 Presence, Numbers, and Distribution of Exchanges

Physically nonconnected homologs (termed “univalents”; Figure 3.5a) in oocytes or spermatocytes that are asynaptic and not stabilized by at least one

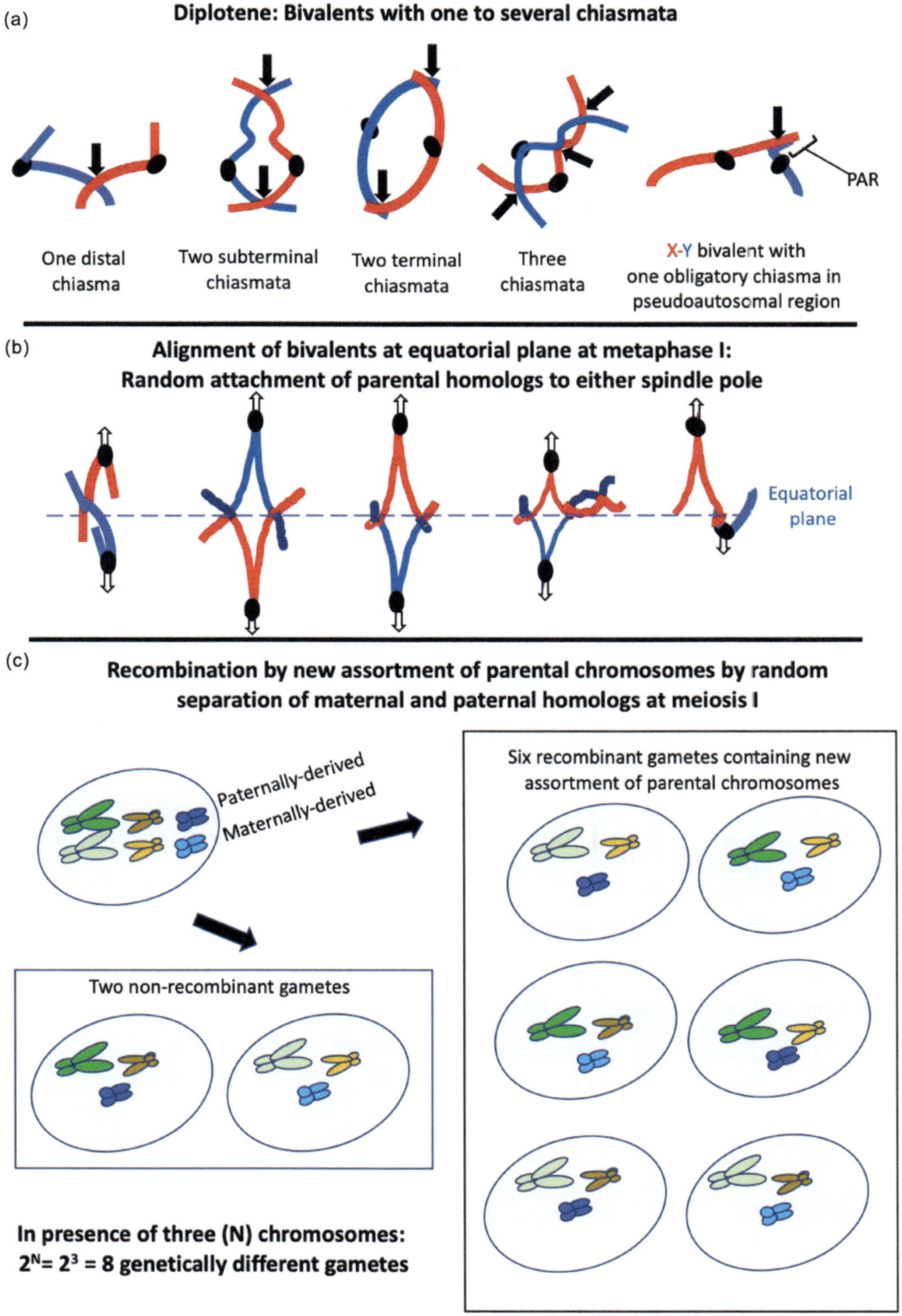

(a) **Diplotene: Bivalents with one to several chiasmata**

PAR

One distal chiasma

Two subterminal chiasmata

Two terminal chiasmata

Three chiasmata

X-Y bivalent with one obligatory chiasma in pseudoautosomal region

(b) **Alignment of bivalents at equatorial plane at metaphase I: Random attachment of parental homologs to either spindle pole**

Equatorial plane

(c) **Recombination by new assortment of parental chromosomes by random separation of maternal and paternal homologs at meiosis I**

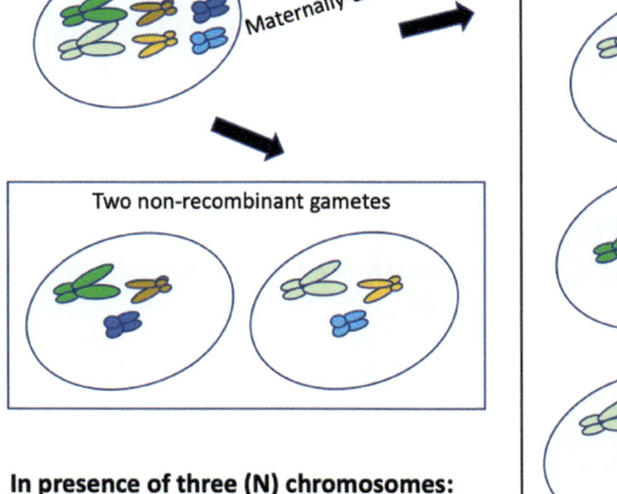

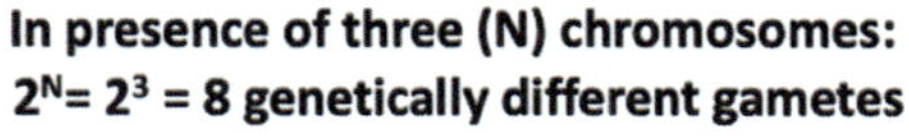

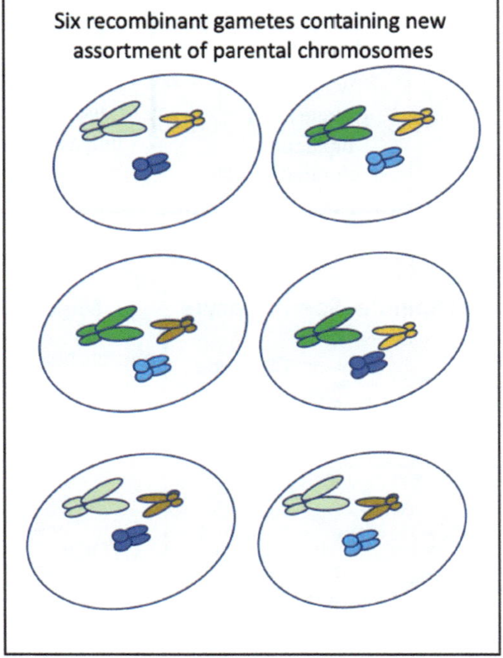

Figure 3.4 Random segregation of homologs at meiosis. Bivalents with distal, single, or multiple chiasmata (a, solid arrows) congress to the spindle equator with first chiasma at the equatorial plane (b) and random attachment of paternally and maternally derived homologs to either spindle pole at metaphase I (b, open arrows) yielding gametes with recombinant (c, right side) or non-recombinant (c, left side) assortment of maternally (dark colors) or paternally (light colors) derived chromosomes. For further explanation, see text.

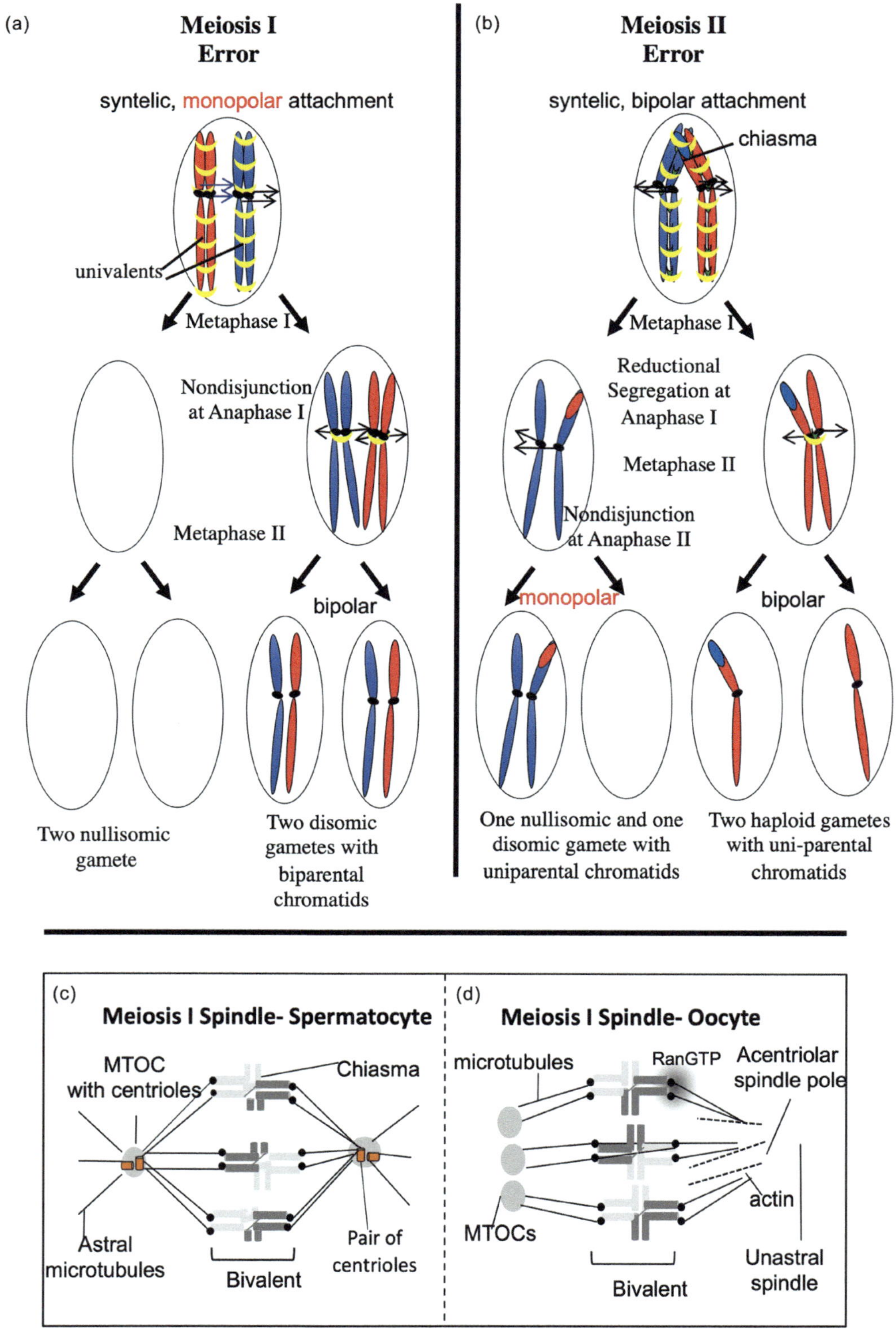

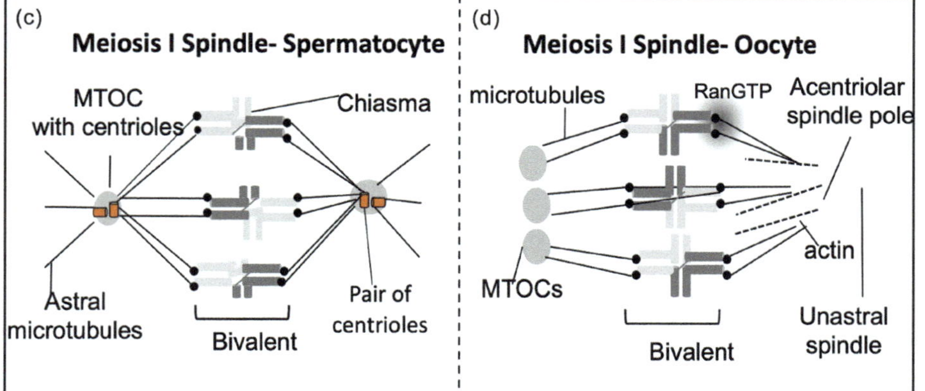

Figure 3.5 Typical errors in chromosome segregation (nondisjunction) at meiosis I in the presence of univalents and monopolar syntelic attachment of homologs to only one spindle pole (a, arrows at centromeres). Errors at meiosis II involving monopolar attachment of

chiasma have a high risk of undergoing meiotic errors at meiosis I [15,16]. They are more likely to attach randomly to only one spindle pole (indicated by arrows in Figure 3.5; syntelic monopolar attachment). Through nondisjunction one of the products from first meiosis will be nullisomic and the other disomic for the respective homolog. Short chromosomes that usually contain only one crossover, particularly those with a single distal chiasma close to the telomere, are more likely to undergo nondisjunction due to the absence or the precocious resolution of a chiasma. This can also be due to reduced constraints favoring wrong attachments and reverse segregation such as in the short acrocentric chromosomes 13, 14, 15, 21, and 22 in the human. When the two univalents of the daughter cell from first meiosis segregate their chromatids equationally at meiosis II, two disomic gametes arise, each containing two instead of one copy of the chromosome derived originally from father and mother (biparental origin) (Figure 3.5a, right side). In the case of chromatids of homologous chromosomes that have already separated at meiosis I instead of meiosis II (termed "reverse segregation," not shown), the random segregation of the two physically unconnected sister chromatids in the daughter cells at meiosis II can also generate disomic gametes with two uniparental chromatids plus a nullisomic gamete [15,16].

After a normal reductional first meiosis, a typical second meiotic error may occur when centromeres of both sister chromatids fail to separate at meiosis II, for instance when centromeres of sister chromatids attach to only one spindle pole, or randomly become attached to the same pole due to a precocious loss of cohesion between the centromeres of sister chromatids (i.e. precocious separation of sister chromatids, PSSC; indicated by arrows in Figure 3.5b, left side). This will generate one nullisomic and one disomic gamete containing two monoparental chromatids (blue in Figure 3.5b). After fertilization these will generate a monosomic or trisomic embryo, respectively.

Apart from presence of univalents, an excess of recombination or recombination at unusual sites (e.g. single exchanges in distal, telomeric parts of chromosomes [9,15,16] or in close proximity to the centromere [15,17]) can also predispose to meiotic errors. Furthermore, nonallelic recombination, erroneous repair after double-strand DNA breaks [18], replication errors, and retrotransposition can contribute to disturbances in meiosis and genomic rearrangements in the human germline [19].

3.4.2 Sexual Dimorphism in Spindle Formation

Similar to mitotically dividing cells, spermatocytes possess two pairs of centrioles, barrel-shaped organelles with two perpendicularly arranged cylinders formed from microtubules that are surrounded by amorphous material (Figure 3.5c). Together with pericentriolar proteins, this serves as the major microtubule organizing center (MTOC) that promotes microtubule polymerization. Prior to nuclear membrane breakdown the centriolar MTOCs have already separated and thus promote the formation of a bipolar spindle after diplotene in the comparatively small cytoplasm of spermatocytes [20,21]. The spindle in spermatocytes possesses characteristically fusiform poles (Figure 3.5c) and astral microtubules that anchor the spindle centrally in the cytoplasm (astral spindle). During prometaphase I, the sister chromatids of each homolog within bivalents attach to spindle fibers, and the bivalents migrate on the spindle until a stable bipolar attachment and alignment at the spindle equator is achieved (Figure 3.5c). This facilitates division into two equal-sized daughter cells upon first and second meiotic division in males.

Oocytes do not possess centrioles. Within the large ooplasm a unique and small acentriolar spindle is formed upon resumption of meiosis (Figure 3.5d) [20]. In some mammals like the mouse multiple acentriolar MTOCs initially organize a central array of microtubules after germinal vesicle breakdown on which the bivalents become located (termed "circular bivalent stage"). Upon prometaphase I the MTOCs then gradually assemble at the flat spindle poles of a

Figure 3.5 *(cont.)* chromatids to the same spindle pole at metaphase II (b, left side) can yield disomic or nullisomic gametes. Sexual dimorphism in spindle formation (c, d) impacts probability of meiotic errors. Spermatocytes (c) possess pairs of centrioles (orange) associated with microtubule organizing centers (MTOCs, gray) that facilitate bipolar spindle organization, whereas oocytes do not possess centrioles and form unastral acentriolar spindles in which either acentriolar MTOCs contribute to organization of microtubules (mouse) or primarily a chromosome-mediated RanGTP gradient is involved in microtubule polymerization and spindle formation (human) and actin filaments contribute to spindle stabilization and unequal division (d). For further explanation, see text.

bipolar, barrel-shaped, unastral spindle (Figure 3.5b) [20,21]. The small RanGTP and TPX2 have been recognized as major drivers of a mainly chromatin-dependent microtubule assembly in oocyte spindle formation. Thus, there is no preexisting bipolar structure in oocytes as in spermatocytes. Rather, meiosis in human oocytes is unique in that a gradient of RanGTP from chromosomes generates microtubule assembly and together with motor proteins and microtubule and centromeric regulatory factors organizes spindle formation (Figure 3.5d). Initially, spindles are therefore frequently multipolar rather than bipolar and bivalents often reorient, turn, and oscillate on spindle fibers. Centromeres detach and reattach to microtubules for extended periods of time during a fairly long prometaphase I of oogenesis that takes several hours before a bipolar spindle is generated (Figure 3.5d). During the extended prometaphase I centromeres of sister chromatids or a single chromatid in one homolog frequently attach to both poles (termed “merotelic attachment”) before becoming bipolarly attached. Finally, tension from bipolar attachments to spindle fibers results in alignment of bivalents at the spindle equator (Figure 3.5d) [4,20–22].

Usually, the release of checkpoint proteins from unattached chromosomes can cause mitotic and meiotic arrest [4,21]. Because of diffusion, this checkpoint may be more permissive in the large cytoplasm of an oocyte with a small spindle compared with spermatocytes and can therefore result in progression to anaphase I in the presence of wrongly attached chromosomes. Furthermore, spindle formation, chromosome segregation, and the unequal division in oocytes involve the actin cytoskeleton, which stabilizes the spindle microtubules in female meiosis and is required for unequal division at polar body formation [22]. The meiotic spindle in oocytes is also acting as a sink for the local accumulation of regulatory factors such as mRNAs, which are essential for high cytoplasmic competence and early embryonic development, and is surrounded by mitochondria. Disturbances in the spindle, for instance by prolonged aging of the metaphase II arrested oocyte, can therefore have profound effects on developmental competence apart from inducing aneuploidy. Whereas the barrel-shaped spindle in rodent oocytes is oriented parallel to the oolemma and turns 90° upon anaphase progression, the human oocyte spindle is attached with only one more fusiform spindle pole to the oolemma before anaphase I and anaphase II.

The complex processes in chromatin-dominated spindle formation can contribute to high susceptibility to meiotic errors in female meiosis compared with male meiosis, especially in aged females (for further discussion see Ottolini et al. [9], Capalbo et al. [16], Eichenlaub-Ritter [20], Lane and Kauppi [21] and Mogessie et al. [22].

3.4.3 Age and Meiotic Errors

Aneuploidies from meiotic nondisjunction are almost always lethal. Thus, monosomy of chromosome X is the only monosomy that is compatible with survival in the human (Turner syndrome). Depending on the chromosome involved and the gene dosage effect of chromosomal imbalance, aneuploidy predisposes to developmental arrest of the preimplantation embryo, implantation failure, spontaneous abortion, or congenital genetic disease in offspring (e.g. trisomy 21 or Down syndrome; trisomy 18 or Edwards syndrome; trisomy 13 or Patau syndrome). Inefficient crossover maturation appears related to increased susceptibility to aneuploidy in human oogenesis compared with spermatogenesis (see also below) [18]. Trisomies rise exponentially in spontaneous abortions with advanced maternal age and depletion of the follicle pool [15,16].

In order to discover the origin of age-related susceptibility to meiotic errors, single cell analysis of human oocytes and their first and second polar bodies or trios of PB1, PB2 and female pronuclear DNA, for instance by karyomapping (Meio-Mapping), have recently been used [4,9,11]. Aneuploidy in oogenesis exhibits a U-shaped curve [4,15]. Increased aneuploidy in oocytes of young females appears predominantly associated with unusual recombination patterns [10,11,16]. By contrast, the exponential rise in segregation errors at advanced maternal age in human oogenesis predominantly appears to involve deterioration of sister chromatid cohesion, precocious separation of sister chromatids, and the equatorial separation of sister chromatids instead of homologs at meiosis I (reverse segregation) that predispose to first and second meiotic errors [15,16]. Male meiotic errors are rare and especially affect the sex chromosomes [23].

3.4.4 Mutations and Rearrangements Affecting Meiosis

Models from mouse to yeast have revealed that the recombination genes and proteins are highly conserved. The presence of alleles or mutations leading to altered expression of the genes for components of the SC or those for DNA recombination and repair can cause sex-specific infertility or subfertility in animal models and in the human to meiotic abnormalities, aneuploidy, and cell death (see Chapters 7–9). Nonhomologous recombination can also contribute to disturbances in chromosome segregation and formation of chromosomally unbalanced gametes [2,19].

Heterozygosity for balanced translocations can also cause meiotic errors and chromosomal imbalance in gametes [24,25]. For instance, when there is a crossover in a loop formed during homolog pairing at prophase I of meiosis between the normal homolog and the one with a pericentric inversion (inversion includes the centromere; Figure 3.3d, bracket on left side), this may lead to the formation of gametes with duplications or deletions after resolution of the chiasma (recombinants; R in Figure 3.3d) containing two copies of one allele plus a deletion of another one (Figure 3.3d, allele d or a, respectively) when homologs and chromatids separate at meiotic divisions (gray or black chromatids in Figure 3.3d). Paracentric inversions, in which a crossover occurs in the inverted segment outside of the centromere, can result in the formation of an acentric fragment or chromatids with more than one centromere at anaphase I. Heterozygosity for a Robertsonian translocation (RB) can also increase the risks for aneuploidy by segregation of the RB chromosome together with one acrocentric from the trivalent to one spindle pole and the other acrocentric to the other pole. The presence of a trivalent may also influence the behavior of other chromosomes (so-called "interchromosomal effect") predisposing to meiotic errors [24,25].

3.4.5 Sex-Specific Differences in Entry and Completion of Meiosis

There are distinct sex-specific differences in the meiotic processes in gametogenesis between males and females that are of relevance for the sensitive time windows associated with high susceptibility to meiotic disturbances, for example to exposures during meiotic prophase, periods of meiotic arrest, or when reductional chromosome segregation occurs (Figure 3.6).

Initiation of germ cell formation is regulated by the sex of the somatic cells in the fetal gonads and not by the genetic constitution of the primordial germ cells themselves [26]. An ovary and oocytes are constitutionally initiated after migration and colonization of sexually indifferent primordial germ cells to the genital ridges in the female fetus. In absence of a Y chromosome that contains sex-determining genes [27] mitotic divisions generate nests of oogonia (Figure 3.6a, left side) that become primary oocytes entering premeiotic S-phase and develop from leptotene to diplotene stage within the fetal ovary. After diplotene chromosomes decondense within the germinal vesicle (dictyate stage) prior to birth and oocytes become meiotically arrested for the first time. Meiotic arrest can last for decades (Figure 3.6a, red arrow, left side). Primary oocytes in dictyate stage become surrounded by a layer of granulosa cells forming primordial follicles that are meiotically quiescent from around the fourth month of gestation. The primary oocytes entering meiosis in the fetal ovary thus comprise the pool of all the oocytes able to become competent to mature and generate an egg throughout the female's reproductive period. After puberty, primordial follicles become periodically recruited and start to grow in the sexually adult female. Most of them become atretic and degenerate. Only a few primordial follicles develop to primary follicles and secondary follicles with more than one layer of granulosa cells. Some develop further to secondary, preantral, and antral stage, and become hormone-sensitive. In tertiary follicles a fluid-filled space, the antrum, separates the layers of outer mural granulosa cells from inner cumulus granulosa cells that are in intimate contact with the growing oocyte [28]. In the human, for meiotic progression only one fully grown oocyte within a large preovulatory antral follicle (Graafian follicle) resumes meiosis and matures to metaphase II at each natural ovulatory stimulus. Meiotic resumption of the oocytes with progression to metaphase II and first polar body formation requires the acquisition of high maturational and developmental competence of the oocyte during the extensive growth phase, in cross-talk with and support by granulosa cells. Thus, gene products that are necessary for maturation and early embryonic development are acquired [28,29]. Ovulated oocytes with a first polar body surrounded by mucified cumulus cells arrest again meiotically in metaphase II until fertilization triggers anaphase II and completion of meiosis II, second polar body emission, and female and male pronuclear formation

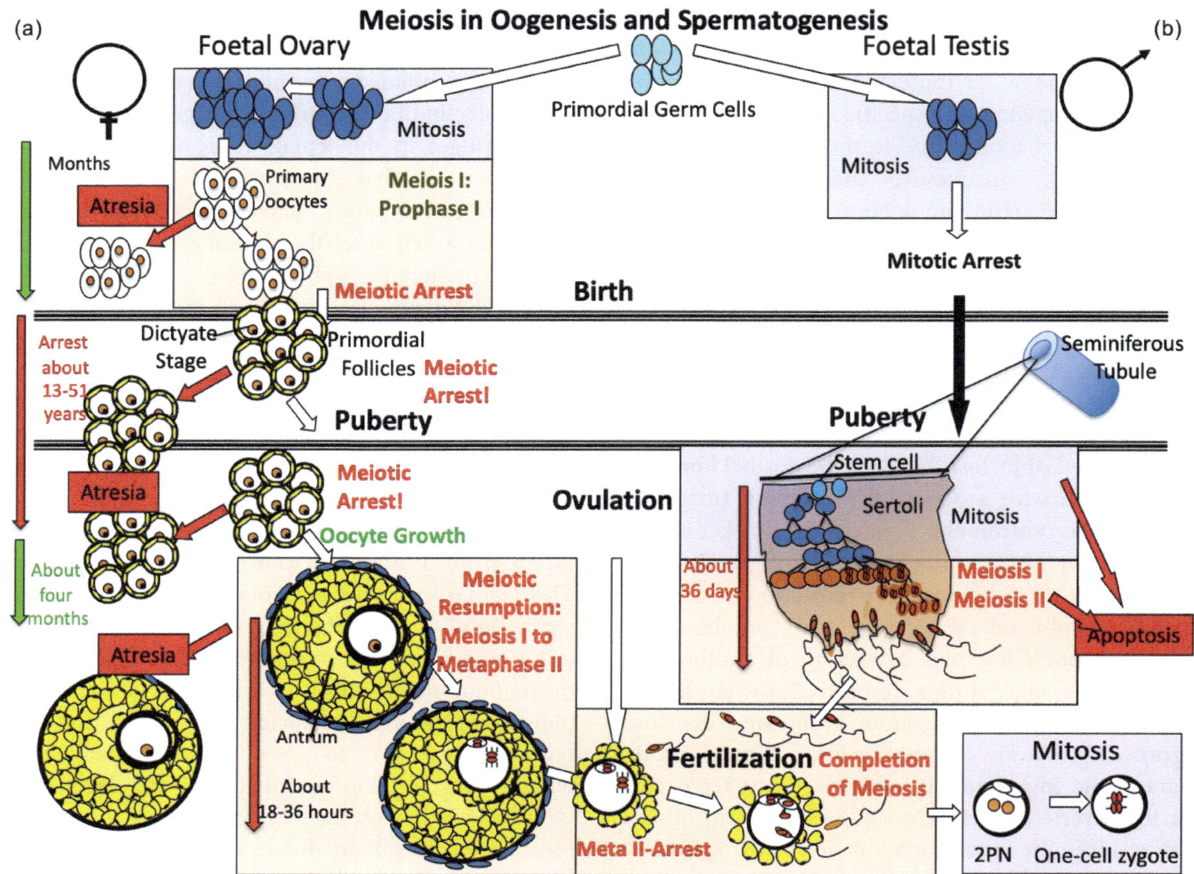

Figure 3.6 Germ cell formation. Timing of meiosis in oogenesis (a, left side): beginning of meiosis in fetal ovary and completion decades later, after prophase I in fetal ovary, long meiotic arrest in dictyate stage, growth of primary oocyte within follicle, followed by resumption of oocyte maturation and ovulation in metaphase II and progression to anaphase II and cytokinesis only after fertilization. Timing of spermatogenesis (b, right side): multiple mitotic divisions and mitotic arrest in fetal testis followed by generation of spermatogonia from division of spermatogonial stem cells within the seminiferous tubules at puberty to advanced paternal age, and rapid progression of spermatocytes from meiosis I to meiosis II followed by spermiogenesis for release of mature sperm into the lumen of the seminiferous tubules. Follicular recruitment and atresia diminish the pool of follicles and oocytes throughout the reproductive life until the pool is critically diminished, whereas apoptosis can lead to transient reduction in spermatocytes but by continuous divisions of stem cells spermatogonia, spermatocytes, spermatids and sperm can be generated until old age in the male. For further explanation, see text.

(Figure 3.6). The end of reproductive life in the female is reached when the follicle pool reaches a critical size and is emptied by demise of follicles due to follicular recruitment and atresia).

In contrast to the female, in the male fetal testis there are some mitotic divisions and testis cords are formed that differentiate into seminiferous tubules which contain spermatogonial stem cells (Figure 3.6b, right side). By mitotic divisions the stem cells become renewed and generate spermatogonia that further divide and differentiate into B-type spermatogonia embedded in Sertoli cells of the seminiferous tubules in the postpubertal male. The spermatogonia can finally enter premeiotic S-phase, and meiosis proceeds uninterrupted from S-phase to meiosis II in the spermatocytes of the testis of the adult male (Figure 3.6). Chromatin transiently decondenses after first meiosis before chromosomes become condensed again and spermatocytes progress through second meiosis, forming haploid spermatids that are still in contact with each other by cytoplasmic bridges connecting gametes from mitotic and meiotic divisions. Meiosis is followed by spermiogenesis during which chromatin becomes highly condensed and packed with specialized proteins (protamines) in the sperm nucleus. Maturation of sperm further involves formation of a sperm tail, mitochondrial fusion, and formation of a midpiece and assembly of an acrosome. Meiosis up to the generation of mature sperm takes only about 26 days, in contrast to oogenesis where dictyate arrest can last from about 13 to 51 years, and folliculogenesis about 4 months. Meiosis in males is therefore much shorter (Figure 3.6). Progression to meiosis and spermatogenesis within the testis occurs in waves and relies on a tightly regulated balance between self-renewal of spermatogonial stem cells and their differentiation into spermatogonia and meiotically dividing spermatocytes (Figure 3.6). Because of the presence and mitotic divisions of spermatogonial stem cells, males remain fertile to old age whereas depletion of the follicle pool restricts female fertility.

Since oocytes are among the most long-lived cells in the body, extrinsic and intrinsic processes can affect oocyte quality and genetic integrity, especially during the long meiotic arrest. Maternal aging is thus the primary risk factor of aneuploidy in female meiosis [9–11]. However, susceptibility to meiotic errors is also especially high at a very young age [15,16]. Exposures during pregnancy can influence the fertility of the females in the next generation whose oocytes become exposed in the maternal fetal ovary. Mutations and exposures prior to and after birth causing extensive atresia and follicular and/or oocyte demise can lead to a small pool size and to premature ovarian insufficiency [30]. Spermatocytes are derived from the dividing and differentiating stem cell niche and the mitotically dividing spermatogonia and meiosis can take place from puberty until old age. Apart from transient effects on male meiosis, damage to the stem cell niche, for instance by chemotherapeutic agents or adverse exposures, can also affect male fertility and reproductive lifespan or contribute to subfertility or infertility. Furthermore, mutations may increase during the increased numbers of mitotic divisions during male aging. Stem cells with certain rare mutations can clonally expand and thus increase mosaicism of mutant spermatogonia in the testes. However, there is only a minor, though clinically relevant, effect of paternal age on aneuploidy [23].

It is suspected that oxidative damage, mitochondrial insufficiencies, and altered gene expression, particularly those affecting chromosome constitution, cell cycle, spindle formation and chromosome cohesion, contribute to meiotic errors and risks for increases in aneuploidy with advanced maternal age [4,20,21]. Since chromosome cohesion is established at premeiotic S-phase and appears to remain unchanged during the long meiotic arrest in oogenesis, the loss of cohesion between sister chromatid arms and centromeres may lead to precocious chiasma resolution, presence of univalents, reverse segregation, and increases in meiotic errors [3,4,10,16]. Postovulatory aging can also impact oocyte quality, competence, fertilization, and genetic integrity. Disturbances in centromere function, spindle formation, and cell cycle control can contribute to meiotic abnormalities especially in female meiosis [4,17,20,31].

3.4.6 Constitutional Conditions, Exposures, and Lifestyle

Studies in human and animal models have shown that maternal aging but also exposures to therapeutic drugs, namely those affecting DNA integrity or spindle formation (e.g. microtubule depolymerizing drugs), as well as to occupational agents like pesticides that influence meiosis directly or indirectly may lead to aneuploidy, chromosomal instability, and subfertility or infertility [32]. Genetic variants and mutations that influence recombination, mitochondrial homeostasis, or spindle formation and cell cycle control are some of the multiple factors that can independently or synergistically with aging affect meiosis and fertility.

Analysis of chromosomal constitution, spindle morphology, and chromosome congression in human oocytes and embryos as well as in animal models have also suggested that certain stimulation protocols and exposures to environmental chemicals, as well as lifestyle factors (e.g. smoking) and diet, may contribute to abnormalities or arrest of meiosis in males and females. Furthermore, there is emerging evidence that exposures of adult males or females in utero to endocrine-disrupting chemicals that might alter epigenetic marks and pairing and recombination between homologous chromosomes can contribute to meiotic errors and affect fertility directly or transgenerationally, but much more research is needed to confirm this [32].

3.5 Conclusions and Future Perspectives

Formation of euploid functional germ cells is essential for maintaining fertility and genetic integrity from one generation to the next. Recent years have provided a wealth of information on conserved genes and processes in germ cell formation and meiosis. Furthermore, new and refined methods for analysis of the genome before and after conception have been established [3]. Gender-specific differences in meiotic timing are known to influence windows of highest susceptibility to meiotic disturbances. Our species appears to be exceptionally prone to nondisjunction and germ cell aneuploidy in the female. Certain chromosome constitutions appear also particularly critical, for instance those inducing nonallelic homologous recombination, erroneous repair of DNA breaks, replication errors, or complex rearrangements that can result from chromosome breakage and nonhomologous reannealing ("chromothripsis") [19].

Overall, all studies have confirmed that unperturbed meiosis is essential for fertility, and crossing-over serves several purposes in achieving the formation of euploid germ cells:

1. On the cellular level, the presence of chiasmata together with chromosome cohesion physically attaches the two homologs such that they can stably align on the spindle at first meiosis, with the centromeres of each homolog's sister chromatids attaching to only one spindle pole, thus supporting high fidelity in the reductional chromosome segregation at anaphase I.
2. The random orientation and segregation of the originally paternally and maternally derived homologs in the bivalents at the anaphase I transition results in recombination by new assortments of parental chromosomes in the gamete, the major mechanism for shuffling and mixing of parental genes to the next generation [34].
3. Intrachromosomal recombination between originally paternally and maternally derived alleles on nonsister chromatids to form recombinant chromatids in gametes contributes to genetic diversity, evolution, and survival.

While meiotic errors that occur at first division can occasionally be compensated at second meiosis or mitotic division in the embryo (secondary nondisjunction) so that a balanced constitution arises in all or some of the blastomeres of the embryo, uniparental disomy can arise if both chromosomes in the embryo are exclusively from the mother or father. When genes on such chromosomes are only expressed from the paternal or maternal allele due to imprinting, this may result in a gene dosage effect and aberrant expression that causes genetic disease.

The advent of novel, sensitive, and reliable methods for analysis of the genome before and after fertilization and gene expression and chromosomal constitution in germ cells and embryos is promising to detect disturbances, and provide information on genetic and etiologic factors that affect meiosis and genomic integrity. This can be used to inform patients and hopefully develop strategies for prevention and treatment, or selection of germ cells or euploid embryos with high developmental potential for assisted reproduction. Raising awareness in either gender of the processes involved in normal meiosis and germ cell formation is important for avoiding adverse exposures and making informed decisions. For the clinician, embryologist, and geneticist it is crucially important to develop best strategies that are minimally invasive, cost- and time-efficient, and of greatest reliability in obtaining euploid embryos and minimizing risks for abnormal meiosis in order to improve the benefits of treatment for patients who want to conceive a healthy child.

References

1. Alberts B, Johnson A, Lewis J, et al. (eds.) Meiosis. In *Molecular Biology of the Cell*, 4th ed. New York: Garland Science, 2002.
2. Bolcun-Filas E, Handel MA. Meiosis: the chromosomal foundation of reproduction. *Biol Reprod* 2018;99(1):112–26.
3. Ishiguro K. The cohesin complex in mammalian meiosis. *Genes Cells* 2019;24(1):6–30.
4. Greaney J, Wei Z, Homer H. Regulation of chromosome segregation in oocytes and the cellular basis for female meiotic errors.*Hum Reprod Update* 2018;24(2):135–61.
5. Gray S, Cohen PE. Control of meiotic crossovers: from double-strand break formation to designation.*Annu Rev Genet* 2016;50:175–210.
6. Dunne OM, Davies OR. Molecular structure of human synaptonemal complex protein SYCE1. *Chromosoma* 2019;128 (3):223–36.
7. Hassold T, Sherman S, Hunt P. Counting cross-overs: characterizing meiotic recombination in mammals. *Hum Mol Genet* 2000;9(16):2409–19.
8. Poriswanish N, Neumann R, Wetton JH, et al. Recombination hotspots in an extended human pseudoautosomal domain predicted from double-strand

break maps and characterized by sperm-based crossover analysis. *PLoS Genet* 2018;4(10):e1007680.

9. Ottolini CS, Newnham L, Capalbo A, et al. Genome-wide maps of recombination and chromosome segregation in human oocytes and embryos show selection for maternal recombination rates. *Nat Genet* 2015;47(7):727–35.
10. Handyside AH, Montag M, Magli MC, et al. Multiple meiotic errors caused by predivision of chromatids in women of advanced maternal age undergoing in vitro fertilisation. *Eur J Hum Genet* 2012;20:742–7.
11. Ottolini CS, Capalbo A, Newnham L, et al. Generation of meiomaps of genome-wide recombination and chromosome segregation in human oocytes. *Nat Protoc* 2016;11:1229–43.
12. Kong A, Thorleifsson G, Gudbjartsson DF, et al. Fine-scale recombination rate differences between sexes, populations and individuals. *Nature* 2010;467 (7319):1099–103.
13. Baudat F, Imai Y, De Massy B. Meiotic recombination in mammals: localization and regulation. *Nat Rev Genet* 2013;14 (11):794–806.
14. Manhart CM, Alani E. Roles for mismatch repair family proteins in promoting meiotic crossing over. *DNA Repair (Amst)* 2016;38: 84–93.
15. Gruhn JR, Zielinska AP, Shukla V, et al. Chromosome errors in human eggs shape natural fertility over reproductive life span. *Science* 2019;365(6460):1466–9.
16. Capalbo A, Hoffmann ER, Cimadomo D, Ubaldi FM, Rienzi L. Human female meiosis revised: new insights into the mechanisms of chromosome segregation and aneuploidies from advanced genomics and time-lapse imaging. *Hum Reprod Update* 2017;23 (6):706–22.
17. Nambiar M, Smith GR. Pericentromere-specific cohesin complex prevents meiotic pericentric DNA double-strand breaks and lethal crossovers. *Mol Cell* 2018;71(4):540–53.e4.
18. Wang S, Hassold T, Hunt P, et al. Inefficient crossover maturation underlies elevated aneuploidy in human female meiosis.*Cell* 2017;168(6):977–89.e17.
19. Hattori A, Fukami M. Established and novel mechanisms leading to de novo genomic rearrangements in the human germline. *Cytogenet Genome Res* 2020;160(40):167–76. DOI 10.1159/000507837
20. Eichenlaub-Ritter U. Oocyte ageing and its cellular basis. *Int J Dev Biol* 2012;56(10–12):841–52.
21. Lane S, Kauppi L. Meiotic spindle assembly checkpoint and aneuploidy in males versus females.*Cell Mol Life Sci* 2019;76 (6):1135–50.
22. Mogessie B, Scheffler K, Schuh M. Assembly and positioning of the oocyte meiotic spindle. *Annu Rev Cell Dev Biol* 2018;34:381–403.
23. Colaco S, Sakkas D. Paternal factors contributing to embryo quality. *J Assist Reprod Genet* 2018;35(11):1953–68.
24. Machev N, Gosset P, Warter S, et al. Fluorescence in situ hybridization sperm analysis of six translocation carriers provides evidences of an interchromosomal effect. *Fertil Steril* 2005;84:365–73.
25. Mateu-Brull E, Rodrigo L, Peinado V, et al. Interchromosomal effect in carriers of translocations and inversions assessed by preimplantation genetic testing for structural rearrangements (PGT-SR). *Assist Reprod Genet* 2019;36(12):2547–55.
26. Nicholls PK, Schorle H, Naqvi S, et al. Mammalian germ cells are determined after PGC colonization of the nascent gonad. *Proc Natl Acad Sci USA* 2019;116 (51):25677–87.
27. Spiller C, Koopman P, Bowles J. Sex determination in the mammalian germline. *Annu Rev Genet* 2017;51:265–85.
28. Conti M, Franciosi F. Acquisition of oocyte competence to develop as an embryo: integrated nuclear and cytoplasmic events. *Hum Reprod Update* 2018;24(3):245–66.
29. Richani D, Dunning KR, Thompson JG, Gilchrist RB. Metabolic co-dependence of the oocyte and cumulus cells: essential role in determining oocyte developmental competence. *Hum Reprod Update* 2021;27(1):27–47.
30. Veitia RA. Primary ovarian insufficiency, meiosis and DNA repair. *Biomed J* 2020;43 (2):115–23.
31. Zielinska AP, Bellou E, Sharma N, et al. Meiotic kinetochores fragment into multiple lobes upon cohesin loss in aging eggs. *Curr Biol* 2019;29 (22):3749–65.e7.
32. Pacchierotti F, Eichenlaub-Ritter U. Environmental hazard in the aetiology of somatic and germ cell aneuploidy. *Cytogenet Genome Res* 2011;133(2–4):254–68.
33. Harper JC, Aittomäki K, Borry P, et al. Recent developments in genetics and medically assisted reproduction: from research to clinical applications. *Hum Reprod Open* 2017;2017(3):hox015.
34. Veller C, Kleckner N, Nowak MA. A rigorous measure of genome-wide genetic shuffling that takes into account crossover positions and Mendel's second law. *Proc Natl Acad Sci USA* 2019;116 (5):1659–68.

Chromosomes in Early Human Embryo Development: Incidence of Chromosomal Abnormalities, Underlying Mechanisms, and Consequences for Diagnosis and Development

Esther B. Baart, Effrosyni A. Chavli, and Diane Van Opstal

4.1 Introduction

Reproduction in humans is considered to be a relatively inefficient process, as the chance of achieving a spontaneous pregnancy after timed intercourse is only approximately 30%. This is much lower than the 70–90% estimated for other species, such as the rhesus monkey, the captive baboon, or rodents and rabbits. The inefficiency of human reproduction is mainly explained by the high incidence of preclinical losses, an estimated 40–60% of all conceptions. Early pregnancy loss is mainly explained due to the occurrence of chromosome abnormalities, which have been identified in most spontaneous abortion samples investigated.

The introduction of assisted reproductive technology (ART) and specifically in vitro fertilization (IVF) has allowed better insight into human early embryo development and it has become clear that chromosomal abnormalities identified in abortion material from in vivo conceptions are also frequently identified in preimplantation embryos generated by IVF. This indicates that chromosome instability is an inherent feature of human conceptions. So far, maternal age remains the only etiologic risk factor identified for chromosomal aneuploidy [1]. However, increasing evidence has made it clear that most preimplantation embryos do not have a uniform chromosomal constitution in all cells and are said to be mosaic as a result of postmeiotic errors in chromosome segregation. Although the mechanisms underlying this phenomenon and its consequences for developmental potential of the preimplantation embryo are still poorly understood, intensified research efforts over the last few years have begun to shed some light into this "black box" of early pregnancy.

This chapter aims to provide insight into the type and frequency of chromosomal abnormalities commonly found in human embryos before and after implantation. It explores molecular mechanisms that may contribute to the observed high error rate, and to what extent these errors could be induced by IVF procedures. The developmental capacity of mosaic embryos is investigated by analyzing in vitro data on blastocyst development. Furthermore, by giving an overview of the type and frequency of chromosome abnormalities observed after implantation, we aim to investigate the implications of chromosomal mosaicism for implantation and further embryonic development.

4.2 Chromosome Abnormalities in Human Preimplantation Embryos

4.2.1 Contribution of Meiotic Errors to Embryo Aneuploidy and Consequences for Embryo Development

Despite the high frequency and clinical importance of human aneuploidy, the underlying mechanisms leading to an abnormal chromosome constitution remain poorly understood. As detailed in Chapter 3, the incidence of aneuploidy is strongly related to maternal age and is also dependent on the stage of development: only 0.3% of newborns are aneuploid, which increases to 4% in stillbirths, and 35% when spontaneous abortions are investigated. Interestingly, the types of abnormalities that are observed also differ. Among newborns and stillbirths, the most common abnormalities are trisomy of chromosome 13, 18, or 21, or sex-chromosomal aneuploidies (i.e. 45,X, 47,XXX, 47,XXY, and 47,XYY). In contrast, trisomies of all chromosomes have been described in spontaneous abortions, with the most common being trisomy 15, 16, 21, and 22. With the exception of trisomy 21, these aberrations are lethal early in pregnancy, and only allow fetal survival beyond the first trimester of pregnancy if present in mosaic form. In contrast, trisomy 13, 18, and 21 and sex-chromosomal aneuploidy can be tolerated and may lead to the birth of an affected child. The only significant monosomy observed at the different stages of development is 45,X and this condition accounts for at least 10% of all spontaneous abortions [2].

By using analysis of DNA polymorphisms in trisomic cases, the origin of the extra chromosome can be examined. Results from such studies reveal that, depending on the chromosome investigated, there is variation in the parental origin (maternal or paternal) and also in the meiotic division where the segregation error occurred. For example, paternal meiotic errors are observed to account for 30–50% of cases of 47, XXY and trisomy 2, but they are rarely observed in other trisomies. Trisomy 16 is almost exclusively maternal in origin and seems to be solely derived from errors during meiosis I, whereas trisomy 18 frequently arises due to an error during maternal meiosis II. However, despite the variation, errors during maternal meiosis I overall explain most of the abnormalities observed [2].

These observations in clinically recognized pregnancies were confirmed by analyses of oocytes and sperm cells. Aneuploidy was found in only 1–4% of sperm, but varied between 10 and 70% in human oocytes, depending on maternal age and method of investigation. Interestingly, although abnormalities involving all chromosomes have been observed in oocytes, the most common aneuploidies were those involving chromosomes 15, 16, 18, 21, and 22 [2]. Thus findings in oocytes correlate well with findings in liveborn infants and pregnancy losses. From this it can be concluded that meiotic aneuploidy originates mostly during maternal meiosis I, arises in an age-dependent manner, and involves particular chromosomes more frequently.

However, as the application of IVF made human preimplantation embryos available for chromosome analysis, it soon became clear that the contribution of meiotic errors to preimplantation embryo aneuploidy is exceeded by errors arising during the first cleavage divisions.

4.2.2 Incidence of Aneuploidy and Mosaicism in Preimplantation Embryos

Most of our early knowledge concerning the chromosomal constitution of human preimplantation embryos came from the analysis of cleavage-stage embryos by preimplantation genetic testing for aneuploidy performed 3 days after fertilization, when embryos are usually composed of 6–10 blastomeres (Figure 4.1). At that time, molecular cytogenetic analysis of interphase nuclei by fluorescence in situ hybridization (FISH) was the most frequently used technique for the analysis of chromosomal abnormalities in human embryos. These abnormalities can arise from an error during meiosis, resulting in a uniform abnormality present in all cells, or from segregation errors occurring during the first mitotic divisions. The latter event results in chromosomal mosaicism, defined as the coexistence of karyotypically distinct cell lineages derived from a single zygote. Mosaic embryos can be composed of a

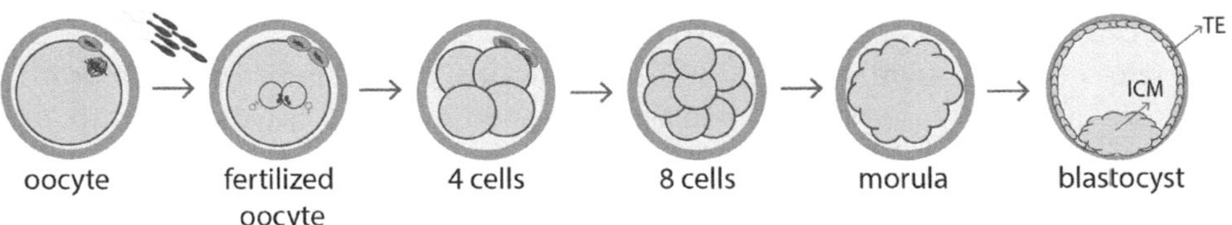

Figure 4.1 Schematic representation of human embryo development in vitro. Following fertilization, the human zygote undergoes eight to nine rounds of cell division before implantation. During this process, maternal RNA transcripts are degraded, the zygotic epigenome undergoes reprogramming, and the embryonic genome is activated. After the cleavage divisions, the embryo undergoes compaction and then the first lineage specification results in formation of the blastocyst, comprising an outer layer of polarized epithelial cells, the trophectoderm (TE); a compact inner cell mass (ICM); and a fluid-filled cavity, the blastocoel.

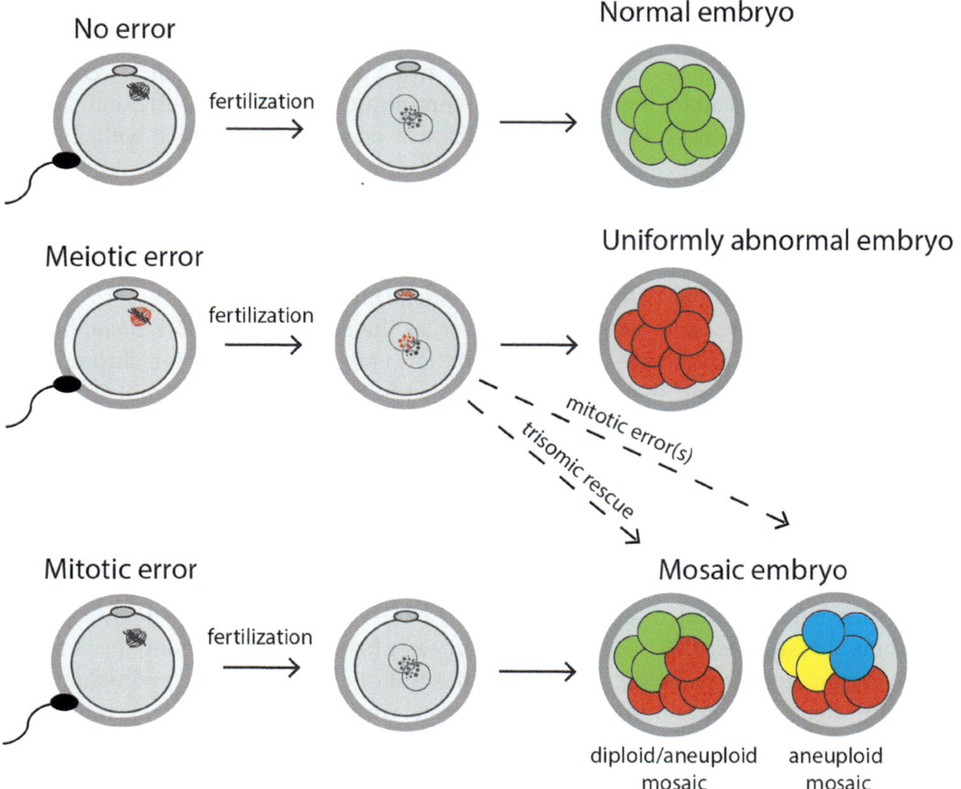

Figure 4.2 Schematic representation of human cleavage stage preimplantation embryos classified according to their chromosomal constitution. If no chromosome segregation errors occur during male or female meiosis and the subsequent cleavage divisions, the embryo will be uniformly normal. If an error occurs during either male (sperm) or female (oocyte) meiosis, one pronucleus will be abnormal in the zygote, resulting in a uniformly abnormal embryo if no further errors occur. If an error or errors occur during the cleavage divisions, the embryo will be diploid/aneuploid mosaic with one or more abnormal cell lines. If an error occurs during meiosis, followed by one or more further errors during the cleavage divisions, the embryo will be aneuploid mosaic with two or more abnormal cell lines. If the mitotic error results in correction of the meiotic error (e.g. loss of the chromosome involved in a trisomy), this is referred to as trisomic rescue.

mixture of chromosomally normal and abnormal cells, referred to as diploid–aneuploid mosaic. Sometimes normal cells occur together with two or more different abnormal cells, referred to as complex mosaic. Mosaic embryos can also be composed of abnormal cells only, but with different abnormalities, referred to as aneuploid mosaic or chaotic, if all cells show different and complex chromosomal abnormalities (Figure 4.2). The rate of mosaic embryos described in the literature is highly variable (15–90%). This depends on the method of chromosome analysis, the number of chromosomes analyzed, the developmental stage of the embryo, and the definition of mosaicism. A systematic review and meta-analysis of studies analyzing the chromosomal constitution of all cells of 815 human embryos at the cleavage stage showed 22% to be uniformly diploid, while only 5% were either uniformly abnormal (i.e. aneuploid, haploid, polyploid) or complex abnormal [3]. Chromosomal mosaicism was found in 73% of the embryos. Within the group of mosaic embryos, 80% were found to be diploid–aneuploid mosaic, with the remainder showing aneuploid mosaicism, the result of a mitotic segregation error in a zygote already carrying a meiotic error.

The first reports on chromosomal mosaicism in human embryos were much debated. It was hypothesized that the observed high error rate was due to technical artifacts of the FISH procedure, potentially leading to overlapping or nonspecific signals and hybridization failure, as proper control material was lacking. However, several studies using novel comprehensive chromosome screening (CCS) techniques that allow the screening of all chromosomes, including

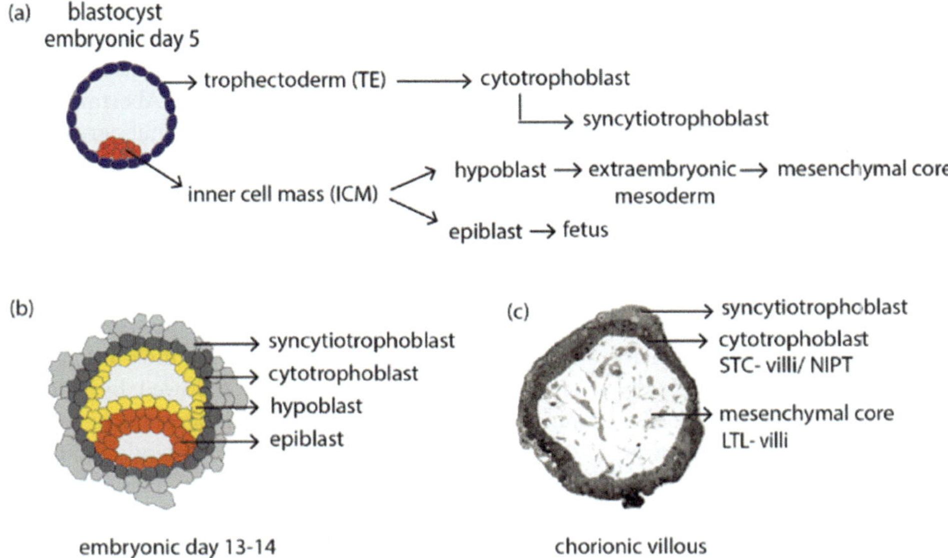

Figure 4.3 Schematic representations of lineage differentiation during early human development before and after implantation. (a) The first lineage specification results in formation of the blastocyst, comprising an outer layer of polarized epithelial cells, the trophectoderm (TE), and a compact inner cell mass (ICM). The extraembryonic TE develops into the two tissues of the fetal part of the placenta (syncytiotrophoblast and cytotrophoblast), while during the second lineage specification the ICM will form the epiblast and the hypoblast. The epiblast later gives rise to the fetus, whereas the hypoblast gives rise to the extraembryonic mesoderm from which the mesenchymal core of chorionic villi is derived. (b) Schematic representation of the different germ layers in a postimplantation human embryo 13–14 days after fertilization. (c) Cross-section through a chorionic villus during the first trimester of pregnancy. Short-term cultured villi (STC-villi), as well as noninvasive prenatal testing using cell-free (cf)DNA screening (NIPT), investigate the cytotrophoblast of the chorionic villi (derived from TE). The long-term preparation method (long-term cultured villi, LTC-villi) investigate the mesenchymal core of chorionic villi (derived from extraembryonic mesoderm).

quantitative polymerase chain reaction (qPCR), array comparative genomic hybridization (aCGH), single nucleotide polymorphism microarrays (SNP array) and next generation sequencing (NGS), have confirmed the high incidence of chromosomal mosaicism at the cleavage stage. In a large set of more than 28 000 biopsied day 3 embryos analyzed with SNP array, about half of the embryos were found to be diploid–aneuploid mosaic, and 15 % to be aneuploid mosaic [4,5].

Improved culture media formulations and technologies providing more stable culture conditions have made it possible to safely culture human IVF embryos until day 5 after fertilization. Four days after fertilization, the embryo undergoes compaction to form a morula, and one day later a fluid-filled cavity forms and the blastocyst stage is reached (Figure 4.1). Novel procedures for cryopreservation made it possible to cryopreserve embryos at this stage. In a blastocyst, the first two embryonic lineages have formed, consisting of a compact inner cell mass (ICM) and outer layer of polarized epithelial cells, the trophectoderm (TE). The TE will give rise to the placenta, while during the second lineage specification the ICM will give rise to the epiblast and the hypoblast that will lead to the fetus and yolk sac, respectively (Figure 4.3). Early studies using surplus embryos donated for research reported a decrease in the incidence of aneuploid and mosaic embryos as development continues to the morula and blastocyst stages and the proportion of chromosomally abnormal cells present within the embryo declines [4,6]. More recent studies using CCS methods confirm these early observations, although the detection of chromosomal mosaicism is not always straightforward (as discussed in Section 4.3.3). Studies investigating multiple biopsies obtained from embryos donated for research reported a mosaicism rate ranging from 14 to 50%, indicating that a significant proportion of blastocysts is still aneuploid or chromosomally mosaic [5].

Alongside numerical chromosome abnormalities, CCS techniques also revealed a high prevalence of segmental aneuploidies in human preimplantation embryos, where parts of a chromosome are duplicated or deleted. These partial aneuploidies are thought to be the result of DNA double-strand breaks followed by erroneous repair in the zygote or the two-cell embryo. It seems that chromosomes harbor fragile sites that are more frequently involved in these breakages. Studies analyzing blastocysts with high resolution techniques (NGS, aCGH) estimate the presence of segmental aneuploidy in approximately 15% of the cases. Around 70% of these segmental aneuploidies exhibit a mosaic profile after CCS analysis, making it technically challenging to reliably distinguish these two biological phenomena [5]. The clinical significance of segmental aberrations is not well understood yet.

In conclusion, although errors during female meiosis are thought to explain most of the aneuploidy observed in stillbirths and liveborns, this is not the case for the aneuploidies observed in preimplantation embryos. Detailed analysis of preimplantation embryos from young women undergoing IVF indicated that only 10–20% of the abnormalities observed can be explained by a chromosome segregation error during meiosis [4]. The remaining proportion of abnormalities observed in embryos can only be explained by errors occurring during the first cell divisions, indicating that chromosome segregation in human preimplantation development is especially error-prone.

4.2.3 Mitotic Mechanisms Leading to Chromosomal Abnormalities

As stated above, most of the abnormalities observed in human cleavage-stage embryos are the result of postmeiotic events. As illustrated in Figure 4.4, losses and gains of whole chromosomes during mitosis can be explained in several ways. Mitotic nondisjunction occurs when during mitosis the sister chromatids do not separate properly or have separated prematurely. Both defects can result in aberrant segregation, where one daughter cell ends up with both chromatids and the other with the reciprocal loss. Alternatively, nondisjunction can also be the result of defective attachment of the chromosome to the spindle in combination with a weakened mitotic checkpoint that fails to prevent anaphase before all chromosomes are properly attached to the spindle in a bipolar fashion (Figure 4.4). Anaphase lagging occurs when improper attachments of the chromosome to the spindle have formed, in which one sister chromatid is simultaneously attached to microtubules emanating from both poles. This will result in one chromatid being pulled correctly to one spindle pole, whereas the other chromatid is pulled from both directions and remains at the spindle midzone. With ingression of the cleavage furrow it can be lost, so one daughter cell will have the normal chromosome complement, whereas the other daughter cell has a chromosome loss. Aberrant cytokinesis can also cause chromosome malsegregation. Cells can skip cytokinesis or undergo cell fusion, resulting in polyploidization. This can lead to the formation of a multipolar spindle in a subsequent division, giving rise to complex abnormalities in the daughter cells.

Analysis of the abnormalities observed within mosaic embryos shows that they are most often consistent with anaphase lagging, followed by nondisjunction. Both mechanisms can also occur within the same embryo and are also frequently found in combination with meiotic errors [4]. This can result in a mitotic error that in fact rescues a cell with a trisomy if the extra chromosome is lost through anaphase lagging (trisomic rescue) (Figure 4.2). Alternatively, a cell with a monosomy can be corrected if a nondisjunction event for the same chromosome occurs. Both events can result in both homologs to stem from the same parent, a condition termed uniparental disomy (UPD; see also Section 4.5.4).

4.3 Molecular Mechanisms Regulating Chromosome Segregation during Preimplantation Embryo Development

Research into the origin of embryo aneuploidy has for a long time focused on the contribution of the oocyte, and a link between maternal age and an increased risk for meiotic errors has been well established. Meiotic errors in oogenesis are currently explained by a "two-hit" hypothesis: homologous chromosomes can form more susceptible crossover configurations during the meiotic pairing process, resulting in segregation errors as the result of multifactorial perturbations induced by aging in the oocyte's spindle and checkpoint function [1,2]. However, this model does not explain the high rate of chromosomal mosaicism in human embryos, which can occur independently of meiotic segregation errors. Moreover, no indications have been found that the incidence of mitotic segregation errors during the cleavage divisions increases with maternal age [4].

An abnormal chromosome number is extremely detrimental to the cell, and intricate mechanisms have evolved to segregate chromosomes equally to the

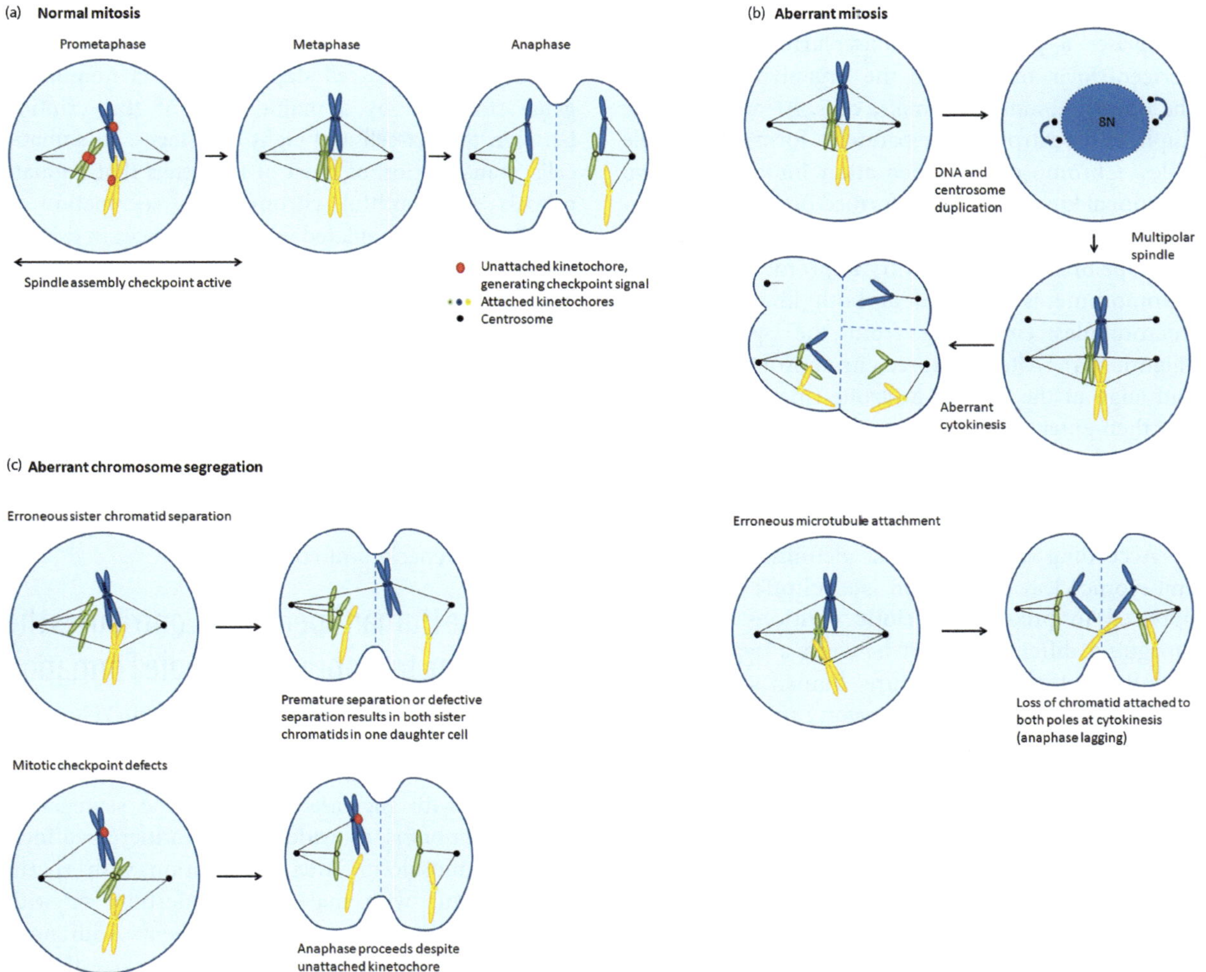

Figure 4.4 (a) Schematic representation of normal chromosome alignment and segregation. At prometaphase, the spindle assembly checkpoint is activated at each unattached kinetochore. Microtubule capture of both kinetochores of a chromosome results in silencing of the checkpoint and chromosome alignment at the metaphase plate. Anaphase is not initiated until all chromosomes have achieved bipolar attachment to the spindle and congression to the spindle midzone. During anaphase, the sister chromatids are pulled apart. After invagination of the cell membrane and completion of cytokinesis, the sister chromatids are equally separated to the two daughter cells. (b) Losses and gains of whole chromosomes during mitosis can be explained in several ways. Aberrant mitosis can cause chromosome loss. Too many centrosomes can result from errors in centrosome duplication or skipping cytokinesis, producing polyploidization (8N). Multiple centrosomes induce multipolar spindles and chromosome malsegregation. (c) Mitotic nondisjunction results in reciprocal loss and gain of a chromosome in the two daughter cells. This can be caused by erroneous sister chromatid separation (chromatids separate prematurely or not at all) or mitotic checkpoint defects that fail to prevent anaphase before all chromosomes are properly attached. Anaphase lagging results in loss of a sister chromatid from one daughter cell and can be the result of erroneous microtubule attachment (merotelic attachment).

daughter cells. Normal somatic cells have a very low segregation error rate and the importance of this is illustrated by the fact that it takes about 3×10^{13} cell divisions to give rise to a human. That chromosome segregation is extremely error-prone in the early human embryo is largely counterintuitive, as these early divisions are crucial for embryo survival and reproductive success. In the following sections we explore novel insights into chromosome segregation regulation in early mammalian embryos, and how it differs from somatic cells, in order to understand the biological basis for these errors.

4.3.1 The First Embryonic Cleavage Divisions Differ from Normal Mitotic Divisions

The formation of the bipolar spindle is a crucial event for cell division, and ensures correct chromosome segregation to daughter cells. The centrosome, which comprises a pair of centrioles surrounded by the pericentriolar material, is the organizing center of the mitotic spindle in somatic cells. At prophase, the duplicated centrosomes separate to form two spindle poles. Chromosome condensation initiates and two functional kinetochores are formed on the sister chromatids at the primary constriction. After nuclear envelope breakdown and entry to prometaphase, each chromosome is captured at both kinetochores by microtubules emanating from the spindle poles (Figure 4.4). After capture, chromosomes congress and align at the spindle midzone (metaphase). Cells can then enter anaphase where sister chromatids are separated and equally pulled to the daughter cells. The chromatin decondenses (telophase) and the nuclear envelope reforms while cytokinesis is completed.

According to the classic definition, the first embryonic cleavage division is a mitotic division, as opposed to preceding meiotic divisions. However, intriguing differences exist between a zygote and a somatic mitotic cell. Mature human oocytes are arrested at metaphase of the second meiotic division (MII). Upon fertilization by a sperm cell, the oocyte completes the second meiotic division and the highly condensed chromatin of the sperm nucleus decondenses, resulting in the formation of the haploid paternal pronucleus. The parental pronuclei are maintained physically separate in the ooplasm during the subsequent G1, S, and G2 phases of the first embryonic cell cycle. Upon entry into mitosis, maternal and paternal chromatin is condensed into chromosomes that align for the first time at a common metaphase plate.

The cleavage divisions that subsequently follow are termed "reductive." This means that the cells do not undergo cellular growth after the division but decrease in size as embryo development continues toward the blastocyst stage (Figure 4.1).

4.3.2 Fertilization and the Transition from Maternal to Embryonic Control

Oocytes and early embryos are transcriptionally inactive. During oogenesis, mammalian oocytes store an abundance of mRNAs, proteins, and macromolecular structures in the ooplasm to sustain the first cell cycles following fertilization and to facilitate the transition from maternal to embryonic control, a process called embryonic genome activation (EGA). In humans, EGA was believed to occur at the four- to eight-cell stage, but recent reports suggest that it starts as early as the two-cell stage for a select number of genes, followed by a major wave of transcription between the six-cell and eight-cell stages. In somatic cells, transcription of most of the genes that regulate mitosis and faithful chromosome segregation is tightly cell cycle regulated and disturbances in expression levels and timing cause chromosome instability. The quality and quantity of this maternal store of transcripts and proteins in the oocyte is therefore likely to be crucial for chromosome segregation regulation during the first embryonic divisions. In line with this, recent studies using single-cell mRNA profiling in blastomeres in relation to embryo aneuploidy and developmental potential have reported differential expression of genes related to cell cycle regulation and mitotic checkpoint control.

4.3.3 The Human Sperm Cell Contributes the Centrosome to Embryonic Spindle Formation

Compared with meiotic errors in the oocyte, aneuploidy rates in sperm are very low. However, compromised spermatogenesis has been reported to be associated with increased chromosome segregation errors in embryos, as evidenced by an increased incidence of mosaicism after fertilization with sperm from patients with male factor infertility or with testicular retrieved sperm from patients with non-obstructive azoospermia. This raises the question of whether the sperm contributes to chromosome malsegregation in the early embryo. Interestingly, mammalian oocytes, except for mouse oocytes, lack the centrosome needed to form a bipolar spindle. During fertilization in humans, the centrosome is contributed by the sperm cell and contributes to spindle assembly in the zygote (Figure 4.5).

Centrioles form the core of the centrosome and the number of centrioles is critical for proper chromosome segregation. Just like chromosomes, their number is reduced during gametogenesis. Centrosomal components are remodeled during spermiogenesis and this is necessary to transmit a functional centrosome to the embryo. In humans, this reduction results in each mature spermatozoon carrying two remodeled centrioles that combine with proteins provided by the oocyte to generate a reconstituted centrosome in the zygote. It was recently shown in mouse that the centrosome

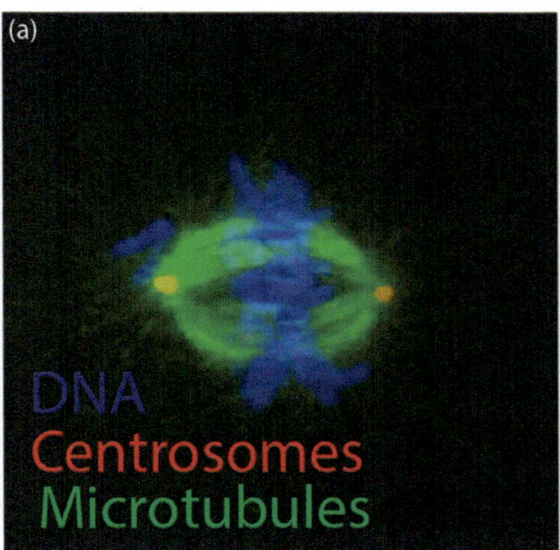

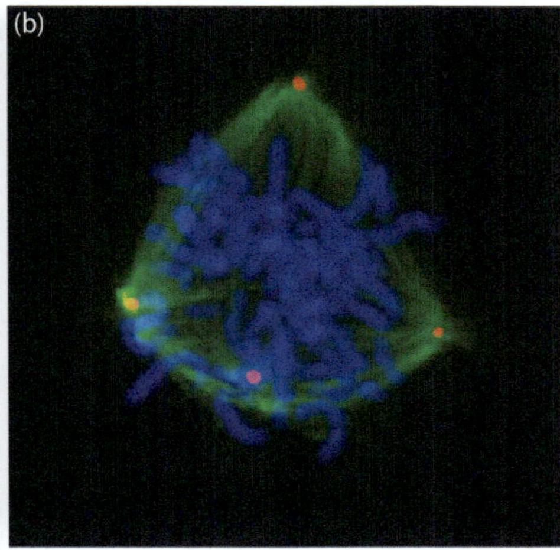

Figure 4.5 Confocal images of human tripronuclear zygotes at prometaphase, showing chromosomes (DNA, blue) together with immunolocalization of tubulin protein in the mitotic spindle (α-tubulin, green) and pericentrin, a protein present in the centrosome (red/yellow). Z-stack images were merged into a single image. (a) A tripronuclear zygote fertilized by one sperm cell followed by failure to extrude the second polar body (digynic zygote), showing a spindle with two centrosomes and two spindle poles. (b) A tripronuclear zygote fertilized by two sperm cells simultaneously (diandric zygote) with chromosomes in the process of congressing on a spindle that displays multiple poles formed by four centrosomes. Source: images courtesy of Kaylee Holleman.

reduction process is not completed as sperm cells leave the testis but continues during transit through the epididymis. Thus, testicular sperm cells or sperm cells from patients with disturbances in spermiogenesis may harbor centrosomes that have undergone an incomplete remodeling process.

Dispermic human zygotes are fertilized by two sperm cells simultaneously and inherit an extra centrosome delivered by the second sperm cell (Figure 4.5). An increased incidence of mosaicism is observed in dispermic human zygotes compared with monospermic or digynic embryos. Extra centrosomes have been shown to induce transient multipolar spindles. Although these poles can fuse to reform a bipolar spindle, this can result in chromosomes where one sister kinetochore attaches to both spindle poles, resulting in anaphase lagging. Thus, the transfer of an incompletely reduced centrosome from the sperm to the oocyte is likely to contribute to spindle formation disturbances and therefore to chromosome segregation errors in the embryo. Multipolar spindles are frequently observed at all stages of human preimplantation development and the presence of multiple or defective centrosomes may thus contribute to the incidence of complex aneuploidies (Figure 4.4).

Mouse zygotes do not inherit the centrosome from the sperm cell. Intriguing experiments using novel live imaging techniques in mouse zygotes have shown that two bipolar spindles form just before the pronuclei break down [7]. Each spindle then captures and aligns either the maternal or paternal chromosomes. These two spindles fuse their poles before anaphase to become one spindle. Defects in alignment of the two spindles were shown to increase chromosome segregation errors. Spindle formation in human zygotes has so far not been studied in such detail, but such studies are needed to further understand the origin of chromosome segregation errors.

4.3.4 Compromised Functionality of the Spindle Assembly Checkpoint?

The cell's most important defense mechanism against aneuploidy is the so-called spindle assembly checkpoint (SAC). This is a complex surveillance mechanism that delays anaphase until biorientation of all chromosomes at the metaphase plate. The SAC is composed of Mad1, Mad2, and Bub3 proteins and BubR1, Bub1, and Mps1 kinases. When activated, it generates a biochemical inhibitory signal that stops cells from entering anaphase. Once all pairs of sister kinetochores are properly captured by the spindle microtubules and are under tension, the SAC is silenced, resulting in sister-chromatid separation and cells being able to enter anaphase (Figure 4.4).

The SAC improves chances of a successful mitosis by delaying anaphase to provide the cell with an opportunity to correct attachment errors

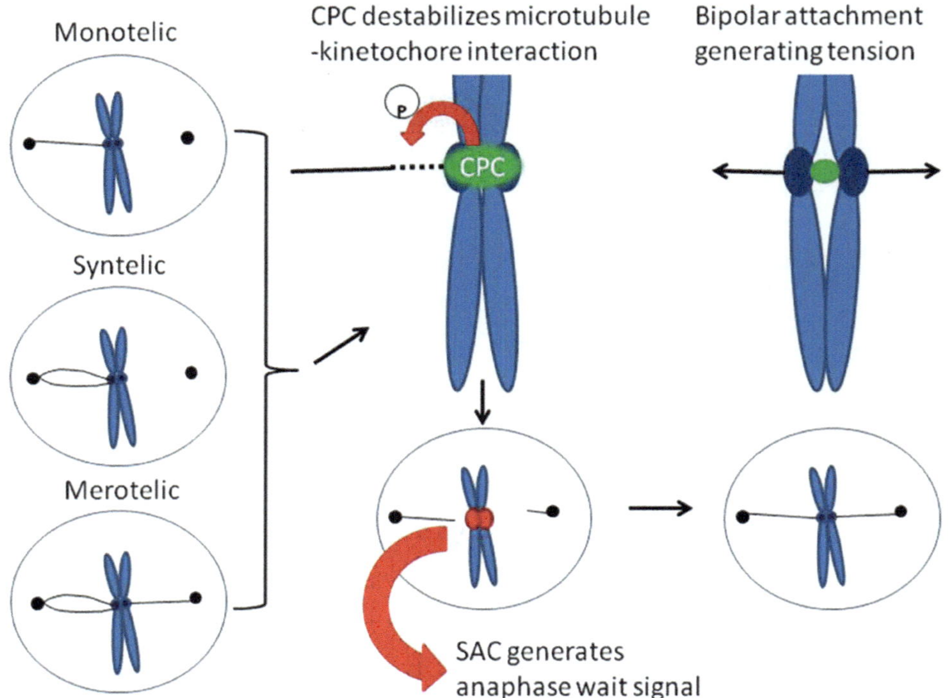

Figure 4.6 Schematic representation of erroneous, nonbipolar, attachments and how they are corrected by joint action of the chromosomal passenger complex (CPC) and the spindle assembly checkpoint (SAC). The CPC localizes to the region between the kinetochores and phosphorylates several proteins present in the kinetochore, resulting in destabilization of microtubule–kinetochore attachments. The CPC thereby generates unattached kinetochores that activate the SAC. The SAC in turn prevents anaphase, providing the chromosome the opportunity to form new attachments. As soon as bipolar attachment is achieved, tension is generated across the kinetochores, pulling them apart. The CPC can no longer reach its targets, enabling microtubule–kinetochore attachments to become stable and resulting in deactivation of the SAC.

(Figure 4.6). However, it is not necessary for the completion of mitosis. The high incidence of chromosome aneuploidies and mosaicism during the first cleavage divisions in human embryos has been proposed to result from a dysfunctional SAC, allowing embryonic cells to divide while not all chromosomes have achieved proper attachment. However, human embryos readily arrest at prometaphase of the cell cycle when microtubule–kinetochore interactions are disturbed by treatment with spindle poisons [8]. Although this is a strong indication of an active SAC, the checkpoint can in fact be weakened. The biochemical signal produced by each unattached kinetochore can be quantitatively reduced, resulting in lower concentrations of signal-producing components. This creates a situation in which more than one unattached kinetochore is needed to produce enough signal to inhibit anaphase onset, and in which chromosome separation can occur with unaligned chromosomes. In mammalian oocytes, this was shown to be caused (in part) by the large cytoplasmic volume that characterizes the oocyte, effectively diluting the signal [9]. As early embryos have a similarly large cytoplasmic volume and rely on the maternal store of proteins and mRNAs, the dilution of kinetochore signaling to the SAC may be a property inherited from the oocyte.

Experimental evidence supported this "cytoplasmic volume" hypothesis in embryos from several nonmammalian species. However, a careful examination in mouse embryos where the cytoplasmic volume was experimentally diminished demonstrated that functioning of the SAC was not related to volume [9]. Unattached kinetochores were readily detected by the SAC, but this did not delay anaphase regardless of the cytoplasmic volume. Weak SAC activity thus seems to be an inherent feature of mouse embryos, although this rarely results in aneuploidy. It stands to reason that human embryos also suffer from weakened SAC activity, possibly in combination with other problems that in the end do result in aneuploidy. However, this remains to be directly assessed.

The observation that the SAC recognizes unattached chromosomes, but fails to delay anaphase, may implicate the malfunction of some proteins in the SAC pathway in early embryos. Bub1 kinase was shown to be an essential component of the SAC, functioning to generate the anaphase halt signal after the detection of unattached kinetochores by Mad1 and Mad2. Interestingly, experimental studies in rhesus macaque oocytes have demonstrated oocyte quality related differences in Bub1 expression. Bub1 and Bub3 mRNAs were also reported to be downregulated in human zygotes of reduced developmental potential (see McCoy [4] and references therein). This illustrates that oocytes may not have acquired an optimal maternal store of relevant transcripts during oocyte growth and maturation, and this can be expected to have an impact on chromosome segregation in the embryo. We can easily imagine that such a phenomenon may arise during ovarian stimulation, as explained in more detail below, and/or that some subfertile women will be more prone to produce such oocytes with reduced mitotic competency.

Although having a weakened SAC at this early stage of life seems entirely counterproductive, there is a possible explanation. From time lapse studies it has become clear that human embryos with the highest implantation potential show highly synchronized cleavage divisions. If each cell required a different amount of time to correct chromosome attachment errors and complete mitosis, synchronization would be rapidly lost. Thus, it is feasible that an accurate chromosome segregation error control mechanism was sacrificed in favor of cell cycle synchronization.

4.3.5 Distinct Constitution of the Chromosomal Passenger Complex during the Cleavage Divisions

Kinetochore capture by the mitotic spindle is a trial-and-error process. While unattached chromosomes activate the SAC to delay anaphase until biorientation is achieved, erroneous attachments cannot induce an SAC response. Therefore, non-bipolar microtubule–kinetochore interactions are destabilized by the chromosomal passenger complex (CPC) to allow attachment correction (Figure 4.6). This complex normally consists of Aurora B kinase, survivin, borealin, and INCENP. To be able to function properly, targeting of this complex to the region between the kinetochores is crucial. One of the kinases involved in the SAC, Bub1, is essential for correct localization of the CPC by providing one of the recruiting marks on the chromatin underlying the centromeric region. In turn, the CPC plays an important role in proper activation of the SAC. Once correct microtubule–kinetochore connections are formed, the CPC normally stops destabilizing them.

In human zygotes, Aurora C is the main active subunit in the CPC as opposed to Aurora B in normal mitotic cell division [8]. Aurora C was first identified in the testis and oocytes and has been coined the meiotic counterpart of Aurora B. It has been shown to compensate for Aurora B to support mitotic progression in somatic cells and mouse embryos. However, structural differences between the proteins exist, with so far unknown implications for CPC localization and/or function. Experiments in oocytes have demonstrated that both Aurora kinases are necessary for correct completion of the meiotic divisions, and that their activities need to be balanced. Interestingly, research in oocytes also shows that Aurora kinase B/C activity does not stop once stable attachments are formed, but that it remains actively destabilizing microtubule connections until the oocyte completes the meiotic division, increasing chances of chromosome segregation errors. If Aurora kinase B/C activity functions similarly in human cleavage-stage embryos, it is likely to exacerbate chromosome segregation errors, especially in combination with a weakened SAC and spindle formation abnormalities.

4.4 Clinical Implications of the High Rate of Chromosomal Abnormalities in IVF Embryos

4.4.1 Impact of IVF Procedures on Chromosome Abnormalities and Mosaicism

Our knowledge of the high rate of chromosomal abnormalities is almost exclusively derived from embryos generated by IVF. It is therefore possible that these abnormalities are induced by ovarian stimulation and/or suboptimal in vitro culture conditions. Following recruitment into the growing pool, the oocyte expands from 35 to 120 μm in diameter, which represents a 100-fold increase in volume over a period of several months. Oocyte growth and maturation are linked to follicle development, and bidirectional signaling occurs between oocytes and granulosa cells. Oocytes must achieve both nuclear and cytoplasmic maturity in order to sustain the early stages of embryonic development. Control of the late stages of follicle development lies primarily with follicle-stimulating hormone (FSH) and luteinizing hormone (LH), but granulosa cell-derived growth factors play a major role in modulating the action of these gonadotropins during follicle development. The orderly expression of these somatic and oocyte-derived factors allows the follicle to acquire specific properties at a precise time and in a specific sequence.

So far, it is largely unknown how ovarian stimulation and the development of multiple preovulatory follicles interfere with these intraovarian factors. Experimental evidence from studies in mice shows that disturbances in the complex interplay of signals that regulate folliculogenesis alter the late stages of oocyte growth, increasing the risk of chromosome malsegregation in subsequent meiotic divisions [2]. Human IVF studies confirm that ovarian stimulation affects chromosomal competence, and a prospective comparison of different stimulation regimens suggests that this mainly affects the incidence of postzygotic errors at the cleavage stage. Significant variation in rates of mosaicism was also reported after a comparison of four different IVF laboratories, including a change in mosaicism rates within one center after switching the ovarian stimulation protocol. However, more recent comparisons after preimplantation genetic testing for aneuploidies (PGT-A) at the blastocyst stage reported no differences related to the ovarian stimulation protocol [5]. Nevertheless, if the protocols used have a similar effect, this does not exclude an effect of ovarian stimulation on oocyte competence. Because IVF procedures without any form of hormonal stimulation to time oocyte pick-up are exceedingly impractical, the impact of ovarian stimulation is difficult to investigate.

Ovarian stimulation may also not be the only contributing factor, as a high rate of mosaicism is also observed in embryos obtained from young women undergoing natural cycle IVF [4]. The differences in the incidence of chromosomal mosaicism may also reflect variation in culture conditions, such as pH and composition of the culture medium, temperature differences, air quality, and oxygen concentration. Supraphysiologic oxygen levels (i.e. atmospheric oxygen levels) have been shown to increase malsegregation in cleavage-stage in vitro cultured embryos from a nondisjunction-prone mouse model. However, a study investigating PGT-A results from in vivo produced blastocysts obtained after ovarian stimulation, intrauterine insemination and uterine lavage from fertile women reported no differences in aneuploidy and mosaicism rates to in vitro generated blastocysts [10].

Chromosomal mosaicism is not exclusively observed in human embryos. Incidences of 20–40% of chromosomal mosaicism have also been described in bovine, equine, porcine, and nonhuman primate embryos. Interestingly, although mosaicism was observed in embryos from these species produced both in vitro and in vivo, the proportion of mosaic embryos and the proportion of aberrant cells was significantly lower in the in vivo produced embryos (see references in Avo Santos et al. [8]). From this we can conclude that, in humans and other nonrodent mammals, chromosome segregation in cells of the preimplantation embryo is predisposed to errors and that this can be exacerbated by (suboptimal) IVF procedures.

4.4.2 Implications of Chromosomal Mosaicism for Preimplantation Genetic Testing for Aneuploidies

As an increasing body of evidence suggested that the incidence of chromosomal abnormalities in IVF embryos was extremely high, it also became clear that good embryo morphology did not necessarily exclude an abnormal chromosomal constitution. It was hypothesized that screening of IVF embryos for chromosomal abnormalities before transfer would increase pregnancy rates after IVF. This led to the introduction of PGT-A, where biopsy of day 3 embryos and the removal of one or two blastomeres was combined with FISH analysis to determine the chromosome copy number of selected chromosomes. However, the observed high rate of diploid–aneuploid mosaicism undermined the reliability of the PGT-A diagnosis, as in this case the blastomere biopsied is not representative of the remaining embryo. If a diploid cell is biopsied from a diploid–aneuploid mosaic embryo, it will be transferred, while the remaining

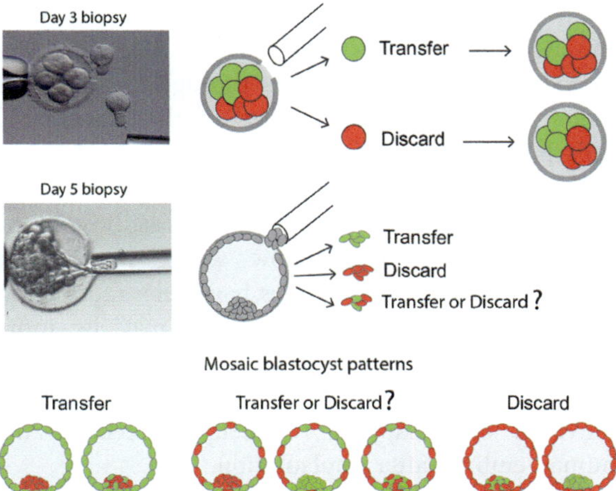

Figure 4.7 Impact of chromosomal mosaicism on the reliability of a PGT-A diagnosis when performed at the cleavage stage (3 days after fertilization) or the blastocyst stage (5 days after fertilization). If a normal cell is biopsied from a mosaic embryo at the cleavage stage, the decision will be to transfer, while in fact the proportion of abnormal cells will increase. Vice versa if an abnormal cell is biopsied, and the decision to transfer or discard. At the blastocyst stage, embryos can still be chromosomally mosaic, with various possible patterns. Mosaicism can also be detected within the biopsied sample, complicating the decision to transfer or discard.

embryo in fact contains aneuploid cells. Moreover, the proportion of diploid cells decreased as a result of the biopsy. If an aneuploid cell is biopsied, the embryo is not transferred, while the proportion of diploid cells has in fact increased (Figure 4.7). Furthermore, the limited number of chromosomes that could be screened by FISH analysis constituted an additional factor undermining the reliability of PGT-A. Although initially positive effects on implantation and ongoing pregnancy rates were reported in retrospective studies, a meta-analysis of nine randomized controlled trials in 2011 clearly demonstrated the lack of a positive effect on IVF outcomes after PGT-A on day 3 embryos [11], after which this approach was largely abandoned.

New developments regarding embryo culture and the development of CCS techniques made it worthwhile to revisit preimplantation genetic testing. For PGT-A at the blastocyst stage, 5–10 cells can safely be removed from the TE. The possibility of analyzing more than one or two cells in combination with CCS techniques has greatly reduced the technical error rate. Despite the elimination of technical issues, chromosomal mosaicism remains an issue at the blastocyst stage and our knowledge about the diagnostic efficacy and accuracy of PGT-A is still limited.

The assessment of the true incidence of chromosomal mosaicism at the blastocyst stage remains extremely complicated and undermined by technical and biological limitations. Although all CCS techniques enable the genomic screening of all chromosomes, each of them has a different detection threshold regarding mosaicism. NGS reportedly has the greatest resolution and sensitivity in identifying chromosomal abnormalities in a small group of cells and has become the most common approach for PGT-A at the blastocyst stage [5] (see also Chapter 13 on PGT). Nevertheless, the simultaneous analysis of a multiple cell biopsy undermines the detection of intra-sample mosaicism. From a biological perspective, abnormal cells are not always equally distributed in the blastocyst. The timing of the segregation error defines the degree of mosaicism and the affected cell lineages, resulting in diverse mosaic patterns (Figure 4.7).

The representability of a TE biopsy for the remaining embryo, and especially the ICM that will later form the fetus, has been investigated in embryos donated for research by NGS analysis of the chromosomal constitution of both cell lineages. A combined analysis of different blastocyst disaggregation studies including 338 blastocysts showed the ICM and TE to be concordant in 77% of the embryos [5]. From these, 21% were found to be uniformly normal, and 67% uniformly abnormal. The remaining 12% showed diverse mosaic patterns in both TE and ICM. From the embryos with discordant results, in 23 embryos the TE was solely affected whereas in seven embryos chromosomal abnormalities were involved only in the ICM. Five embryos consisted of an abnormal TE and a mosaic ICM or vice versa, while seven embryos showed a mosaic pattern in both cell layers. As most of these embryos were previously diagnosed as abnormal by PGT-A, the average concordance rate between ICM and TE might be biased.

The prevalence of chromosomal mosaicism at the blastocyst stage can undermine the diagnostic accuracy of PGT-A in two ways. Firstly, as a single TE biopsy does not always accurately represent the chromosomal constitution of the entire embryo, PGT-A might result in an erroneous classification. When the biopsied TE cells seem to be affected, the rest of the embryo might be normal. In such a case, an embryo with possible developmental potential will be discarded. Inversely, a normal TE sample does not guarantee the absence of chromosomal aberrations in the ICM (Figure 4.7). Secondly, due to intra-sample mosaicism some embryos are characterized as mosaic. The destination of such an embryo is questionable, and the decision to transfer or discard depends on the clinical management of each IVF center.

4.5 Impact of Chromosomal Mosaicism on Embryo Development and Implantation

4.5.1 Development from the Cleavage Stage to the Peri-implantation Blastocyst

Our knowledge about the fate and impact of abnormal cells during early embryo development remains limited. Early studies showed that the proportion of human embryos that can be classified as mosaic declines from over 80% of day 4 embryos to less than 60% of day 5 developing blastocysts and subsequently decreases even further to 42% of day 8 postimplantation embryos [6]. Also, the proportion of chromosomally abnormal cells within an embryo declines. An early model proposed by Evsikov and Verlinsky in 1998 suggested the self-elimination of the whole embryo if the number of aneuploid cells at the morula stage reaches a certain threshold level, while embryos under this threshold of aneuploid cells can reach the blastocyst stage. Embryo arrest, cell arrest, apoptosis, and preferential proliferation of normal cells are the proposed explanations for the depletion of abnormal cells in embryos.

In recent years, efforts have been made to elucidate whether there is a definite selection against specific chromosomal abnormalities during human embryo development. Investigation of the chromosomal constitution of more than 28 000 day 3 blastomere biopsies and more than 6000 TE biopsies showed varying profiles of whole chromosome abnormalities between cleavage and blastocyst stage embryos [4]. Complex chromosomal abnormalities were less frequently observed at the blastocyst stage, indicating an early arrest of complex abnormal embryos. Furthermore, the coexistence of multiple chromosomal losses was largely eliminated from day 5 samples, suggesting a selection against embryos carrying monosomic cell lines.

Recent work in mouse embryos, where an aneuploid cell line was experimentally introduced by forming a chimera, demonstrated that these induced mosaic embryos are able to implant [12]. The aneuploid cells are largely depleted through the course of embryo development. The fate of these aneuploid cells has been shown to depend on the embryonic lineage: aneuploid cells in the fetal lineage (ICM) are actively eliminated by apoptosis, whereas those in the placental lineage (TE) show severe proliferative defects. As no viable trisomies have been reported in mice, mouse embryo development may be more intolerant to aneuploidy. It therefore remains unknown to what extent these findings are representative of what happens to aneuploid cells in the human embryo after implantation.

To shed some light on this question, novel developments in culture techniques have enabled in vitro implantation conditions, allowing embryo culture until 12 days after fertilization. This system was used to culture blastocysts donated for research after PGT-A diagnosis. After extended culture and in vitro implantation the chromosomal constitution was reanalyzed. Interestingly, 70% of the embryos diagnosed as mosaic on day 5 were found to be chromosomally normal on day 12. Moreover, embryos carrying trisomies, duplications, or mosaic aberrations were more likely to survive until day 12 in comparison to embryos with monosomies, deletions, or complex chromosomal constitutions [13].

The global uptake of PGT-A is increasing, and more clinics proceed to the transfer of embryos diagnosed as mosaic. Based on the limited information that is available, the transfer of mosaic embryos can result in healthy live births. On the other hand, transfer of these embryos is associated with decreased implantation and pregnancy potential, as well as increased risk of genetic abnormalities and adverse pregnancy outcomes. This may be related to the meiotic or mitotic origin of the aberration, the proportion of abnormal cells within the embryo, the abnormality involved, the chromosome(s) involved, parental origin of the aberration, or the presence of aberrant cells within the TE or ICM lineage.

Efforts are being made to reach a consensus to define which mosaic embryos could qualify for transfer. For instance, mosaic embryos consisting of chromosomal abnormalities associated with a risk for clinically significant fetal UPD (i.e. both copies of a chromosome pair derived from one parent only), severe intrauterine growth retardation, or liveborn syndromes should be avoided [14]. However, this approach remains ambiguous and the fate of mosaic blastocysts after implantation remains uncertain. To explore this question, in the following sections we provide an overview of what is known about the incidence of chromosomal mosaicism after implantation and during further embryonic and fetal development.

4.5.2 The Incidence of Chromosomal Mosaicism in Chorionic Villi and Amniotic Fluid during Prenatal Diagnosis

Chromosomal mosaicism is a well-known phenomenon in chorionic villi. Although fetus and placenta originate from the same zygote, their chromosomal constitution can be different; this was discovered soon after the introduction of chorionic villus sampling (CVS) for prenatal diagnosis in the first trimester of pregnancy. There are numerous reports of abnormal karyotypes in chorionic villi that were later not confirmed in fetal cells. Chorionic villi can be processed for classical cytogenetic studies in two ways: the direct or semi-direct technique, also called short-term cultured villi (STC-villi) and long-term cultured villi (LTC-villi). These techniques differ in the origin of the cells that are investigated in the cytogenetic preparations: the cells in STC-villi are derived from the cytotrophoblast of chorionic villi and those of LTC-villi are predominantly of the mesenchymal core. Cytotrophoblast and mesenchymal core have different embryonic origins: the cytotrophoblast is derived from the TE of the blastocyst, whereas the mesenchymal core is derived from the ICM, and more particularly the hypoblast (see Bianchi et al. [15] and references therein) (Figure 4.3). Chromosome analysis of both cell types in fact represents the investigation of two different embryonic compartments that may be chromosomally different because of postzygotic mitotic division errors. Because of its embryonic origin, the karyotype of the mesenchymal core better represents the karyotype of the fetus than that of the cytotrophoblast.

Conventional prenatal cytogenetic diagnosis in first-trimester chorionic villi revealed chromosomal mosaicism in about 2% of the samples. The time and place of a postzygotic mitotic error during embryonic development, as well as the mitotic or meiotic origin of the chromosome aberration, will determine the pattern of mosaicism. While early division errors of an originally normal zygote may cause generalized mosaicism involving both placental and fetal

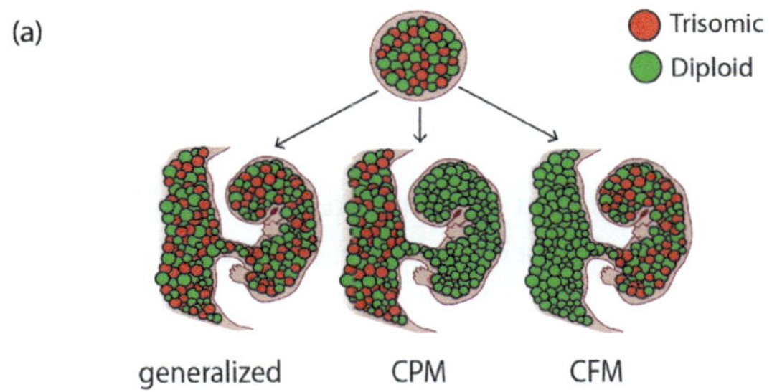

(b)

	Cytotrophoblast	Mesenchymal core
CPM type 1	(mosaic) abnormal	normal
CPM type 2	normal	(mosaic) abnormal
CPM type 3	(mosaic) abnormal	(mosaic) abnormal

Figure 4.8 (a) Types of chromosomal mosaicism as identified during prenatal diagnosis. In cases of generalized mosaicism the chromosomal abnormality is found in both placenta and fetus. In confined placental mosaicism (CPM) the chromosomal abnormality is limited to the placenta and in confined fetal mosaicism (CFM) to the fetus. (b) CPM can be classified into three subtypes depending on the cell lineage(s) affected.

compartments, later errors, affecting specific cell lineages, will lead to confined placental mosaicism (CPM) and, more rarely, confined fetal mosaicism (CFM) (Figure 4.8a). According to which cell line is affected, three types of CPM are distinguished: CPM I, with a chromosome aberration restricted to the cytotrophoblast; CPM II, with the chromosome aberration present in only mesenchymal core; and CPM III, when both cell lineages are affected (Figure 4.8b).

Nowadays, chromosomal microarray analysis (CMA) of DNA isolated from uncultured chorionic villi has in many cases replaced conventional cytogenetic techniques. CMA may involve aCGH, enabling the detection of copy number variations (CNVs) below the resolution of conventional karyotyping (<10 Mb), and genome-wide SNP arrays, which use a combination of genomic dosage and genotyping data. The dual property of this tool has proven to be very useful for detection and quantification of chromosomal mosaicism even below 5%, among other advantages [16]. CMA has the added advantages of high-throughput analysis, ability to automate, rapid turnaround time, and the avoidance of culturing fetal cells and therefore prevention of cell selection in favor of normal cells but also of culture artifacts leading to pseudomosaicism. In most cases, DNA is isolated from the whole chorionic villus, representing a mixture of cytotrophoblast and mesenchymal core or from mesenchymal core alone. The current preference for CMA seems to have led to the abandonment of the separate analysis of cytotrophoblast and mesenchymal core, hampering the comparison of results on the incidence of mosaicism between studies using conventional and molecular techniques. Based on a higher sensitivity of CMA for detection of mosaicism, it could be expected that the detection rate of chromosomal mosaicism would increase. Three studies have specifically evaluated the incidence of mosaicism using CMA by adapting routine settings for CNV detection with array in order to improve mosaicism detection. In comparison to the 2% reported after karyotyping, CMA studies reported frequencies of chromosomal mosaicism of 1.78%, 3%, and 4.1%, respectively (see Lund et al. [17] and references therein). In conclusion, these studies using the most sensitive cytogenetic technologies with adapted detection criteria did not reveal a much higher incidence of chromosomal mosaicism than karyotyping studies.

If chromosomal mosaicism is encountered in chorionic villi, it mostly represents CPM with a fetal confirmation rate of about 10–28%, primarily depending on the cell type in which the chromosome abnormality is seen (cytotrophoblast and/or mesenchymal core) and on the type of chromosome abnormality [18,19]. CPM I has a low risk for fetal involvement, and therefore follow-up cytogenetic investigations in amniotic fluid cells to verify the fetal chromosome constitution may be skipped. On the other hand, CPM II or III, with an affected mesenchymal core originating from the ICM of the blastocyst, as does the fetus, has a higher risk of being confirmed in the fetus as well. However, with the introduction of CMA the separate analysis of both cell layers seems to have been abandoned by most groups. This may lead to unnecessary follow-up amniocentesis in some pregnancies in which only the cytotrophoblast is affected. Additionally, true fetal mosaicism that is present in the mesenchymal core may go undetected due to dilution with a normal cytotrophoblast.

Confined placental mosaicism involving a trisomy, which accounts for about 50% of all CPM cases, may result from a somatic duplication of a whole chromosome in placental progenitor cells originating from a diploid zygote, or from a trisomic conceptus with loss of the extra chromosome in embryonic but not placental progenitor cells. This phenomenon is called trisomic zygote rescue and is associated with a theoretical one in three risk of UPD in the normal cell line [20]. The incidence of UPD in a series of 24 unselected cases of CPM involving a trisomy was 4%, confirming the mitotic nature of most CPM cases of trisomy [21]. Although the majority of pregnancies with CPM proceed uneventfully, CPM is associated with an increased risk for intrauterine growth retardation, fetal loss, or poor perinatal outcome. This is potentially caused by disturbed placental function, UPD (if an imprinted chromosome is involved or if homozygosity of a recessive allele leads to a recessive disease), or hidden fetal mosaicism with the abnormal cell line in fact also present in fetal tissues. CPM III is especially associated with pregnancy complications. The aberration involved in CPM III is most likely to have a meiotic origin, as the chromosome aberration is present in both cytotrophoblast and mesenchymal core. The normal karyotype of the fetus is likely the result of trisomic rescue (Figure 4.2) [22].

Chromosomal mosaicism in amniotic fluid cell cultures, as determined with classical cytogenetic techniques as well as CMA, is even rarer than it is in chorionic villi. The frequency ranges from 0.1 to 0.3% with a fetal confirmation rate of about 70% [23].

4.5.3 Noninvasive Detection of Chromosomal Mosaicism Using Cell-Free DNA Screening

For many years CPM could only be detected by CVS or placental studies after birth. However, nowadays noninvasive prenatal testing (NIPT) for detecting fetal chromosome aberrations in cell-free DNA (cfDNA) of maternal plasma may also reveal CPM. NIPT investigates cfDNA fragments of which the majority has a maternal origin and studies have shown that the fetal cfDNA fraction is derived from the apoptotic cytotrophoblast of chorionic villi. Therefore, the NIPT result in fact reflects the cytogenetic result from STC-villi (Figure 4.3). This means that NIPT can detect both CPM I and III. In fact, chromosomal mosaicism is the major cause of discordant NIPT results and the reason that NIPT is a screening and not a diagnostic test, requiring diagnostic follow-up investigations such as CVS or amniocentesis in the case of an abnormal result. Additionally, when the

NIPT is normal, there is a small risk for false-negative results due to CFM (Figure 4.8a).

NIPT mostly involves a targeted test for detection of the common aneuploidies, trisomy 13, 18 and 21, with or without the sex chromosomes. However, some groups use a genome-wide approach with NGS allowing investigation of other chromosome aberrations as well, such as structural chromosome aberrations, but also trisomies of chromosomes other than 13, 18, and 21. These trisomies are called rare autosomal trisomies and these are mostly confined to the placenta (CPM). The incidence of these rare trisomies in 10 genome-wide NIPT studies (N = 196 662 cases) was found to vary between 0.14 and 1.03%, depending on the referral reason, with a mean of 0.32%. This was significantly lower than the incidence of 0.41% in 57 538 CVS trophoblast samples [24]. The authors of this study concluded that the sensitivity of NIPT for detection of CPM I and III is significantly lower than when using CVS. Another study found a higher proportion of mosaicism for the common aneuploidies with NIPT as compared to conventional karyotyping, indicating a higher sensitivity of NIPT for placental mosaicism detection [25]. The use of study cohorts with probably different a priori risk figures for CPM may explain the conflicting results of both studies. Based on first-trimester chorionic villi and term placental cytogenetic studies, there is evidence that NIPT is probably more sensitive for detection of CPM involving the cytotrophoblast than CVS. It is assumed that the entire placental trophoblast sheds cfDNA into the maternal circulation and that a CPM restricted to smaller placental areas may be detected by NIPT and missed by CVS, although this will depend on factors such as fetal fraction (e.g. fetal part of the cfDNA), which will determine the sensitivity of the test [26]. However, more studies will be necessary to gain insight into the representation of the placental chromosomal constitution in the cfDNA fraction in cases of CPM and about the sensitivity of NIPT to detect it. Nevertheless, based on these data it is predicted that the prevalence of CPM as detected with cfDNA will not significantly exceed the 1–2% established with prenatal diagnosis in chorionic villi.

4.5.4 Chromosomal Mosaicism in the Term Placenta

For confirmation of abnormal NIPT results that are not confirmed in the fetus, placental cytogenetic studies are indicated. Some of these studies revealed several striking observations previously made in preimplantation embryos, but hardly described in prenatal or postnatal clinical cytogenetics. In 10 cases where NIPT revealed a trisomy, the trisomic cell line could be confirmed in the mosaic placenta in all cases [27]. Moreover, in 3 of 10 cases, there were two different normal cell lines, apart from the trisomic cell line: one showing a UPD of the involved (trisomic) chromosome and the other a normal biparental disomy (BPD). UPD/BPD mosaicism is extremely rare, and has only been described in a few postnatal cases of Prader–Willi and Silver–Russell syndromes. Nevertheless, in term placentas of cases in which NIPT revealed a rare autosomal trisomy, it seems to be rather common. This type of mosaicism indicates that trisomic rescue may involve multiple rescue events during early embryogenesis.

Moreover, in a few cases in which NIPT showed a structural chromosome aberration, SNP array analysis of different placental biopsies revealed complex chromosomal mosaicism involving multiple cell lines with different structural chromosome aberrations. This seemed to demonstrate the mitotic chromosome instability seen in cleavage-stage embryos. It is possible that the placenta reflects the cytogenetic chaos in the cleavage-stage embryo and functions as a litter basket for chromosomally abnormal cell lines that arose during early embryonic development. However, these placental studies were conducted in cases with abnormal NIPT results, indicative for the presence of a chromosome aberration in the placenta. On the contrary, an investigation of 111 placentas from 49 IVF and 62 natural random pregnancies with CMA only revealed 15 large (>100 kb) de novo CNVs, of which only three involved mosaic (partial) trisomies [28]. This led the authors to conclude that the high levels of chromosomal mosaicism in IVF cleavage-stage embryos are not preserved at later stages of prenatal development.

4.5.5 Chromosomal Mosaicism in Chorionic Villi and Placentas of Spontaneous versus IVF Pregnancies

As discussed above, it is important to realize that the high incidence of mosaicism may partly be the result of IVF procedures, but whether this high frequency is also found with in vivo-generated embryos is not known. Therefore, it could be hypothesized that the incidence of chromosomal mosaicism found during prenatal diagnosis is higher in IVF compared with non-IVF pregnancies. A few studies have investigated the incidence of CPM as detected with karyotyping in first-trimester chorionic villi in IVF/intracytoplasmic sperm injection (ICSI) pregnancies as compared to that in spontaneous conceptions and found no significant difference [29]. This observation was recently extended to first-trimester missed abortions, studied with SNP array, in which no differences were found in the incidence of chromosomal mosaicism between ART (3.6%) and natural conception group (4%) [30]. Also no difference was found in the incidence of de novo (sub)microscopic DNA imbalances with CMA on fetal and placental tissues collected after birth of naturally and IVF-conceived liveborn neonates [28]. Thus, mosaicism associated with ART persists beyond the preimplantation embryo at a rate similar to that associated with pregnancies conceived spontaneously. By the end of the first trimester and at term no difference is found in the rate of mosaicism between IVF and non-IVF pregnancies. Whether a difference exists during early pregnancy remains to be investigated.

4.5.6 Chromosomal Mosaicism in Spontaneous Abortions

About 15% of clinically recognized pregnancies end up in a first-trimester spontaneous abortion, mostly due to a chromosome aberration. In a recent review of studies reporting cytogenetic findings with conventional karyotyping, an incidence of chromosome aberrations of 45% was found [31]. Mosaicism accounted for less than 6% of the chromosome abnormalities in this study. Because of the high incidence of culture failures and maternal cell contamination, which may account for about 30% of all 46,XX abortion samples [31], researchers have searched for other techniques that are able to determine cytogenetic abnormalities in uncultured abortion tissue. A few studies on tissues of spontaneous abortions used interphase FISH and these concluded that at least 50% of spontaneous abortions with an abnormal karyotype have mosaic forms of chromosome abnormalities [32]. However, this high incidence has not been confirmed by numerous studies using aCGH and SNP array, which revealed chromosomal mosaicism in about 4%, comparable to conventional karyotyping studies (see Li et al. [30] and references therein).

4.5.7 Do We Underestimate Chromosomal Mosaicism beyond the Preimplantation Embryo Stage?

It is important to mention that the incidence of chromosomal mosaicism as determined by classical cytogenetic techniques may be underestimated for several reasons. These figures are mostly determined in the course of cytogenetic diagnosis by using karyotyping of banded chromosomes. This usually involves the analysis of a limited number of metaphases. For instance, for prenatal diagnosis about 10 metaphases are usually investigated, excluding a level of mosaicism of 26% with 95% confidence. This means that low-level mosaicism (<26%) is disregarded during routine chromosome studies, probably leading to an underestimation of the real frequency of mosaicism in chorionic villi and amniotic fluid. In addition, classical cytogenetic techniques make use of cell cultures, because they depend on metaphases for chromosome analysis. In vitro cell cultures may change the in vivo level of mosaicism in favor of normal cells and therefore mosaicism can go undetected. Moreover, in cases of abortion tissue, there is a high culture failure rate and maternal cell contamination may lead to false-negative results. Finally, the resolution of karyotyping is limited to 5–10 Mb and therefore mosaicism of some structural chromosomal rearrangements that appear to be common in preimplantation embryos can easily be missed during prenatal diagnosis if smaller than 10 Mb.

However, the routine introduction of more sensitive cytogenetic techniques such as aCGH and especially SNP arrays in prenatal and postnatal diagnosis did not substantially increase the incidence of chromosomal mosaicism. One of the reasons might be the use of routine detection criteria. One study on spontaneous abortions was able to detect mosaicism in 10%, but only by lowering the algorithm thresholds [33]. But lowering the algorithm thresholds means that specificity also decreases. Therefore, if mosaicism is not suspected, or is not searched for, a laboratory will not use these specific criteria, potentially leading to underestimation of the true incidence of mosaicism. Another reason might be that chromosomal mosaicism may be tissue-limited and may remain undetected if the affected tissues are not investigated. For karyotyping of products of conception, mostly cultured chorionic villi are used, which involves cytogenetic analysis of predominantly the mesenchymal core of the villi that is derived from the compartment of the extraembryonic mesoderm [15]. By doing so, the karyotype of cytotrophoblast and the embryo itself is left undetermined. Sometimes only direct preparations of chorionic villi have been investigated, determining the karyotype of the cytotrophoblast while just a few studies investigated both placental cell lineages. Unfortunately, the embryo itself is frequently not investigated. The existence of discordant cytogenetic results between direct and long-term cultured chorionic villi, cultured amniotic membrane, and chorionic plate was detected in 20% of 54 miscarriages [34]. Thus, by analysis of only one compartment in most spontaneous abortion studies, mosaicism can easily go undetected. Depending on type of chromosomal trisomy, the timing of the mitotic segregation error, and the mitotic or meiotic origin of the trisomy, different tissue-specific compartmentalization patterns can be found for the different autosomal trisomies.

Despite this knowledge, in most recent studies using aCGH or SNP array for prenatal diagnosis in chorionic villi or for cytogenetic diagnosis in products of conception or term placenta, DNA is isolated from the whole chorionic villi instead of from cytotrophoblast and mesenchymal core separately. This potentially dilutes the percentage of abnormal cells to fall beneath the detection level of both techniques, which may contribute to an underestimation of the real incidence of mosaicism. Therefore, more cytogenetic studies investigating all placental and embryonic compartments with techniques able to detect (low-level) mosaicism will be necessary to determine the actual incidence of chromosomal mosaicism. Perhaps eventually it will be shown that the extensive chromosomal mosaicism seen at the preimplantation stage largely disappears before a pregnancy is recognized.

4.6 Summary and Conclusions

Screening of human IVF-derived embryos for chromosomal aneuploidies before transfer has revealed that most embryos are chromosomally abnormal, with the bulk of abnormalities originating from errors during the first mitotic divisions of early preimplantation development, resulting in chromosomally mosaic embryos. This phenomenon undermines the reliability of PGT-A at the cleavage stage, but also affects PGT-A at the blastocyst stage. The high error rate in chromosome segregation is observed in most mammalian embryos, and it is likely that mechanisms normally ensuring a low error rate are sacrificed in order to meet the demands of synchronized cell cycle control in response to developmental cues.

The extent of chromosomal mosaicism observed in preimplantation embryos has so far not been reported in clinically recognized pregnancies. This indicates that at least some of the mosaic embryos are likely to arrest development. Indeed, the incidence of embryos carrying chromosomally abnormal cells has been found to decrease during development from the eight-cell to the peri-implantation blastocyst stage. This indicates that a proportion of mosaic embryos undergoes developmental arrest between compaction and blastocyst formation.

However, not all mosaic embryos will elicit a full embryonic arrest and a significant number of mosaic embryos continue to develop beyond implantation if transferred. Through the growth advantage of normal cells or elimination of abnormal cells, some of the mosaic embryos may even become normal or will show limited mosaicism that allows further development. Their fate may then depend on the abnormality involved or on yet unknown factors, such as distribution of abnormal cells over the different compartments of placenta and fetus or on their shape which may be irregular due to different growth velocities of normal and abnormal cells. Subsequently, a (pre)clinical pregnancy loss may occur or mosaicism (CPM or generalized) may be encountered prenatally in chorionic villi or amniotic fluid cells.

The incidence of chromosomal mosaicism in postnatal clinical cytogenetics is estimated to be about 1%. However, little is known about the presence of chromosomal mosaicism in different organs in children and adults, as only blood is routinely investigated in postnatal cytogenetic diagnosis. For instance, chromosomal mosaicism is increasingly recognized to play a role in brain development and disease [35]. However, the biopsy of different tissues is not routine practice in clinical cytogenetics. Perhaps the new technology of cfDNA screening with NGS may further contribute to the evaluation of chromosomal mosaicism in health and disease. Cell-free DNA screening in the context of NIPT has already revealed maternal constitutional chromosomal mosaicism while genomic DNA from blood was normal. The cfDNA in blood plasma in fact may originate from every organ that is in contact with the blood circulation. Therefore, through cfDNA screening organs can become more accessible for cytogenetic evaluation without the need for invasive procedures. Because of the high incidence encountered in preimplantation embryos, chromosomal mosaicism may be underestimated and therefore important biological questions remain concerning the origin and impact of chromosomal mosaicism in embryo development.

References

1. Gruhn JR, Zielinska AP, Shukla V, et al. Chromosome errors in human eggs shape natural fertility over reproductive life span. *Science* 2019;365:1466–9.
2. Nagaoka SI, Hassold TJ, Hunt PA. Human aneuploidy: mechanisms and new insights into an age-old problem. *Nat Rev Genet* 2012;13:493–504.
3. van Echten-Arends J, Mastenbroek S, Sikkema-Raddatz B, et al. Chromosomal mosaicism in human preimplantation embryos: a systematic review. *Hum Reprod Update* 2011;17:620–7.
4. McCoy RC. Mosaicism in preimplantation human embryos: when chromosomal abnormalities are the norm. *Trends Genet* 2017;33:448–63.
5. Popovic M, Dhaenens L, Boel A, Menten B, Heindryckx B. Chromosomal mosaicism in human blastocysts: the ultimate diagnostic dilemma. *Hum Reprod Update* 2020;26:313–34.
6. Santos MA, Teklenburg G, Macklon NS, et al. The fate of the mosaic embryo: chromosomal constitution and development of Day 4, 5 and 8 human embryos. *Hum Reprod* 2010;25(8):1916–26. http://dx.doi.org/10.1093/humrep/deq139.
7. Reichmann J, Nijmeijer B, Hossain MJ, et al. Dual-spindle formation in zygotes keeps parental genomes apart in early mammalian embryos. *Science* 2018;361:189–93.
8. Avo Santos M, Van De Werken C, De Vries M, et al. A role for Aurora C in the chromosomal passenger complex during human preimplantation embryo development. *Hum Reprod* 2011;26(7):1868–81. http://dx.doi.org/10.1093/humrep/der111.
9. Vázquez-Diez C, Paim LMG, FitzHarris G. Cell-size-independent spindle checkpoint failure underlies chromosome segregation error in mouse embryos. *Curr Biol* 2019;29(5):865–73.e3.
10. Munné S, Nakajima ST, Najmabadi S, et al. First PGT-A using human in vivo blastocysts recovered by uterine lavage: comparison with matched IVF embryo controls. *Hum Reprod* 2020;35(1):70–80. http://dx.doi.org/10.1093/humrep/dez24210.1093/humrep/dez242
11. Mastenbroek S, Twisk M, van der Veen F, Repping S. Preimplantation genetic screening: a systematic review and meta-analysis of RCTs. *Hum Reprod Update* 2011;17:454–66.
12. Bolton H, Graham SJL, Van der Aa N, Kumar P, Theunis K, Fernandez Gallardo E, Voet T, Zernicka-Goetz M. Mouse model of chromosome mosaicism reveals lineage-specific depletion of aneuploid cells and normal developmental potential. *Nat Commun* 2016 Mar 29;7:11165. doi: 10.1038/ncomms11165. PMID: 27021558; PMCID: PMC4820631.
13. Popovic M, Dhaenens L, Taelman J, et al. Extended in vitro culture of human embryos demonstrates the complex nature of diagnosing chromosomal mosaicism from a single trophectoderm biopsy. *Hum Reprod* 2019;34:758–69.
14. Grati FR, Gallazzi G, Branca L, et al. An evidence-based scoring system for prioritizing mosaic aneuploid embryos following preimplantation genetic screening. *Reprod Biomed Online* 2018;36:442–9.
15. Bianchi DW, Wilkins-Haug LE, Enders AC, Hay ED. Origin of extraembryonic mesoderm in experimental animals: relevance to chorionic mosaicism in humans. *Am J Med Genet* 1993;46:542–50.
16. Conlin LK, Thiel BD, Bonnemann CG, et al. Mechanisms of mosaicism, chimerism and uniparental disomy identified by single nucleotide polymorphism array analysis. *Hum Mol Genet* 2010;19:1263–75.
17. Lund ICB, Becher N, Christensen R, et al. Prevalence of mosaicism in uncultured chorionic villus samples after chromosomal microarray and clinical outcome in pregnancies affected by confined placental mosaicism. *Prenat Diagn* 2020;40:244–59.
18. Pittalis MC, Dalprà L, Torricelli F, et al. The predictive value of cytogenetic diagnosis after CVS based on 4860 cases with both direct and culture methods. *Prenat Diagn* 1994;14:267–78.
19. Battaglia P, Baroncini A, Mattarozzi A, et al. Cytogenetic follow-up of chromosomal mosaicism detected in first-trimester prenatal diagnosis. *Prenat Diagn* 2014;34:739–47.
20. Wolstenholme J. Confined placental mosaicism for trisomies 2, 3, 7, 8, 9, 16, and 22: their incidence, likely origins, and mechanisms for cell lineage compartmentalization. *Prenat Diagn* 1996;16:511–24.
21. Van Opstal D, Van den Berg C, Deelen WH, et al. Prospective prenatal investigations on potential uniparental disomy in cases of confined placental

trisomy. *Prenat Diagn* 1998;18:35–44.

22. Toutain J, Horovitz J, Saura R. Type 3 confined placental mosaicisms excluding trisomies 16 are also associated with adverse pregnancy outcomes. *Genet Med* 2020;22:446–7.
23. Hsu LY, Perlis TE. United States survey on chromosome mosaicism and pseudomosaicism in prenatal diagnosis. *Prenat Diagn* 1984;4(Spec No):97–130.
24. Benn P, Malvestiti F, Grimi B, et al. Rare autosomal trisomies: comparison of detection through cell-free DNA analysis and direct chromosome preparation of chorionic villus samples. *Ultrasound Obstet Gynecol* 2019;54(4):458–67.
25. Brison N, Neofytou M, Dehaspe L, et al. Predicting fetoplacental chromosomal mosaicism during non-invasive prenatal testing. *Prenat Diagn* 2018;38:258–66.
26. Van Opstal D, Eggenhuizen GM, Joosten M, et al. Non-invasive prenatal testing as compared to chorionic villus sampling is more sensitive for the detection of confined placental mosaicism involving the cytotrophoblast. *Prenat Diagn* 2020;40 (10):1338–42. http://dx.doi.org/10.1002/pd.5766
27. Van Opstal D, Diderich KEM, Joosten M, et al. Unexpected finding of uniparental disomy mosaicism in term placentas: is it a common feature in trisomic placentas? *Prenat Diagn* 2018;38:911–19.
28. Zamani Esteki M, Viltrop T, Tšuiko O, et al. In vitro fertilization does not increase the incidence of de novo copy number alterations in fetal and placental lineages. *Nat Med* 2019;25:1699–705.
29. Jacod BC, Lichtenbelt KD, Schuring-Blom GH, et al. Does confined placental mosaicism account for adverse perinatal outcomes in IVF pregnancies? *Hum Reprod* 2008;23:1107–12.
30. Li G, Jin H, Niu W, et al. Effect of assisted reproductive technology on the molecular karyotype of missed abortion tissues. *Biosci Rep* 2018;38(5):BSR20180605. http://dx.doi.org/10.1042/BSR20180605.
31. van den Berg MMJ, van Maarle MC, van Wely M, Goddijn M. Genetics of early miscarriage. *Biochim Biophys Acta* 2012;1822:1951–9.
32. Lebedev I. Mosaic aneuploidy in early fetal losses. *Cytogenet Genome Res* 2011;133:169–83.
33. Scott SA, Cohen N, Brandt T, et al. Detection of low-level mosaicism and placental mosaicism by oligonucleotide array comparative genomic hybridization. *Genet Med* 2010;12:85–92.
34. Kalousek DK, Barrett IJ, Gärtner AB. Spontaneous abortion and confined chromosomal mosaicism. *Hum Genet* 1992;88:642–6.
35. Vorsanova SG, Yurov YB, Iourov IY. Dynamic nature of somatic chromosomal mosaicism, genetic–environmental interactions and therapeutic opportunities in disease and aging. *Mol Cytogenet* 2020;13:16.

DNA Is Not the Whole Story: Transgenerational Epigenesis and Imprinting

Ashwini Balakrishnan and J. Richard Chaillet

5.1 Introduction

There are two types of genomic information accurately inherited or maintained following DNA replication: the entire genomic DNA sequence and epigenetic information in the form of patterns of CpG methylation on a subset of the genome. These DNA methylation patterns are crucially important for mammalian development, primarily because they regulate gene transcription. There are cyclical declines and increases in DNA methylation during gametogenesis and embryogenesis, and the oscillations in methylation occurring across generations must depend on de novo methylation and demethylation processes to rearrange the existing methylation patterns. Within any single reproductive cycle, changes in genomic methylation are the outcome of a linked sequence of active and passive processes that rearrange genomic methylation to generate epigenetic milestones, each with a defined future role. For example, genomic imprints are established during gametogenesis by de novo methylation to generate mature gametes with complete complements of paternal methylation imprints in sperm and maternal methylation imprints in oocytes. The establishment of a collective set of imprints within a sperm and an oocyte is an epigenetic milestone, whose purpose after sperm–oocyte fusion is to ensure monoallelic expression of imprinted genes during fetal development. We postulate that another essential epigenetic milestone is found in the blastocyst-stage embryo; this milestone is achieved at the generation, through the poorly understood process of epigenetic reprogramming, of pluripotent embryo stem cells, whose role is to contribute to the development of the conceptus. Primordial germ cells (PGCs), largely devoid of genomic methylation and poised to differentiate into gametes with sex-specific imprints, and adult stem cells, poised to differentiate into organs, would be other possible epigenetic milestones. The developmental locations of

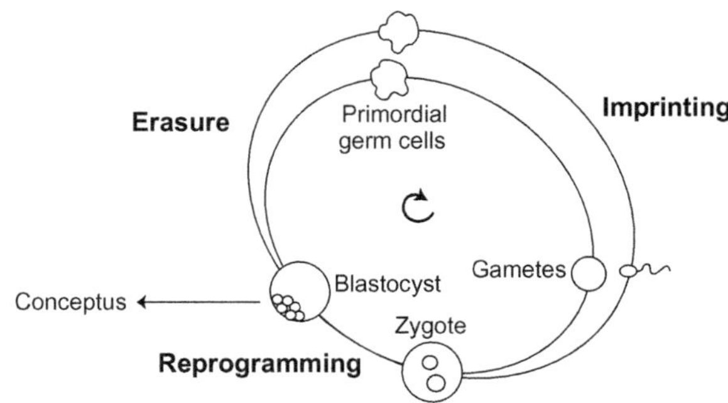

Figure 5.1 Depiction of the developmental locations of the fundamental epigenetic milestones. Erasure, imprinting, and reprogramming are three processes involving major rearrangement of CpG methylation patterns in the genome.

three fundamental milestones (gametes, blastocyst, and PGCs), and the epigenetic processes by which they are crafted, are depicted in Figure 5.1.

In this chapter, we describe the molecular processes underlying generation of epigenetic milestones, and discuss how these processes define much of what we know about transgenerational epigenesis. We will focus on genomic imprinting because much is known about the molecular mechanisms of genomic imprinting, and these mechanisms involve two critically important milestones, the gamete and the blastocyst. We will also discuss abnormalities in the processes of generating epigenetic milestones, and how these abnormalities can reduce or possibly even enhance transgenerational epigenesis.

5.2 DNA Is Not the Whole Story

The nonequivalence of the genome inherited from the mother versus the father was convincingly demonstrated in the early 1980s. Pronuclear transfer experiments in mice were performed to create embryos that contained either two copies of the maternal genome (gynogenotes) or two copies of the paternal genome (androgenotes) [1]. Both gynogenetic and androgenetic uniparental embryos were not viable and had opposing phenotypes. Gynogenetic embryos displayed more advanced embryonic development than androgenetic embryos whereas androgenetic embryos emphasized well-developed extraembryonic tissues. These observations indicated that the chromosomes inherited from the male and female parents are functionally different, and importantly, since this difference is independent of the embryo's sex, the genome differences must reside in the autosomes.

Mouse embryos with uniparental disomies, in which each conceptus inherits both copies of a particular chromosome from just one parent, beautifully validated the theory that functional differences exist between maternal and paternal versions of one or more autosomes. For example, mouse embryos with uniparental disomies of chromosome 11 had very different phenotypes depending on whether the chromosome was inherited maternally or paternally. Maternal disomic mice for chromosome 11 (and nullisomic for paternal chromosome 11) were smaller than their chromosomally normal littermates while paternal disomic mice for chromosome 11 (and nullisomic for maternal chromosome 11) were larger than their normal littermates [2]. This sex-specific inheritance effect indicated that the maternal alleles of one or more genes on chromosome 11 function differently than the opposite paternal alleles. We now know that the functional difference between maternal and paternal alleles of particular genes on chromosome 11 is because one allele of each of these genes is expressed (transcribed) and the other is not expressed. Assuming that the DNA sequence of these genes was not reversibly altered after passage through spermatogenesis and oogenesis, the two parental alleles of each gene must be epigenetically distinguished so that one parental allele is expressed and the other allele is silent. Many genes whose parental alleles function differently due to a parent-specific epigenetic footprint have now been definitively identified. They are known as imprinted genes and they are located on those chromosomes found to exhibit developmental abnormalities when inherited as maternal and/or paternal uniparental disomies. Moreover, the parental alleles exhibit unambiguous differences in DNA methylation, a heritable form of genome modification. A list of mouse imprinted genes and their chromosomal location can be found on the MouseBook website [3].

5.3 Establishment of Genomic Imprints

5.3.1 Nature of an Imprint

If functions of the two haploid genomes during embryogenesis are truly strictly determined by their parental origins, then imprints should have certain inalienable features. They must mark two genetically identical alleles differently in mature gametes, and following fertilization imprints should be maintained long enough to affect development of the conceptus. Furthermore, because maternal chromosomes become paternal chromosomes during spermatogenesis and paternal chromosomes become maternal chromosomes during oogenesis, the perpetuation (inheritance) of imprints must be disrupted in every generation before germ cells commit to either the male or female lineage (i.e. in PGCs) to ensure a switch in parental form.

The patterns of DNA methylation observed during development fulfill all of the aforementioned features of the genomic imprinting process. Precise biochemical mechanisms are in place for both the establishment and maintenance of methylation (Figure 5.2), which ensures the accurate formation and then perpetuation of patterns of genomic methylation following DNA replication. In this scheme, unmethylated symmetric CpGs in DNA are converted to fully methylated CpGs by de novo methyltransferases. The symmetric arrangement of methylated CpG dinucleotides on complementary strands of fully methylated DNA in a cell is reproduced in both daughter cells by DNA replication followed immediately thereafter by maintenance methylation, carried out by a maintenance methyltransferase enzyme.

There are tremendous cyclical changes in the average percentage of total genomic CpGs that are methylated during development (Figure 5.3a). There are two prominent hiatuses in methylation, one in undifferentiated PGCs and the other in blastocyst-stage embryos. Between these two low points, pronounced and apparently monotonic increases in methylation occur during gametogenesis and postimplantation embryogenesis. Notably, any given methylation pattern or imprint that was established on just one parental allele during gametogenesis is maintained during the fall in preimplantation genomic methylation (Figure 5.3b). The important genomic locations of inherited imprinted methylation are called differentially methylated domains (DMDs) because one

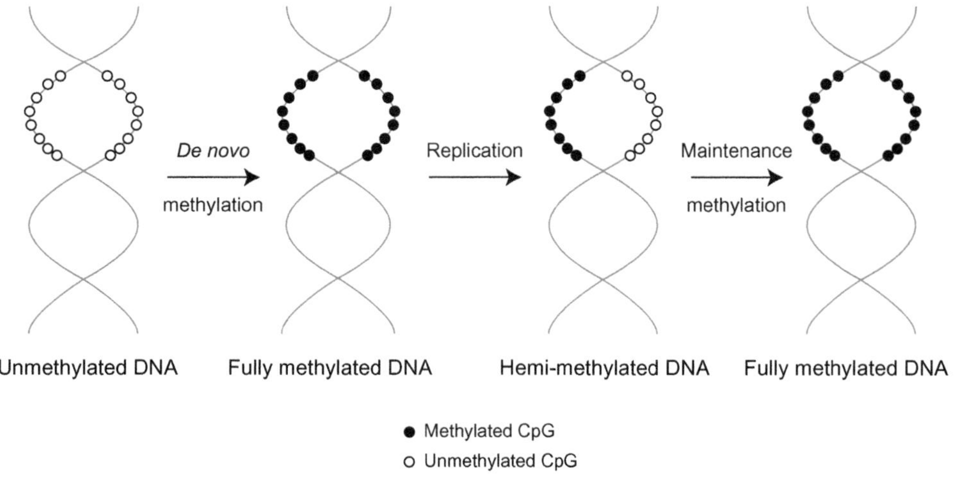

Figure 5.2 Biochemical processes of de novo methylation (conversion of unmethylated DNA to fully methylated DNA) and maintenance methylation (conversion of hemi-methylated to fully methylated DNA) following DNA replication in a CpG-rich region.

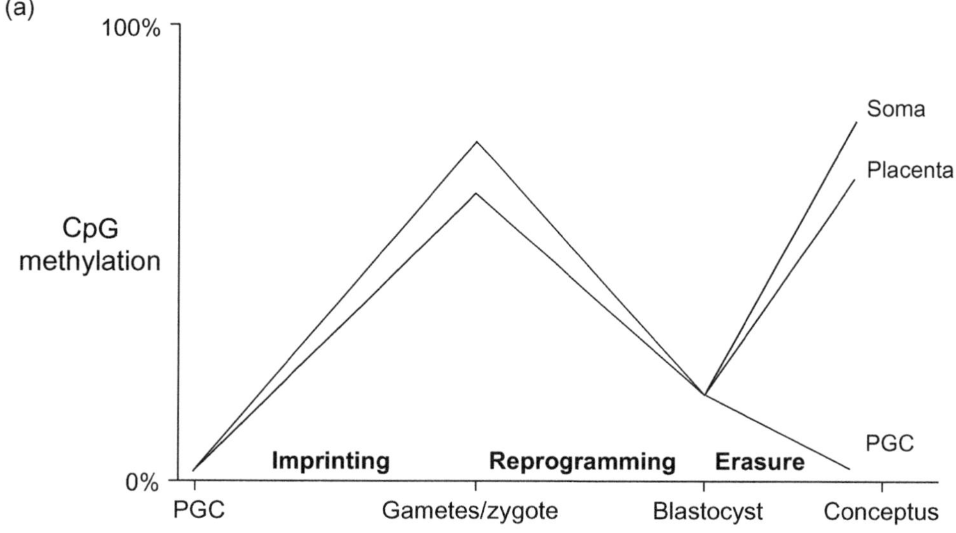

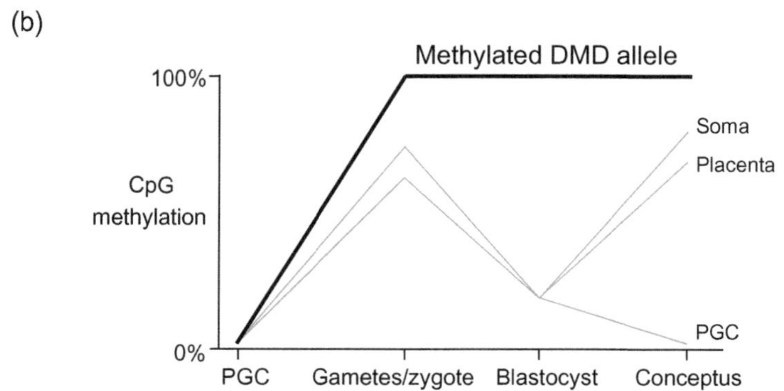

Figure 5.3 (a) Cyclical increases and decreases of genomic DNA methylation during gametogenesis and embryonic development. (b) Inheritance of methylation on differentially methylated domains (DMDs). Methylation is established on one parental allele during gametogenesis and this methylation is maintained (inherited) after fertilization. PGC, primordial germ cell.

parental allele is highly methylated and the opposite allele is unmethylated.

In discussing the important correlations between genomic methylation and genomic imprinting, we will first address the way in which parent-specific methylation is acquired in the gametes. Thereafter, we will focus on the all-important mechanism whereby patterns of DNA methylation are faithfully inherited following fertilization.

5.3.2 Enzymes Involved in Imprint Establishment

A ground state of an unmethylated diploid genome is present in undifferentiated PGCs. Although there are some quibbles about whether the PGC genome is completely stripped of methylation marks, it is clear that all maternal and paternal methylation imprints are absent [4,5]. From this ground state, as the germ cells differentiate through spermatogenesis in males and oogenesis in females, new methylation of a large percentage (50–70%) of genomic CpG dinucleotides occurs (de novo methylation). Within this newly methylated DNA are the DMD sequences. It is important to realize that DMD methylation is a very minor component of methylation acquired during the entirety of both female and male gametogenesis. It is logical that parent-specific imprints are established during gametogenesis, as the two alleles are separated (in two organisms) where different processes specific to either female or male germ cells can act on the two otherwise identical parental alleles, methylating one but not the other. In this way, gametogenesis serves as the developmental stage for the establishment of parent-specific genomic imprints. Hence, during every developmental cycle, imprints are erased in PGCs (see below) and established during maturation of the male and female gametes [4].

What are the enzyme(s) that methylate unmethylated DMD sequences in one parental germline but not in the opposite one and how specific are they? These enzymes are de novo DNA methyltransferases, whose specific function is to convert unmethylated DNA to fully methylated DNA, in which complementary CpG residues on both strands are methylated (Figure 5.2). There are two known mammalian de novo DNA methyltransferases, DNMT3A and DNMT3B (Figure 5.4a). Both are excellent candidates for establishing patterns of methylation because in vitro both enzymes exhibit primarily de novo methyltransferase activity, with little maintenance methyltransferase activity. A series of in vivo studies in the mouse indicated that DNMT3A is the active de novo methyltransferase during both male and female gametogenesis, and appears to be responsible for the acquisition of all DNA methylation during these developmental stages. To elucidate the role of DNMT3A in imprinting, the *Dnmt3a* gene was eliminated specifically in male and female germ cells before embryonic day 14.5 (E14.5), i.e. in PGCs [6]. Embryos from female mice missing germ cell DNMT3A died around E10.5, while males without germ cell DNMT3A showed such impaired spermatogenesis that the effects of missing methylation marks on embryonic development in the next generation could not be studied in this model. Maternal imprints were lost in the embryos from the females lacking DNMT3A in their germ cells while spermatogonia from mutant males lacked DMD methylation at a subset of paternal imprints. DNMT3A is involved in methylation-induced transposon silencing at a subset of repetitive elements specifically in male PGCs [7]. Defective spermatogenesis in male PGCs without DNMT3A may be due to meiotic failure caused by partial retrotransposon reactivation (see also Chapter 6).

In the female germline, genomic methylation occurs during the oocyte growth phase after the meiotic arrest, and a number of different sequence types, not just DMD sequences, become methylated by DNMT3A [6]. In contrast, in the male germline, de novo methylation occurs premeiotically in fetal prospermatogonia and targets repetitive DNA of many different types. Because de novo methylation catalyzed by DNMT3A is established on many more sequences than just germline imprints, this enzyme itself does not provide the machinery for specifically directing methylation to the relevant imprinted regions. A similar approach was used to explore the role of *Dnmt3b* in genomic imprinting. Loss of DNMT3B in PGCs had no effect on the establishment of male or female imprints. Thus, DNMT3A and not DNMT3B is the de novo methyltransferase used to catalyze the addition of methylation to imprinted DMD sequences in germ cells.

5.3.3 Auxiliary Factors

DNMT3L is a protein with homology to DNMT3 family members but lacking a functional catalytic

(a)

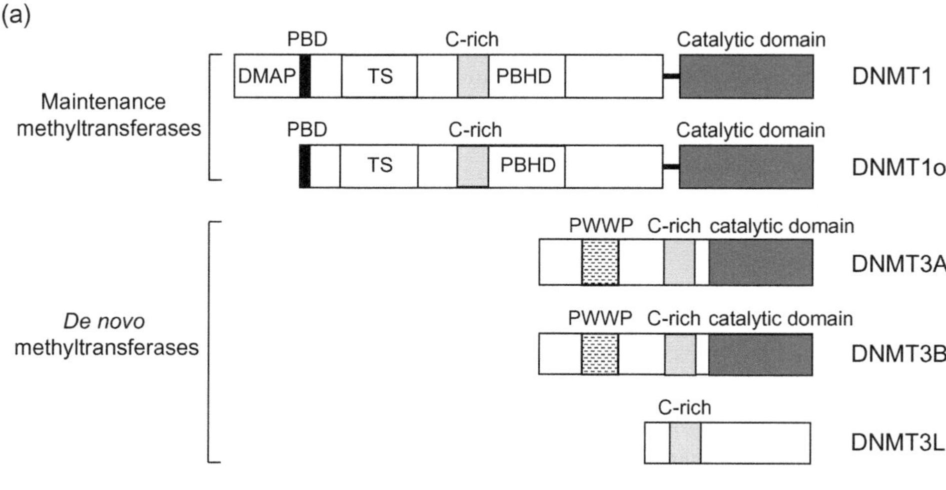

(b)

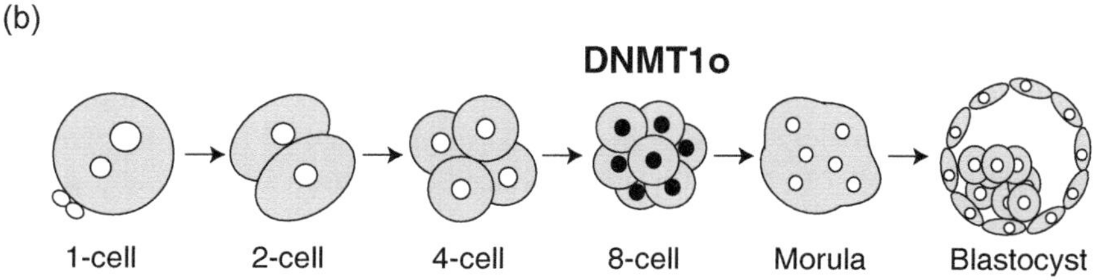

(c)

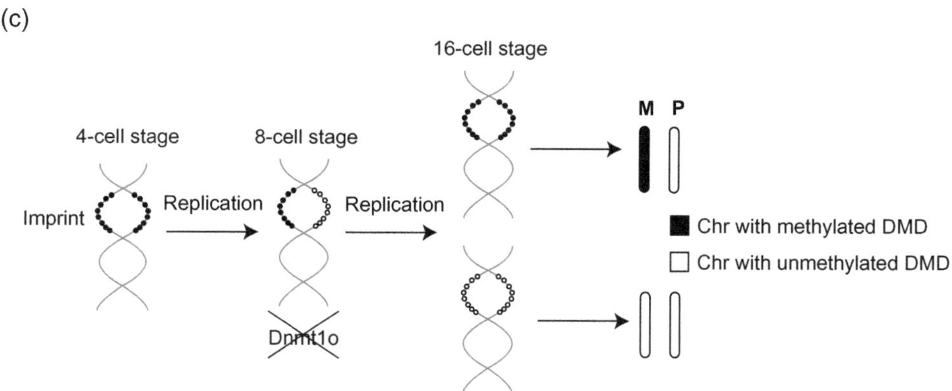

Figure 5.4 (a) Depiction of various domains in de novo and maintenance methyltransferases. Dnmt3L lacks the methyltransferase catalytic domain and DNMT1o (oocyte specific) lacks the N-terminal DNMT1 domain known to interact with DNMT1-associated protein 1 (DMAP1). DMAP, DMAP1-interaction domain; PBD, PCNA-binding domain; PBHD, Polybromo homology domain; PWWP motif, Proline–Tryptophan–Tryptophan–Proline motif; TS, Replication focus targeting sequence. (b) DNMT1o is synthesized in the oocyte and retained in the cytoplasm except at the 8-cell stage when it moves into the nuclei of all blastomeres. (c) Loss of DNMT1o at the 8-cell stage results in an abnormally unmethylated imprint in one-half of cells after the 16-cell stage. A representative maternally (M) imprinted chromosome is shown. The normally methylated (imprinted) and abnormally unmethylated chromosomes will segregate into different cells. DMD, differentially methylated domain.

domain (Figure 5.4a). DNMT3L has been shown to interact with both DNMT3A and DNMT3B [8]. It is the interaction between DNMT3L and DNMT3A that is relevant to the establishment of parental imprints. Concomitant increases in DNMT3A and DNMT3L concentrations are seen in the embryonic testis between E15.5 and E18.5 and their levels are down-regulated shortly after birth. In the ovary, levels of

DNMT3A are relatively constant but the levels of DNMT3L peak sharply in the phase associated with rapid oocyte growth [9]. These results indicate that high levels of both DNMT3A and DNMT3L coincide with acquisition of sex-specific methylation in the ovary and testis. Importantly, these phases are associated with the establishment of functional genomic imprints [10–12].

Genetic experiments confirmed that germ cells are the sites where DNMT3L has biological functions. *Dnmt3l*$^{-/-}$ adult mice obtained from crosses between heterozygous *Dnmt3l*$^{+/-}$ female and male mice are viable and have normal genomic methylation, indicating that DNMT3L is not required for postzygotic de novo methylation. There is, however, a strict requirement for DNMT3L in germ cells because *Dnmt3l*$^{-/-}$ males are sterile and embryos from *Dnmt3l*$^{-/-}$ females are phenotypically abnormal and die around mid gestation (E10.5) [13,14]. The embryos from *Dnmt3l*$^{-/-}$ females show loss of DMD methylation at several maternally imprinted loci indicating that DNMT3L is needed to establish imprints in the oocyte. DNMT3L is also needed for the establishment of methylation in male germ cells. *Dnmt3l*$^{-/-}$ males are sterile with no germ cells present in the adult male. Paternally imprinted sequences are partially demethylated by the lack of DNMT3L in germ cells from 17-day-old male mice. However, there seems to be a complete loss of methylation at retrotransposon elements in the male germline, which are consequently transcribed at a high frequency causing meiotic failure [14]. The importance of DNA methylation as a mechanism of repression of retrotransposon transcription is discussed in Chapter 6.

The simultaneous germ-cell expression of DNMT3A and DNMT3L, as well as the similarities in phenotypes of mice with germ-cell deletions in the *Dnmt3a* and *Dnmt3l* genes, indicate that DNMT3L is an essential, albeit catalytically inactive, auxiliary factor for DNMT3A-mediated de novo methylation in female and male gametogenesis. Crystallographic studies have provided insight into the mechanism of imprint establishment by the DNMT3A–DNMT3L complex. DNMT3L and DNMT3A interact through the carboxy terminal region of DNMT3L and this complex forms a tetramer through interaction between two DNMT3A–DNMT3L dimers [15]. DNMT3L stabilizes the catalytic activity of DNMT3A, and mutations in interacting amino acids between DNMT3L and DNMT3A cause loss of DNMT3A catalytic function. Further experiments illustrated that the DNMT3A–DNMT3L complex methylates CpG residues with an average spacing of 9.5 residues. This interval was observed in maternally imprinted genes, which had an average periodicity of 9.5 base pairs between them, but not in the DMDs of paternally imprinted genes. *Dnmt3a* and *Dnmt3l* knockout experiments also illustrated that maternally imprinted DMDs are more susceptible to the loss of these proteins than paternally imprinted DMDs.

Although there is a dual requirement for DNMT3A and DNMT3L in the establishment of maternal and paternal imprints, these two proteins together do not account for the specific establishment of genomic imprints. What other factors, if any, direct the DNMT3A/DNMT3L complex to sequences on which methylation imprints are deposited? One attribute of genomic structure that appears to be relevant is the chromatin makeup of target regions. DNMT3L binds to genomic sequences devoid of H3K4 methylation through its Plant Homeo Domain (PHD)-like domain (Figure 5.5a). This interaction, as well as other similar ones may refine the targets of the DNMT3A/DMNT3L complex toward imprinted sequences [16]. Of course, there are likely to be limits to the specificity provided by such genome-wide post-translational modifications of histones, and the real specificity lies in some as yet unknown features of the DNA target sequences themselves.

5.3.4 Acquisition at Different Times

The different times at which imprints are established in the maternal and paternal germlines is intriguing. Paternally imprinted DMDs such as the *H19/Igf2* DMD acquire methylation before the onset of meiosis in prospermatogonia during their mitotic proliferation in mice. These paternal methylation imprints are maintained through the remainder of male gametogenesis by maintenance methylation, and are present in haploid sperm. In contrast, maternal imprints are acquired postnatally, after mitotic expansion of oogonia and during the period of oocyte growth and meiotic arrest. The reason for the mitotic acquisition of paternal imprints and the meiotic acquisition of maternal imprints is not known. It is also not known why the acquisition of the different maternal imprints is a highly orchestrated process in which some imprints such as methylation of *Snurf/Snrpn* DMD

(a)

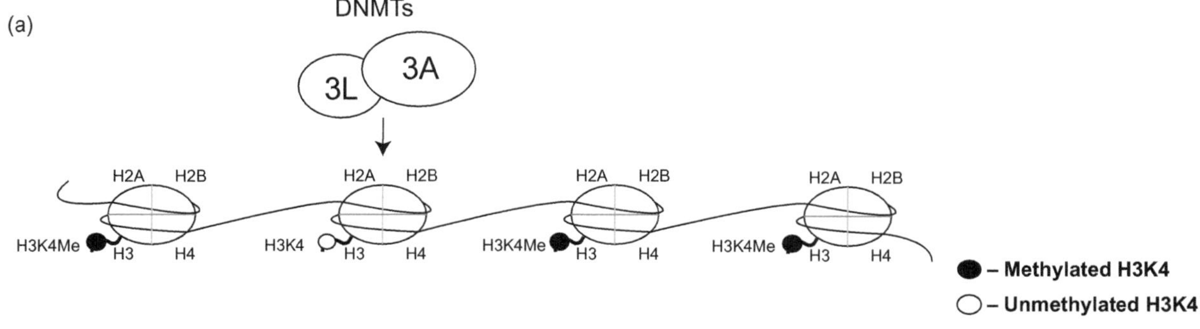

(b)

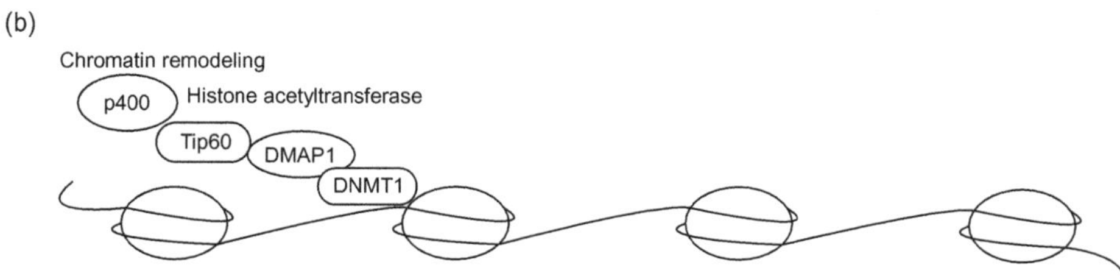

Figure 5.5 (a) De novo DNA methylation is directed to a region devoid of histone H3K4 methylation by DNMT3L, which interacts with unmethylated H3K4. (b) Chromatin remodeling by the Tip60-p400 complex and DNMT1-mediated maintenance methylation may be coordinated by interactions with DMAP1 during preimplantation reprogramming.

sequences are established early in meiotic maturation whereas others such as methylation of *Peg1* DMD sequences are established just before completion of meiosis I in mice [10–12]. In humans, a similar pattern of imprint establishment is observed where paternal imprints are acquired prior to entry into meiosis and maternal imprints are established during oocyte growth after meiotic arrest. The timing of imprint establishment makes oocytes particularly vulnerable to artificial reproduction technologies that accelerate oocyte maturation (such as ovarian hyperstimulation with gonadotrophins) because immature oocytes with incomplete imprints may also be released. Furthermore, in vitro maturation of germ cells and preimplantation embryos can affect imprint establishment or maintenance as described in Chapter 12.

5.4 Inheritance and Maintenance of Genomic Imprints

5.4.1 Enzymes Involved in Imprint Maintenance

Maintenance methylation involves the methylation of cytosines in CpG residues of the newly synthesized DNA strand, but only in CpG dinucleotides complementary to methylated CpGs of the parent DNA strand. Therefore maintenance methyltransferases act on hemi-methylated DNA after each S phase to convert it to fully methylated DNA. The DNMT1 DNA cytosine methyltransferase is the only enzyme known to possess maintenance methyltransferase activity in mammals (Figure 5.4a). DNMT1 has been shown to possess a high affinity for hemi-methylated DNA and is targeted to replication foci during S phase [1].

DNMT1 is required for the maintenance of overall genomic methylation. The catalytic methyltransferase activity of DNMT1 is located in the C-terminal domain, while the N-terminus contains domains specialized for nuclear translocation and other regulatory functions. Embryos developing in the absence of DNMT1 die around the middle of gestation, and have a severe reduction in the methylation of imprinted DMDs and repetitive DNA sequences. Several imprinted genes are expressed abnormally from both parental alleles (biallelic expression), consistent with the loss of parent-specific imprinted methylation. Because *Dnmt1*$^{-/-}$ embryos lack appropriate methylation marks at several sequences, it was not possible to determine the contribution of disrupted imprinted gene expression to the observed embryonic lethality.

DNMT1 has been shown to possess two main protein-coding isoforms – the longer somatic DNMT1s isoform and the short DNMT1o isoform expressed specifically in oocytes and preimplantation embryos (Figure 5.4a). DNMT1s is expressed in all stages of preimplantation development including the pronuclear stage and is found in the nucleus from the two-cell stage of preimplantation development onward [17]. Translation of the DNMT1s protein is initiated from an AUG codon encoded in exon 1s of the *Dnmt1* gene, whereas translation of the oocyte-specific transcript is initiated from an AUG codon encoded in exon 4 of the gene. Mouse DNMT1o is a catalytically active maintenance methyltransferase, lacking the first 118 amino acids of mouse DNMT1s, but otherwise identical. DNMT1o is synthesized solely in the oocyte and is a maternal-effect protein found in high concentrations in the ooplasm and cytoplasm of early preimplantation mouse embryos. Cytoplasmic DNMT1o moves into the nucleus only at fourth embryonic S phase (eight-cell stage) (Figure 5.4b). Homozygous $Dnmt1^{\Delta[1]o/\Delta[1]o}$ female mice and embryos derived from these lack DNMT1o in all cells.

Analysis of DNMT1o has provided important insights into the role of the *Dnmt1* gene in genomic imprinting [18]. Although DNMT1s (zygotically produced) is present in the nucleus at this stage, it is unable to compensate for a loss of eight-cell DNMT1o. Because embryos from $Dnmt1^{\Delta[1]o/\Delta[1]o}$ female mice are missing nuclear DNMT1o at the eight-cell stage, the hemi-methylated DNA at this stage is not converted to fully methylated DNA (Figure 5.4c). After the next round of replication, the hemi-methylated DNA duplex gives rise to a hemi-methylated DNA duplex and an unmethylated DNA duplex. At this stage (16-cell) DNMT1s is present in the nucleus and it converts the hemi-methylated duplex into a fully methylated duplex. The unmethylated duplex, which is not a substrate for DNMT1s, remains unmethylated during future cell divisions. The net result is a reduction to 50% of the normal level of methylation. When E9.5 embryos from $Dnmt1^{\Delta[1]o/\Delta[1]o}$ mothers were examined, the loss of methylation was restricted to the genomic imprints. The global genomic methylation and methylation at repeats such as the intracisternal A particle (IAP) provirus were normal in E9.5 embryos from $Dnmt1^{\Delta[1]o/\Delta[1]o}$ females. These experiments indicate two possibilities: (1) DNMT1o has maintenance methyltransferase activity exclusive to genomic imprints, or (2) both imprinted and nonimprinted (for example IAP) sequences lose methylation in the absence of DNMT1o, but only the imprints are unable to regain methylation in later embryogenesis. Different chromosomes and embryo cells lost DMD methylation in this model depending on the random assortment of individual chromosomes (Figure 5.4c). As a consequence, embryos derived from oocytes lacking DNMT1o showed a mosaic loss of imprints in different cells, causing a wide range of embryonic and placental defects. These findings clearly illustrate that perturbations in normally inherited imprinted DNA methylation can severely impact fetal growth and development.

5.4.2 Features of Imprinted DMDs

More than 100 imprinted genes have been identified in the mouse and they are organized into discrete clusters of two or more genes each [1]. Imprinted clusters can stretch for thousands of kilobases and they can consist of paternally and maternally imprinted protein-coding genes, often interspersed with imprinted untranslated genes, as well as nonimprinted genes. In general, each imprinted cluster contains a primary DMD, which serves as the imprinting control element (ICE) and whose methylation is established in the germ line. The importance of an ICE (primary DMD) can be shown by the effect of an ICE deletion on imprinted gene expression. For example, deletion of the ICE between the *Igf2* and *H19* genes results in abnormal expression of the normally silent paternal allele of the *H19* gene (Figure 5.6a). More details of the molecular function of this *H19/Igf2* ICE will be discussed below. Secondary DMDs can also be present in the cluster, which acquire their methylation after fertilization due to *cis*-directed effects of a nearby primary DMD.

DMDs consist of a high density of CpGs, which are arranged as tandem repeats and imprinted genes require a minimum number of tandem repeats for differential methylation (Figure 5.6b) [19]. Tandem repeats occur mostly in maternally imprinted genes at a frequency of six to nine repeats of 18–170 base pairs in size. Paternally imprinted DMDs such as the *H19/Igf2* DMD do not possess tandem repeats even though they possess a high density of CpG residues. It has also been found that DMDs of maternally imprinted genes contain imprinted promoters while paternally

(a)

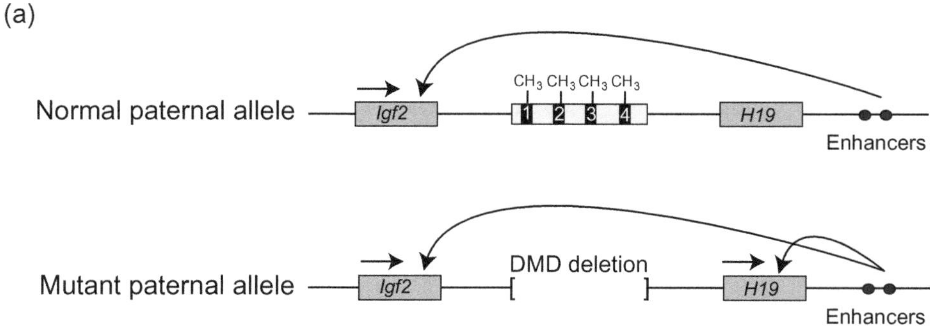

(b)

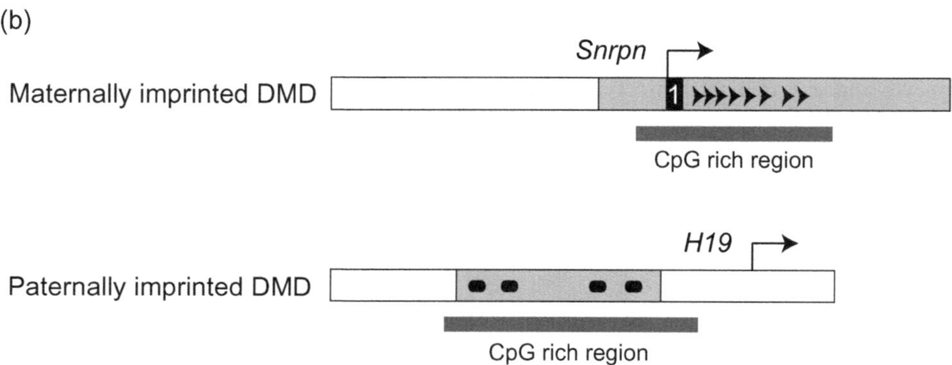

Figure 5.6 (a) Deletion of the differentially methylated domain (DMD) from the paternal allele of the *H19/Igf2* cluster causes paternal *H19* expression and thus abnormal biallelic *H19* expression. (b) Structures of a maternally methylated DMD (overlapping the *Snrpn* gene) and a paternally imprinted DMD (near the *H19* gene) are shown. Gray rectangles represent DMDs; black arrowheads, tandem repeats; arrows, transcription start sites; solid ovals, CTCF binding sites.

imprinted genes lack promoters in the DMD region. These differences indicate that maternally and paternally imprinted DMDs may be regulated differently. Maternally imprinted DMDs are predicted to have similar secondary structure even though they show considerable sequence divergence [1].

Other important features of DMDs were revealed in studies of DMD methylation in mouse embryonic stem (ES) cells. When imprinted sequences in ES cells lose their CpG methylation due to a transient loss of DNMT1, they are unable to recover these marks after DNMT1 is restored [20]. Importantly, this inability to recover DMD methylation occurs despite the expression of both DNMT3A and DNMT3L, two proteins that normally cooperate to establish methylation imprints in germ cells. These experiments indicate that imprinted DMDs have unique inheritance properties. First, they acquire methylation only during gametogenesis. Second, this germline methylation is normally continuously maintained (inherited) in the offspring's cells irrespective of changes in nonimprinted methylation. Lastly, after fertilization, the unmethylated DMD allele does not become methylated. In essence, DMD sequences evolved to maintain or inherit the methylation state established in the gamete, which is highly methylated on one parental allele and unmethylated on the opposite parental allele. Precisely how the sequence makeup of DMDs provides these important inheritance properties is not known.

5.5 Consequences of Inherited Methylation: Monoallelic Gene Expression

Although a normal embryo has two alleles of each imprinted gene, only one parental allele of the gene is expressed. For each imprinted gene, monoallelic expression is determined by the presence of a nearby DMD. Some DMDs overlap a gene's promoter, and the parent-specific methylation directly suppresses transcription. Many regulatory DMDs however are quite distant (>500 kb) from imprinted genes' promoters. How does the methylation of a DMD lead to transcriptional suppression of these genes? There are two main mechanisms that have been characterized to date, which describe how methylation of a DMD can lead to transcriptional suppression of genes in the entire imprinted cluster [21].

5.5.1 Insulator Model: *H19/Igf2* Imprinting

The conserved *H19/Igf2* locus is located on mouse chromosome 7 and human chromosome 11. *H19* is a maternally expressed non-coding RNA, while insulin-like growth factor 2 (*Igf2*) is a paternally expressed gene involved in promoting embryonic growth and development. The expression of these genes is reciprocally controlled by the methylation of the DMD located between them (Figure 5.7a). Hypomethylation of both maternal and paternal DMD alleles leads to biallelic expression of the *H19* gene and corresponding silencing of *Igf2* expression. The mechanism for this regulation was elucidated by identification of binding sites of the insulator protein CCCTC-binding factor (CTCF) in the DMD region. On the maternally inherited chromosome, CTCF binds to the unmethylated DMD and acts as an insulator preventing enhancers 3' of the *H19* gene from activating *Igf2* expression. Instead, the enhancers activate *H19* expression. The DMD in the paternally inherited chromosome is methylated preventing CTCF binding, allowing the enhancer to activate its preferred target *Igf2*.

5.5.2 Noncoding RNA Model: *Kcnq1* Locus

The other generally recognized mechanism of imprinted gene regulation shared by several clusters is through transcriptional suppression by a noncoding RNA (ncRNA). Most imprinted clusters are associated with at least one ncRNA. The *Kcnq1* cluster is regulated by a maternally methylated DMD (KvDMR1), which is located in an intron of the *Kcnq1* gene (Figure 5.7b). When this DMD is unmethylated, which is the normal paternal allele, the long (>60 kb) *Kcnq1ot1* ncRNA, whose promoter lies within KvDMR1, is expressed. This expression in turn leads to the suppression of transcription of

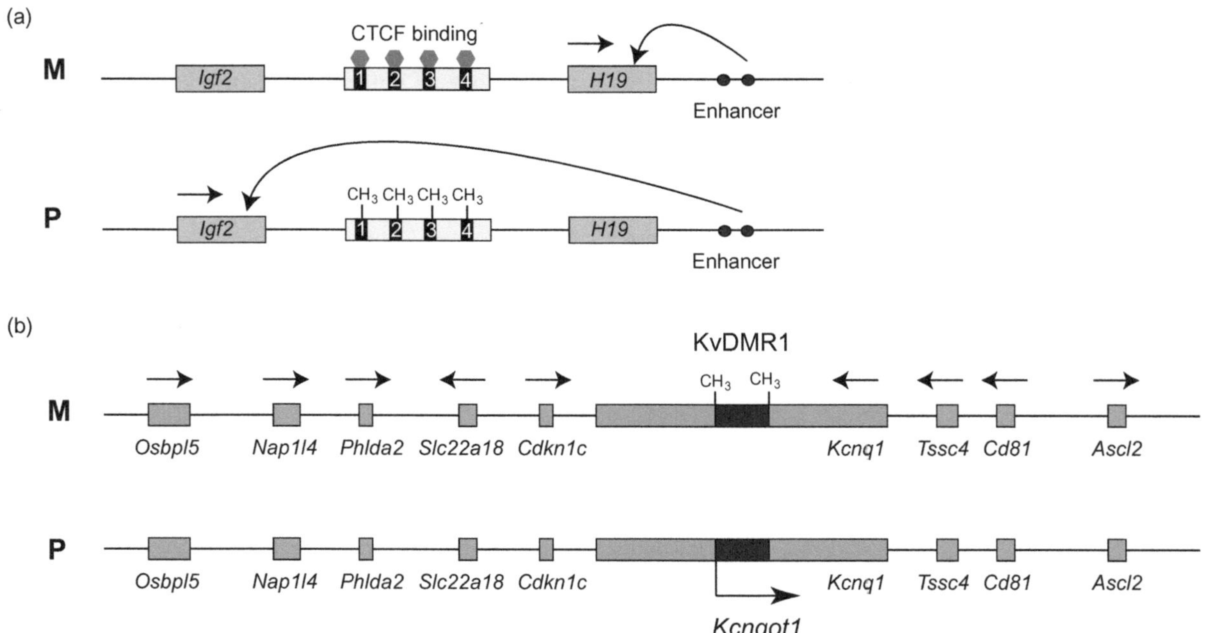

Figure 5.7 (a) Transcriptional regulation of *H19/Igf2* expression by imprinted DMD methylation mediated by CTCF binding at four sites causes selective enhancer activation of *H19*. (b) *Kcnq1* locus regulates monoallelic expression of several genes through expression of noncoding RNA *Kcnq1ot1*. Horizontal arrows indicate transcribed alleles.

surrounding genes on the same chromosome. On the maternally inherited chromosome, the DMD is methylated, suppressing expression of *Kcnq1ot1* and allowing expression of the surrounding genes. The mechanism by which the long ncRNA is able to suppress gene expression is not known. Mechanisms of RNA interference may explain how *Kcnq1ot1* suppresses *Kcnq1* expression but not the expression of genes in the cluster, several kilobases away from KvDMR1. Truncation of the ncRNA *Kcnq1ot1* leads to derepression of the genes on the paternal chromosome and biallelic expression, hence the length of the ncRNA seems to be important for the parent-specific expression. Coating of genes in the cluster by the ncRNA or transcription of the ncRNA itself has been postulated to recruit repressive chromatin marks leading to gene silencing. However, these hypothetical mechanisms have yet to be rigorously tested, and if correct, would not explain how certain genes in the cluster escape imprinting and exhibit nonimprinted, biallelic expression.

5.6 Interplay between Chromatin Modifications and DNA Methylation

Recent studies have identified interactions between DNA methylation and other chromatin modifications that control transcription. Besides regulating ncRNA expression and the function of insulator sequences, DNA methylation in imprinted regions is observed to alter local chromatin states, which further result in suppression or activation of transcription of the surrounding genes. The interplay between DNA methylation and chromatin modifications is complex and interdependent and may vary between different imprinted genes. Studies have reported parent-specific differences in histone acetylation in differentially methylated regions of imprinted genes with the unmethylated allele being associated with a transcriptionally active chromatin state [22]. We will describe two well-documented interactions between DNA methylation and other chromatin features to understand how different chromatin processes may direct DNA methylation and consequently affect transcription.

As discussed in Section 5.3, the DNMT3A–DNMT3L complex mediates de novo methylation of imprinted genes in gametogenesis and loss of DNMT3L in females results in embryos without maternal imprints. DNMT3L was found to interact with histone H3 tails specifically when they were unmethylated on the lysine residue at position 4 (H3K4). Because methylation of H3K4 inhibited this interaction, DNMT3L is specifically directed to regions devoid of H3K4 methylation (Figure 5.5a). This study demonstrates the mechanism by which chromatin modifications could contribute to the targeting of DNA methylation to specific regions [16].

The essential role of the Tip60-p400 chromatin remodeling complex in preimplantation development was shown by the failure to identify homozygous mutant preimplantation embryos from crosses between heterozygous mice that have engineered mutations in any one of a number of genes encoding components of the complex [23,24]. These findings suggest that Tip60-p400 is involved in a process of remodeling the epigenetic makeup of the preimplantation genome, a process possibly associated with epigenetic reprogramming to a pluripotent cell state (see below). Interestingly, the DNMT1-associated protein DMAP1 is one of the essential components of Tip60-p400 [24], further suggesting that preimplantation functions of Tip60-p400 and DNMT1 may be coordinated through DMAP1. If so, we speculate that during preimplantation development, Tip60-p400 chromatin remodeling activity participates in a process of epigenetic reprogramming to rearrange genomic methylation (Figure 5.5b). In this model, by the milestone blastocyst stage, a pattern of genome-wide methylation is established such that the blastocyst's pluripotent stem cells can contribute to normal and complete embryonic development.

5.7 Abnormalities of Imprinting

Genomic imprinting has been shown to be vital for both prenatal and postnatal development and loss of imprinting on certain chromosomes can lead to congenital anomalies [25]. Abnormal imprinting in a region can occur due to deletion or duplication of the DMD region or failure to establish correct methylation patterns. A few of the well-characterized human diseases caused by loss of imprinting result in growth abnormalities or neurodevelopmental disorders. The Beckwith–Wiedemann syndrome is characterized by somatic overgrowth, congenital malformations, and embryonic neoplasia. About 60% of the sporadic Beckwith–Wiedemann syndrome cases have abnormal methylation patterns in a 1-Mb region of chromosomal 11 at position 11p15.5. This region contains both the *H19/IGF2* and *KCNQ1* imprinted clusters, and abnormal CpG methylation results in biallelic expression of either the *IGF2* gene or the *KCNQOT1* transcript or both.

Loss of imprinting can also result in neurobehavioral disturbances in the Prader–Willi and Angelman syndromes. The Prader–Willi syndrome is caused either by maternal uniparental disomy, paternal deletion, or loss of methylation in a 2-Mb region on chromosome 15 at 15q11–13. The shortest region of overlap (SRO) of numerous Prader–Willi syndrome deletions is a 4.3-kb region containing the DMD (ICE) of a large cluster of imprinted genes. Angelman syndrome is due to loss of imprinting and deficiency of the normally maternally expressed gene *UBE3A*. Interestingly, although this gene is embedded in the Prader–Willi syndrome cluster, its imprinting is regulated by a unique ICE.

Genomic imprinting defects can occur during establishment of imprints in germ cells or during the maintenance of imprints in the early embryo. Because assisted reproductive technology (ART) such as intracytoplasmic sperm injection, superovulation of oocytes, sperm cryopreservation, and in vitro fertilization manipulate gametes and early embryos, it is not surprising that ART has been postulated to cause imprinting disorders. In this regard, certain imprinting diseases like the Beckwith–Wiedemann syndrome have been reported to occur at a six to nine times higher frequency in children from ART patients. The effect of various ART procedures on genomic imprinting is further described in Chapter 12.

5.8 Preimplantation Reprogramming

5.8.1 Global DNA Methylation and Pluripotency

The role of the *Dnmt1* gene in maintenance of genomic imprints in preimplantation embryos is complex. After fertilization, both parental genomes undergo a general or global loss of genomic methylation in the presence of nuclear DNMT1 [4]. The notable exception to this demethylation is the continual retention of methylation on imprinted DMD sequences at each and every cleavage (replication) stage of preimplantation development [4,26]. The clear difference between imprinted DMD sequences and global decline in methylation may be the most obvious feature of a complex molecular program whereby genomic methylation is rearranged on one but not the other allele during preimplantation development. A widely held view is that this complex epigenetic rearrangement is a central element in the important biological process of cellular reprogramming. In this process, the totipotent fertilized egg (zygote), a fusion of two highly differentiated gametes, undergoes a series of cleavages to produce a blastocyst containing small, pluripotent inner cell mass (ICM) cells. A programmed reduction in genomic methylation by the blastocyst stage may be required for the global reprogramming to pluripotency of the ICM cells. Following this period of preimplantation reprogramming, during postimplantation development, global methylation levels increase and the cells of the embryo become committed to various cell lineages. The importance of methylation at the blastocyst stage has been nicely illustrated in studies of mice lacking DNMT1; blastocysts lacking all forms of DNMT1 protein develop normally, but following implantation into the uterus, cells of these embryos fail to differentiate and instead undergo apoptosis, leading to embryonic death. Two distinct mechanisms for the programmed reduction in preimplantation methylation have been proposed. One mechanism is the active removal of methylated cytosine from the genome and the other mechanism is the passive loss of genomic methylation due to absence of maintenance methyltransferase activity following replication of the nuclear genome [27].

5.8.2 Model of Active Demethylation

Although both parental genomes lose a significant proportion of methylation during preimplantation development, the paternal genome in the mouse zygote appears to lose its methylation at a faster rate than does the maternal genome [27]. This observation suggests that active demethylation, or removal of methyl groups from cytosine bases outside the replication stage of the cell cycle, occurs during preimplantation. The molecular mechanism for putative demethylation proposes that demethylation occurs through the conversion of methylated cytosine (5mC) to a hydroxyl-methylated cytosine (5hmC) intermediate by a family of dioxygenases called TET (ten-eleven-translocation) proteins [27]. Mouse ES cells express TET1 and TET2, while TET3 is expressed in early preimplantation embryos, and the activity of one or more of these dioxygenases to generate 5hmC might be followed by a mechanism to remove 5hmC and replace it with an unmethylated cytosine, possibly through activity of a base-excision repair pathway.

If active demethylation is responsible for the significant reduction in preimplantation genomic methylation, inherited, imprinted DMD methylation must be blocked from this activity. How might this occur? It might occur by specifically protecting sequences from active demethylation. This protective property has been attributed to the protein STELLA, a maternal-effect protein present in early embryos [27]. When *stella*$^{-/-}$ females are mated to wild-type males, the maternal genome in the resulting embryos shows a decrease in 5mC levels by immunofluorescence, at a stage when the maternal genome may be protected from active demethylation. Bisulfite sequencing analysis of genomic DNA in pronuclear-stage embryos revealed that some maternally imprinted genes, paternally imprinted genes, and IAP repeats show a decrease in methylation when derived from *stella*$^{-/-}$ mothers. The partial loss of methylation on the maternal genome and from imprinted DMDs due to the deficiency in STELLA reveals that they may be multiple molecular components that regulate the methylation status of the maternal genome and imprinted genes, possibly through interaction with DNMTs during various stages of embryonic development. Identification of these proteins and their interaction partners may help better understand the process of epigenetic reprogramming.

5.8.3 Model of Passive Demethylation

An alternative explanation for the loss of genomic methylation accompanying preimplantation development is passive demethylation, or the loss of methylation due to replication of the nuclear genome in the absence of DNMT1-mediated maintenance methylation. The mechanism by which DNMT1 is able to distinguish imprinted DMDs from other genomic CpGs is not known. However, such a discriminating feature of DNMT1 may reside in a recently identified mammal-specific region (amino acids ~190–350), located in the enzyme's N-terminal regulatory domain [20]. Deletion mutagenesis of the mammal-specific region of *Dnmt1* produced a series of mutants that either maintain DMD methylation but not nonimprinted methylation, maintain nonimprinted but not DMD methylation, or maintain neither nonimprinted nor DMD methylation in vitro in ES cells. At present, it is not known whether the mammal-specific region of DNMT1 plays a role in maintaining the high DMD methylation in vivo during the global decrease in methylation after fertilization. A plausible mechanism of passive demethylation would exclude DNMT1 from certain methylated regions of the genome and the regions would consequently become unmethylated over successive rounds of DNA replication.

5.9 Erasure of Parent-Specific Imprints in Primordial Germ Cells

After epigenetic reprogramming to generate blastocyst-stage embryos, the level of global genomic methylation, primarily defined by the large fraction of nonimprinted sequences, rapidly increases in all somatic cells of the conceptus (Figure 5.3). This rise in methylation is due to the combined activities of de novo methyltransferases and DNMT1. Methylation of imprinted DMDs is maintained at a high level during the period of epigenetic reprogramming and thereafter in the developing postimplantation embryo. In contrast, PGCs in the embryo undergo erasure of their imprinted DMD methylation as prelude to the establishment of sex-specific genomic imprints (Figure 5.8). The PGCs migrating to or residing in the fetal gonad are the most logical site of imprint erasure because PGCs are permanently allocated to the germ lineage and this stage precedes establishment of sex-specific and functional imprints. Erasure of imprints completes the cycle of imprint establishment and maintenance allowing for transgenerational inheritance of the imprints established in the new gonad into the somatic cells of the developing embryo in the next generation.

Erasure of both maternal and paternal imprinted methylation patterns on DMD sequences takes place after the migration of the PGCs to the genital ridge, between E10.5 and E13.5 [5]. A few days before this, at E7.25, genomic imprints in PGCs are identical to those found in somatic cells. The timing of erasure differs somewhat for the different DMD sequences. However, analysis of cloned mouse embryos generated from E12.5 or E13.5 PGC nuclei showed biallelic expression of imprinted genes and reduced developmental potential [28], indicating that imprint erasure is complete by day 12.5 of gestation. The enzymes involved in imprint erasure are currently unknown. It is also not known whether imprint erasure is an active demethylating or a passive demethylating process whereby the genome loses its methylation through rounds of DNA replication in the absence of DNMT1 maintenance methyltransferase activity.

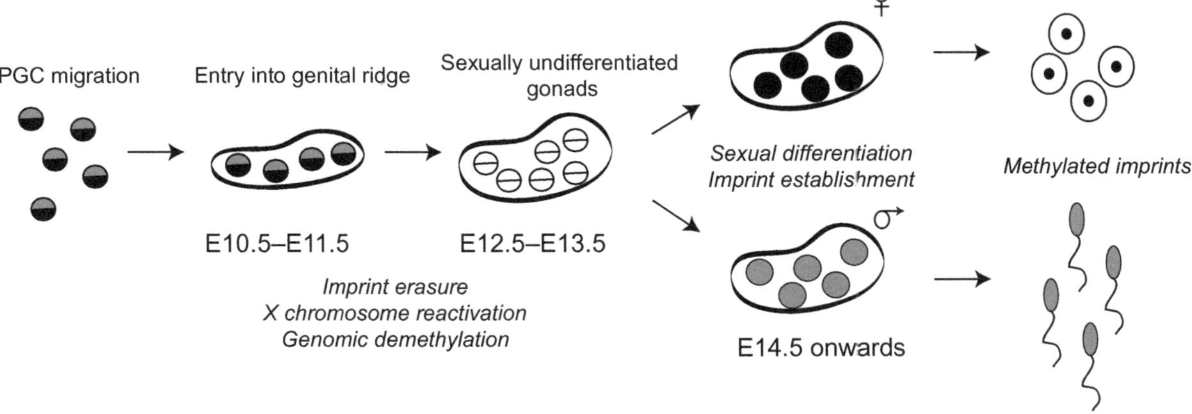

Figure 5.8 Erasure of both paternal (gray portion of circles) and maternal (black portion of circles) imprints in primordial germ cells (PGCs) following migration into the undifferentiated gonad. Imprints are established in maturing germ cells following differentiation of the gonad into an ovary or testis.

Moreover, erasure of methylation is not specific to imprints in the PGCs; methylation is also lost on all categories of nonimprinted DNA sequences. Knowledge of the kinetics of methylation loss on imprinted and nonimprinted sequences would lead to a better understanding of the mechanism of methylation loss – a rapid loss would suggest active demethylation, whereas a slow and steady decline would suggest passive demethylation.

5.10 Conclusions

This chapter illustrates how epigenetic modifications to the DNA sequence in the form of CpG methylation can serve as a heritable mark passed on from generation to generation. Genomic imprints are an excellent example of DNA sequences undergoing transgenerational inheritance of epigenetic modifications. Transmission of genomic imprints consists of an establishment phase mediated by de novo methyltransferases and a maintenance phase by maintenance methyltransferases. The methylation on imprinted DMDs is maintained during epigenetic reprogramming and is erased or reset only in PGCs. Imprinted sequences and DNA methylation are associated with transcription factors and histone modifications, which facilitate their actions on transcriptional control of genes in the imprinted cluster. Defects in imprinted genes have profound effects on development; hence accurate transgenerational inheritance of imprints during ART (see Chapter 12) or natural fertilization is vital to the survival of the embryo.

References

1. Reinhart B, Chaillet JR. Genomic imprinting: *cis*-acting sequences and regional control. *Int Rev Cytol* 2005;243:173–213.
2. Cattanach BM, Kirk M. Differential activity of maternally and paternally derived chromosome regions in mice. *Nature* 1985;315: 496–8.
3. MouseBook: www.mousebook.org/catalog.php?catalog=imprinting
4. Chaillet JR, Vogt TF, Beier DR, et al. Parental-specific methylation of an imprinted transgene is established during gametogenesis and progressively changes during embryogenesis. *Cell* 1991;66:77–83.
5. Hajkova P, Erhardt S, Lane N, et al. Epigenetic reprogramming in mouse primordial germ cells. *Mech Dev* 2002;117:15–23.
6. Kaneda M, Okano M, Hata K, et al. Essential role for *de novo* DNA methyltransferase Dnmt3a in paternal and maternal imprinting. *Nature* 2004;429:900–3.
7. Kato Y, Kaneda M, Hata K, et al. Role of the Dnmt3 family in *de novo* methylation of imprinted and repetitive sequences during male germ cell development in the mouse. *Hum Mol Genet* 2007;16:2272–80.
8. Hata K, Okano M, Lei H, et al. Dnmt3L cooperates with the Dnmt3 family of *de novo* DNA methyltransferases to establish maternal imprints in mice. *Development* 2002;129:1983–93.

9. La Salle S, Mertineit C, Taketo T, et al. Windows for sex-specific methylation marked by DNA methyltransferase expression profiles in mouse germ cells. *Dev Biol* 2004;268:403–15.

10. Kono T, Obata Y, Yoshimzu T, et al. Epigenetic modifications during oocyte growth correlates with extended parthenogenetic development in the mouse. *Nat Genet* 1996;13:91–4.

11. Li JYY, Lees-Murdock DJ, Xu GLL, et al. Timing of establishment of paternal methylation imprints in the mouse. *Genomics* 2004;84:952–60.

12. Lucifero D, Mann MR, Bartolomei MS, et al. Gene-specific timing and epigenetic memory in oocyte imprinting. *Hum Mol Genet* 2004;13:839–49.

13. Bourc'his D, Xu GL, Lin CS, et al. Dnmt3L and the establishment of maternal genomic imprints. *Science* 2001;294:2536–9.

14. Bourc'his D, Bestor TH. Meiotic catastrophe and retrotransposon reactivation in male germ cells lacking Dnmt3L. *Nature* 2004;431:96–9.

15. Jia D, Jurkowska RZ, Zhang X, et al. Structure of Dnmt3a bound to Dnmt3L suggests a model for *de novo* DNA methylation. *Nature* 2007;449:248–51.

16. Ooi SK, Qiu C, Bernstein E, et al. DNMT3L connects unmethylated lysine 4 of histone H3 to *de novo* methylation of DNA. *Nature* 2007;448:714–17.

17. Cirio MC, Ratnam S, Ding F, et al. Preimplantation expression of the somatic form of Dnmt1 suggests a role in the inheritance of genomic imprints. *BMC Dev Biol* 2008;25:8–9.

18. Howell CY, Bestor TH, Ding F, et al. Genomic imprinting disrupted by a maternal effect mutation in the *Dnmt1* gene. *Cell* 2001;104:829–38.

19. Reinhart B, Paoloni-Giacobino A, Chaillet JR. Specific differentially methylated domain sequences direct the maintenance of methylation at imprinted genes. *Mol Cell Biol* 2006;26:8347–56.

20. Borowczyk E, Mohan KN, D'Aiuto L, et al. Identification of a region of the DNMT1 methyltransferase that regulates the maintenance of genomic imprints. *Proc Natl Acad Sci USA* 2009;106(20):806–11.

21. Ideraabdullah FY, Vigneau S, Bartolomei MS. Genomic imprinting mechanisms in mammals. *Mutat Res* 2008;647:77–85.

22. Singh P, Cho J, Tsai SY, et al. Coordinated allele-specific histone acetylation at the differentially methylated regions of imprinted genes. *Nucleic Acids Res* 2010;38:7974–90.

23. Sapountzi V, Logan IR, Robson CN. Cellular functions of TIP60. *Int J Biochem Cell Biol* 2006;38:1496–509.

24. Mohan NK, Ding F, Chaillet JR. Distinct roles of DMAP1 in mouse development. *Mol Cell Biol* 2011;31:1861–9.

25. Paoloni-Giacobino A, Chaillet JR. Genomic imprinting and assisted reproduction. *Reprod Health* 2004;1:6.

26. Tremblay KD, Duran KL, Bartolomei MS. A 5′ 2-kilobase-pair region of the imprinted mouse *H19* gene exhibits exclusive paternal methylation throughout development. *Mol Cell Biol* 1997;17:4322–9.

27. Saitou M, Kagiwada S, Kurimoto K. Epigenetic reprogramming in mouse pre-implantation development and primordial germ cells. *Development* 2012;139:15–31.

28. Lee J, Inoue K, Ono R, et al. Erasing genomic imprinting memory in mouse clone embryos produced from Day 11.5 primordial germ cells. *Development* 2002;129: 1807–17.

Chapter 6

Genes Are Not the Whole Story: Retrotransposons as New Determinants of Male Fertility

Patricia Fauque and Déborah Bourc'his

6.1 Introduction

The release of the human genome sequence has revealed that our genes represent only a minor fraction, with exons making up less than 2% of our DNA. In contrast, transposable elements represent some 45% of the genomic mass. Contrary to other species such as *Drosophila* or plants, these elements are not clustered in specific chromosomal regions in mammals, but are rather scattered throughout the genome, residing between but also inside genes. The transposon landscape of our genome reflects an evolutionary tug-of-war between integration and propagation events orchestrated by these elements, and counteracting defense mechanisms exerted by the host.

By providing raw material for genetic innovation and diversification, transposons have positively participated in genome shaping and speciation during evolution. However, on a short-term basis, transposon activity can adversely modify the function and architecture of the genome [1]. They can induce damaging genetic changes by hopping into new genomic locations, or create some major chromosomal rearrangements by ectopic recombination between nonallelic copies. When occurring in somatic cells, cancer can ensue; when occurring in gametes, heritable mutations can be generated and lead to congenital disorders [2]. Their effect can also be nongenetic, by influencing the expression patterns of adjacent genes through their strong promoter sequences or by inducing alternative splicing or premature transcriptional termination.

To ensure their propagation, transposons need to mobilize in cells destined for the next generation, the germ cells. Recent studies have highlighted the existence of specialized pathways acting in the germline for the lifelong repression of transposons and the protection of the hereditary material [3]. Genetic impairment of this system in mice invariably leads to massive transposon reactivation and complete failure to produce spermatozoa in males [4]. Transposon control therefore appears to be a major safeguard of male fertility in mouse, but also in other animal species. Such a role is highly likely to be conserved in human, but has drawn very little attention so far. The purpose of this chapter is to shed light on these newly appreciated determinants of male reproductive fitness, the transposons. Improper control of these elements, of genetic or environmental origin, could be revealed as a major cause of male infertility.

6.2 Human Transposable Elements

Transposons, also known as mobile DNA elements or "jumping genes," are widespread in nature and comprise an estimated half of the human genome. This is likely an underestimation of the true proportion of genomic transposons, as their repetitive and divergent nature hamper efficient sequencing and annotation. Several classes of transposons have evolved in the human lineage and coexist in the current *Homo sapiens* genome, with some elements having expanded with more success than others (Figure 6.1).

Their mobilizing strategies allow a classification into two main classes, the DNA transposons and the retrotransposons (Figure 6.1). DNA transposons move by a "cut-and-paste" mechanism via an element-encoded transposase. They comprise about 3% of the human genome and are mostly inert remnants of ancient elements that have accumulated deleterious mutations over the course of evolution. Retrotransposons on the other hand still have some active representatives that have retained the ability to duplicate themselves via a "copy-and-paste" mechanism, using an RNA intermediate converted into a

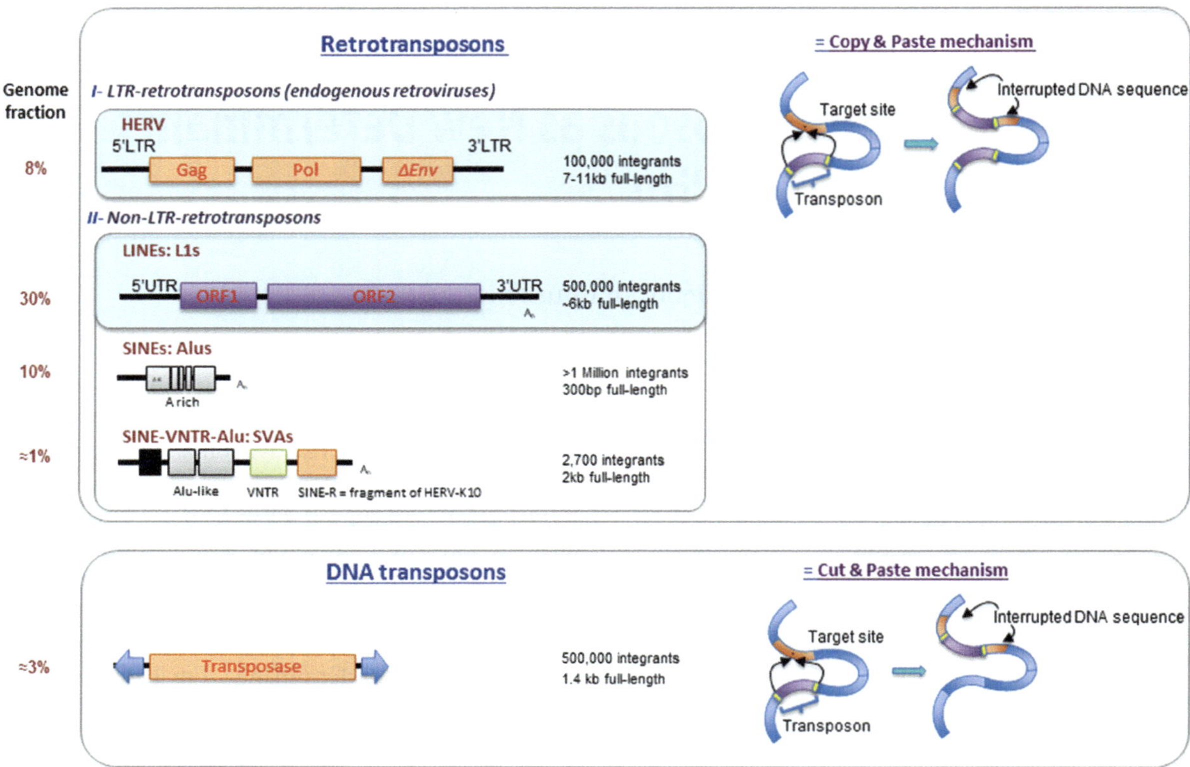

Figure 6.1 Transposable elements in the human genome. The retrotransposon type, the structure of representative retrotransposon elements, and the size, copy number and contribution (percentage) to the human genome as well as the mechanism of mobilization are shown. There are two different classes of transposons: (1) DNA transposons are DNA sequences which move by a cut-and-paste mechanism via an element-encoded transposase; and (2) retrotransposons are DNA sequences that transpose through an RNA intermediate by a copy-and-paste mechanism. Abbreviations for human endogenous retrovirus (HERV): LTR, long terminal repeat; Gag, group specific antigen; Pol, polymerase; Env, envelope protein. For LINE (L1): UTR, untranslated region; ORF1 and ORF2, open reading frames. For *Alu*: A and B, component sequences of the RNA polymerase III promoter; A rich, adenosine-rich segment. For SINE-R/VNTR/*Alu* (SVA): VNTR, variable number of tandem repeats; SINE-R, domain derived from a HERV-K. An, poly(A) tail.

complementary DNA (cDNA) by reverse transcription. While transposons have been largely ignored or even purposely discarded from genomic analyses for a long time, nowadays increasingly technological and computational methods have been specifically developed to allow accurate classification, evaluation of their activity, and mapping of new integration events.

So far three families of retrotransposons have been recognized as competent for mobilization in the human genome: the LINE (Long INterspersed Elements), *Alu*, and SVA (SINE-R/VNTR/*Alu*) elements. They all rely on a combination of host-provided cellular factors and LINEs-encoded proteins to fulfill their life cycle. These mobile families belong to the non-LTR (long terminal repeat) class of retrotransposons. They oppose the LTR class of retrotransposons, also known as human endogenous retroviruses (HERVs), which are considered to be mostly immobile.

6.2.1 LTR Elements

The HERVs comprise 7–8% of the human genome. These elements evolved through the endogenization of infectious retroviruses, which became part of our genetic material after the nonlethal insertion of a viral particle in the germline, and passage of this integrated proviral sequence to the offspring by Mendelian inheritance. Further expansion in copy number occurred by several rounds of either reinfection or intracellular retrotransposition [5]. Among the four known retroviral genera, three have given rise to HERVs, the gamma, the beta, and the spuma retroviruses. Around 7–11 kb long, full-length HERVs resemble their retrovirus ancestors in their structure (Figure 6.1): they are flanked on both 5′ and 3′ sides by LTR sequences that contain binding motifs for cellular transcription factors, while the internal part encodes *gag* and *pol* proteins, which notably encode the reverse transcriptase and integrase activities necessary for retrotransposition. They usually lack a functional envelope (*env*) gene, which prevents most of the HERVs from assembling infectious particles and relegates them to an intracellular existence. Among the 100,000 HERV copies estimated to reside in the human genome, only 3,500 more-or-less complete elements have been identified. Fragments of HERVs are more commonly found, usually in the form of "solo" LTRs, which result from the deletion of intervening sequences by recombination between LTR repeats.

Thirty-one individual groups of HERVs can be distinguished in the human genome, ranging in copy number from one to several thousands. The HERV nomenclature is traditionally based on the host tRNA species they use as a primer for reverse transcription [6]. For example, HERV-K members use a lysine tRNA, while the HERV-H group uses a histidine tRNA. However, this classification cannot apply when the tRNA primer sequence is not known at the time of the discovery of novel elements; in these cases, designation becomes arbitrary and based on neighbor genes, the name of the probe used for cloning, the chromosomal location, or any type of discriminating information. More recently, efforts have been made toward the development of phylogenetic methods to cluster HERV subfamilies according to LTR sequence similarities [7].

Most groups of HERVs are also found in Old World monkey and ape genomes, suggesting that the majority of retroviral endogenization events occurred some 30 million years ago. Nonetheless, some species-specific ERVs have been identified in human and nonhuman primate genomes, demonstrating the continuous incorporation of new elements throughout evolution. Although numerous polymorphic HERVs have been identified in the human population, they mostly consist of contractions from full-length elements into solo LTRs by recombination. There are very few insertional HERV polymorphisms, defined by alleles present in both the pre- and the post-integration forms, which attests to the relative immobility of these elements in the recent era of modern humans. Less than a dozen full-length HERVs are dimorphic, and they are all of the HERV-K type, the most recent retroviral species to have entered our genome. This situation is in striking contrast to the high rate of insertional polymorphisms of LTR retrotransposons in the mouse genome, or the high frequency of non-LTR retrotransposon polymorphisms in the human genome. The relative dormancy of HERVs in the human genome may reflect fast evolutionary erosion and/or the existence of potent restraining pathways acting specifically on these elements.

Because of this apparent immobility, HERVs are often just considered as fossilized versions of ancient retroviruses. Nonetheless, they can be transcriptionally active and produce retroviral proteins that can be co-opted for useful physiologic functions. Syncytin proteins provide a well-characterized example of endogenous retrovirus domestication. Encoded by the *env* gene of HERV-W and HERV-FRD elements, they play a key role in placentation by promoting membrane fusion and syncytium formation [8]. Interestingly, high expression levels of other HERV types, such as HERV-R and HERV-K, have been reported in the human placenta, suggesting an evolutionary, regulatory, or functional connection between this short-lived mammalian-specific reproductive organ and HERV biology. Their altered states may be involved in some human placental pathologies [9].

6.2.2 Non-LTR Elements

The origin of non-LTR retrotransposons is unclear, as they do not resemble infectious retroviruses in their structure or their transposition cycle. They can be stratified into three subgroups: Long INterspersed Elements (LINEs), Short INterspersed Elements (SINEs), and SINE-R/VNTR/*Alu* (SVA) elements. These elements have nothing in common in terms of size and structure. The LINEs represent the most abundant retrotransposon family in humans and in all mammals in general. LINE-1 (L1) elements are considered to be the master retrotransposons, as they account for 30% of the genome, have self-propagating properties, and can provide in *trans* the machinery required for the mobilization of SINEs and SVAs [1].

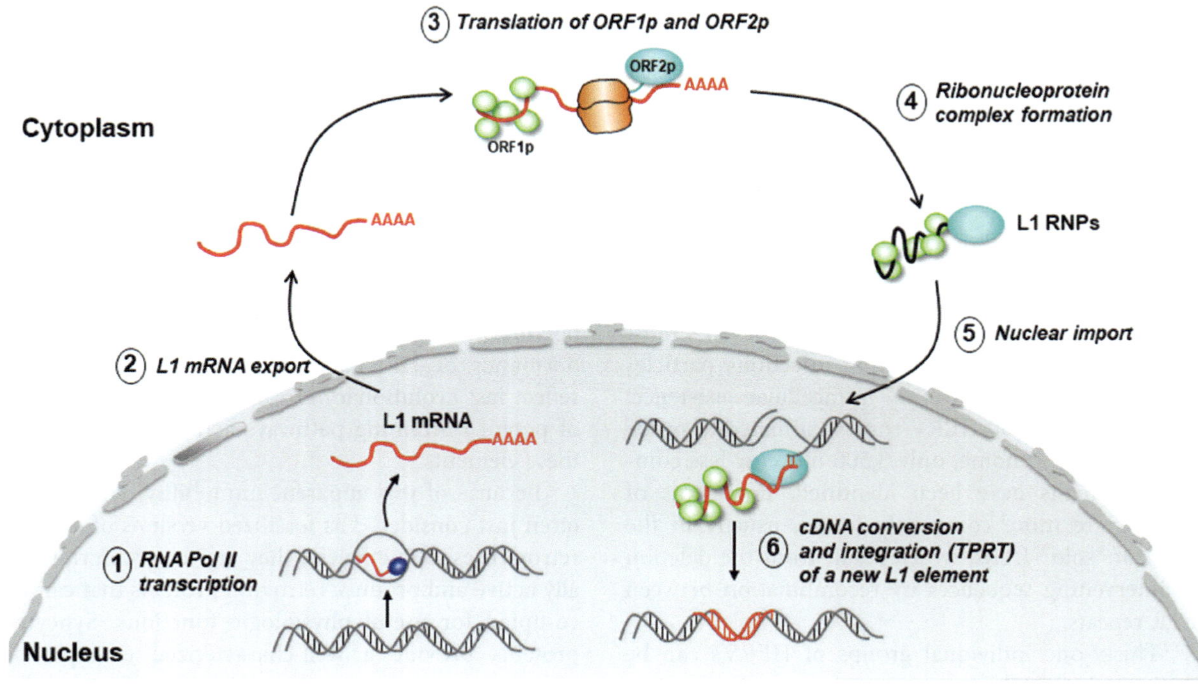

Figure 6.2 L1 retrotransposition cycle. (1) L1 life cycle starts with transcription of the element by RNA polymerase II (RNA Pol II). (2) The L1 mRNA is then exported to the cytoplasm. (3) L1-encoded proteins (ORF1p and ORF2p) are translated and assemble with the L1 mRNA into ribonucleic acid particles (L1 RNPs), which are then imported into the nucleus. (4) The ORF2p-encoded endonuclease nicks genomic DNA in the host genome and is responsible for the reverse transcription of the L1 mRNA. Reverse transcription occurs at the site of integration, in a process referred to as target-primed reverse transcription (TPRT) and is then integrated. The mechanism of second cDNA strand synthesis remains unclear.

Full-length L1 elements are about 6 kb in length and are flanked by 5′ and 3′ untranslated region (UTR) sequences (Figure 6.1). They are transcribed in an RNA polymerase II-dependent manner and translated into two open reading frames, ORF1 and ORF2, which encode the necessary machinery for mobilization. ORF1 is an RNA chaperone, which coats and protects L1 mRNAs, while ORF2 provides the enzymatic activities of an endonuclease and a reverse transcriptase. Contrary to LTR retrotransposons, the reverse transcription of L1 elements is coupled to chromosomal integration, therefore taking place in the nucleus and not in the cytoplasm, in a process known as target-primed reverse transcription (TPRT) (Figure 6.2).

From an evolutionary point of view, L1s are thought to have been part of our genome for hundreds of millions of years, meaning that they predated LTR retrotransposons. Several L1 subfamilies have succeeded each other, each replacing the last as the dominant active member. The majority of L1s are shared by human and chimpanzee genomes. Nowadays, these elements are inert, having accumulated mutations, truncations, and deleterious rearrangements. However, around 1,000 L1s are specific to humans and among them, the Ta (transcribed, subset a) or L1PA1 subfamily has around 100 members, which have intact ORFs, can be expressed, and are competent for mobilization, as shown in cellular retrotransposition assays and their participation in interindividual insertional polymorphisms. Ta elements contain a sequence variant signature within their 3′ UTR, in the form of an ACA instead of a GAG trinucleotide. The L1 Ta members seem to be the only autonomous L1s residing in the human genome, if not the only autonomous retrotransposons of all kind in our genome. Their active machinery allows not only their own retrotransposition but also the *trans*-mobilization of nonautonomous L1 elements that may be transcriptionally active, with an intact promoter but deleted ORF sequences. As mentioned earlier, L1s are also responsible for the mobility of other families of non-LTR retrotransposons, the SINES and the SVAs. Finally, they can promote the mobilization of cellular mRNAs and the creation of intron-less retroposed copies of genes into new genomic locations, which can degenerate into pseudogenes or develop novel functions under selective pressure.

Recent estimation inferred from high-throughput sequencing approaches gave a total L1 retrotransposition rate of 1 in 100 to 1 in 150 births, with 1 in 20 of these births being associated with a disease phenotype. Since the first association of a hemophilia case with a de novo germline L1 insertion in 1988, approximately 95 disease-causing mutations have been attributed to L1 retrotransposition events. They can act as mutagens by directly disrupting exons or inducing misexpression when they insert into introns, or in the proximity of promoter or enhancer regions.

SINEs have no common history with LINES, although they strictly rely on L1-encoded proteins for their mobilization. Among them, *Alu* sequences are the most abundant, with an estimated copy number of 1 million, which accounts for about 10% of human genomic mass [10]. Around 5,500 are specific to humans and not shared by the chimpanzee genome. *Alu* sequences evolved some 65 million years ago from the 7SL RNA, which serves in the ribosomal signal recognition particle for addressing nascent proteins to the endoplasmic reticulum. This sequence may be required for L1-mediated transposition, by placing the *Alu* mRNA in close proximity to the L1 ribonucleoprotein particles. Similar to their precursors, *Alu*s are around 300 nucleotides in length, contain an RNA polymerase III promoter, and end with an A-rich tail (Figure 6.1). Among thousands of *Alu* sequences that have an intact promoter and may be transcriptionally active in the human genome, the *Alu*Ya5/8 and *Alu*Yb8/9 subfamilies account for the vast majority of de novo insertions in humans. Comparisons between individuals in human and primate populations give an estimate of one *Alu* de novo insertion every 20 live births. They provide the highest number of all retrotransposon-derived insertional polymorphisms in the human genome.

The last family of active, though nonautonomous retrotransposons in humans comprises the SVA elements. While LINEs and SINEs have colonized the genome of multiple animal species, vertebrate and invertebrate, SVAs have evolved uniquely in primate genomes, some 25 million years ago. They have a composite structure, combining from 5′ to 3′; an *Alu*-like segment consisting of two inverted *Alu* fragments; a variable number of tandem repeats (VNTR) region, which is made of copies presumably derived from the SVA2 element found in rhesus macaques and humans; a fragment derived from an extinct HERV-K10 element (SINE-R); and a poly(A) tail (Figure 6.1). The events that led to the emergence of this chimeric entity in the hominid lineage are unknown. Although quite convergent in terms of sequence identity, they are very heterogeneous in size, ranging from 700 bp to 4 kb, with a canonical 2-kb long element. SVA elements are likely to be transcribed by RNA polymerase II. They are mobilized by L1-encoded proteins, potentially with the help of their *Alu*-derived sequence, and de novo SVA insertions have been associated with several human diseases. In agreement with their evolutionarily young age, their genomic number is less than the other retrotransposon classes, estimated at 2,700 copies, among which around 800 are specific to humans (subfamilies SVA E and F), while the others are shared with chimpanzees (subfamilies SVA A–D).

6.3 Control Mechanisms of Retrotransposon Activity

As stressed above, derepression of individual retrotransposons can cause diseases by generating insertional mutations or interfering with proper gene control. Massive derepression of retrotransposons has an even more dramatic outcome as it typically leads to lethality and sterility, depending on whether it occurs in the developing embryo or in germline cells, as exemplified in mouse mutant models [3]. To counteract this, the host has evolved various defense mechanisms that maintain retrotransposons under control.

The most potent control of retrotransposon activity is evolution. Inserted retrotransposon copies are subject to host pressures that alter their DNA sequence, either by favoring the maintenance of an empty allele over an allele where the retrotransposon has inserted, or by allowing genetic drift. So, over time, unless they provide an important function for the host, retrotransposons are doomed to become inactive. However, there is still a subset of transcriptionally and retrotransposition-competent elements in our genome. These have to be dealt with on a short-term basis, and the cell uses a plethora of strategies to target them at various stages of their life cycle (Figure 6.2). Retrotransposon families are greatly diverse in sequence, numbers, and evolutionary origins. Accordingly, these restricting pathways are usually based on proteins that are universal and flexible, and are used for other cellular functions, generally related to gene expression control. However, there are also some examples of specialized proteins that are dedicated to either LTR or non-LTR elements.

Currently known host defense mechanisms target (1) transcription of retrotransposons, (2) post-transcriptional processing of retrotransposon mRNAs, and (3) integration of new retrotransposon copies. Most of the corresponding restraining factors have been identified and functionally studied in mammalian model organisms such as mice [3], but the same rules seem to apply to human retrotransposons. While not being directly considered a host defense strategy, it is important to mention that cell division could be a strict requirement for the retrotransposition cycle. In vitro reporter assays notably showed that L1 retrotransposition is strongly reduced in G0-arrested cells [11], implying that nonmitotic cells may be less sensitive to retrotransposon activity. This may be linked to the need for nuclear membrane breakdown to allow the nuclear import of L1 RNPs (Figure 6.2).

6.3.1 Transcriptional Control

Mobilization of a retrotransposon first requires its expression. The primary way to prevent retrotransposition is therefore to block the transcription of genomic copies of retrotransposons. Chromatin modifications, also referred as epigenetic modifications, locally modulate the compaction of the genome and its accessibility to transcription factors. Both the DNA and histone protein components of chromatin are subject to secondary biochemical modifications. Dense DNA methylation assembled at promoter regions is a known potent inhibitor of transcription. The majority of methylated cytosines in human genomic DNA are actually contained in retrotransposon sequences, and indeed it has been proposed that DNA methylation evolved primarily as a defense mechanism against transposable elements [12]. Notably L1, *Alu* and SVA elements have all been shown to be methylated in human somatic tissues. The enzymes responsible for the DNA methylation reaction are the DNA methyltransferases (DNMT) and, among them, DNMT1, DNMT3A and the cofactor DNMT3L cooperate in methylation-dependent transcriptional repression of retrotransposons of the LTR and non-LTR classes in mice. Similarly, chemical impairment of genomic methylation patterns with 5-azadeoxycytidine leads to retrotransposon reactivation in human cellular systems. In addition to members of the DNMT family, proteins that assist the DNA methylation reaction also have a role in retrotransposon repression. Among them, Lsh (lymphoid-specific helicase) is a member of the SNF2 family of chromatin-remodeling ATPases that facilitate the access of DNMTs to DNA, and is required for the methylation and repression of retrotransposons in mice.

Various repressive histone modifications have also been linked to transcriptional repression of retrotransposons. In particular, H3K9, H4K20, and H3K27 methylation are commonly found at 5′-LTR or UTR sequences of retrotransposons, and genetic deletion of the enzymes driving these modifications can lead to retrotransposon reactivation [13]. Proteins that bind H3K9 methylated histones are also required for retrotransposon repression. Notably, in mouse embryonic stem cells, the TRIM28/KAP1 (KRAB-associated protein 1) transcriptional repressor is recruited to and reinforces H3K9 methylation marks at ERVs only, and not at any non-LTR elements. Interestingly, TRIM28/KAP1 also silences infectious retroviruses. This example illustrates that repressors involved in innate immunity against retroviruses have a conserved function on endogenized retroviruses.

6.3.2 Post-Transcriptional Control

Retrotransposons have an RNA-centered mode of replication. Retrotransposon transcripts serve both as messengers to produce retroviral proteins following translation and as templates to generate new genomic copies, following reverse transcription. Accordingly, different defense strategies have been developed against their RNA phase, and at least two forms of RNA alteration, RNA editing and RNA interference, are known to target retrotransposon transcripts.

The term "RNA editing" refers to molecular processes that modify the information content of the RNA molecule, most often by nucleoside deamination. By altering RNAs and therefore promoter strength and/or amino-acid sequence of encoded proteins, RNA editing enzymes can reduce the activity of new copies of retrotransposons. The ADAR family of RNA editases converts adenosine residues into inosines. Analysis of the human transcriptome has revealed that ADARs target double-stranded RNAs that are formed from inverted *Alu* and L1 repeats. The APOBEC proteins form another family that catalyzes the deamination of cytosine residues into uracils and which have greatly expanded in the primate lineage. APOBEC3G was originally shown to reduce HIV replication by inducing the accumulation of uracil mutations on the nascent retroviral cDNA strand and subsequently inactivating the newly integrated copy [14]. Retrotransposition assays have shown that APOBEC3A, 3B, 3C and 3F enzymes are also potent restrictors of different classes of LTR and non-LTR retrotransposons in human and mouse cells. However, retrotransposon restriction triggered by some of the APOBEC3 proteins does not involve C to U conversion. It has been hypothesized that these enzymes mediate cytoplasmic sequestration of L1 RNA and/or L1-encoded proteins, or directly inhibit L1 ORF activity [15]. Considering the master role of L1 in retrotransposition biology, this would not only impact on L1 activity but could also render the L1 machinery inaccessible to nonautonomous retrotransposons of other classes.

RNA interference (RNAi) represents another post-transcriptional mechanism of retrotransposon suppression [16]. In this case, small RNAs operate through homology-based recognition to induce the degradation of complementary retrotransposon transcripts via recruitment of the RNA-induced silencing complex (RISC). The slicing activity in this complex is provided by the Argonaute protein family: the human genome encodes four classical Argonautes (AGO1 to 4), and four germline-restricted members named PIWI (PIWIL1 to 4). Only three *PIWI* genes are present in the mouse genome, which lacks the homolog of PIWIL3.

There are three classes of small RNAs in mammals. The microRNAs (miRNAs) and the endogenous small interfering RNAs (endo-siRNAs) require the DICER protein for their production and associate with the canonical AGO proteins. The piwi-interacting RNAs (piRNAs) are specifically produced in the germline, where they require and associate with PIWIL proteins only, and are produced in a DICER-independent manner. They also differ in size, ranging from 25 to 30 nucleotides, while miRNAs and endo-siRNAs are traditionally 22–24 nucleotides long. So far, all these small RNA types have been linked to retrotransposon biology. DICER deficiency induces the cytoplasmic accumulation of *Alu* RNAs, whose toxic effects are responsible for the development of age-related macular degeneration [17]. PIWIL mutations in mouse compromise piRNA production and lead to global reactivation of LTR and non-LTR retrotransposons in the male germline, which completely compromises sperm production [3,18,19]. Interestingly, the female germline does not use piRNAs to target retrotransposons, but may rather rely on endo-siRNAs. Finally, any component of the RNAi pathway represents a potential candidate for retrotransposon restriction. Among them, the MOV10 RNA helicase, which helps RNA processing within the RISC complex, has been recently shown to severely reduce the retrotransposition rate of L1, *Alu*, and SVA elements in cellular assays.

6.3.3 Integration Control

Finally, the last stage of the retrotransposon life cycle, the integration of the cDNA copy into a new genomic location, is also subject to restriction. The ERCC1/XPF heterodimer complex has endonuclease activity and is involved in DNA repair mostly through the nucleotide excision repair (NER) pathway. Reduction of XPF in human cell lines increases L1 retrotransposition, suggesting that intermediates of the L1 retrotransposition process may be cleaved by this enzyme. Interestingly, other DNA repair enzymes have an inverse effect on retrotransposition: the double-strand break repair protein ATM indeed facilitates L1 integration [20], indicating that various DNA repair pathways may be able to recognize and process retrotransposon integration intermediates in a positive or negative manner.

6.4 Retrotransposons and Fertility: What We Learned from Animal Models

The only opportunity for retrotransposons to propagate is by generating new genomic copies in cells destined for the next generation, i.e. germ cells or early embryonic cells upstream from the formation of the germline. Accordingly, a large number of retrotransposon repressors are specifically expressed in the developing germline. Moreover, functional studies in mice have revealed that these germline factors act as major determinants of reproductive fitness, in particular in males [3,4,18,19]. Genetic, cytologic, and biochemical studies in mouse models

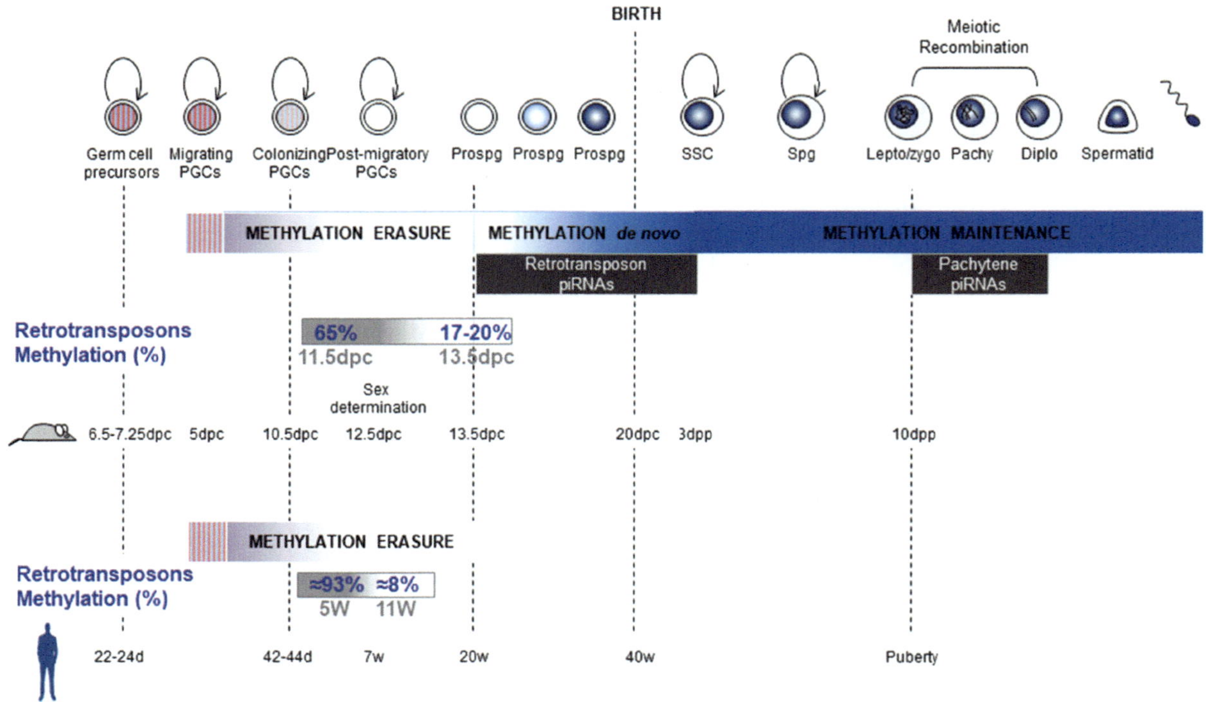

Figure 6.3 Kinetics of spermatogenesis and piRNA/DNA methylation pathway in mice and developmental equivalents in humans. Germ cell emergence occurs around 6.25–7.25 dpc in mouse (22–24 days in human). Global erasure of genomic methylation starts during the migration phase of primordial germ cells (PGCs) and culminates at the time of their incorporation into the genital ridges, at 10.5 dpc. At 12.5 dpc, sexual determination is initiated (7 weeks in human). Retrotransposon methylation strongly decreases during this period (65% to 17–20% at 11.5 dpc and 13.5 dpc in mouse [21–24]; 93% to 8% at 5 weeks and 11 weeks in human [25,26]). At 13.5 dpc, mouse PGCs become prospermatogonia, as they enter mitotic arrest (starts around 20 weeks in human). This cellular change coincides with a wave of *de novo* methylation, which occurs in association with retrotransposon-derived piRNAs, and is completed before birth. Rapidly, the prospermatogonia differentiate into self-renewing spermatogonial stem cells (SSCs) around 3 dpp in mouse. DNA methylation is stably maintained throughout the different phases of postnatal spermatogenesis, until the release of mature spermatozoa. At the time of entry into the prophase of meiosis I, a second population of piRNAs is produced, called pachytene piRNAs, whose function is not related to retrotransposon control. Methylation states are depicted here with increasing intensities of gray in PGCs and then blue from prospermatogonia, representing increasing methylation levels. The timing of methylation erasure, establishment and production of various populations of piRNAs is not precisely known in humans. Dpc, days post coitum; dpp, days postpartum; ProSpg, prospermatogonia; Spg, spermatogonia; Lepto, leptotene; Zygo, zygotene; Pachy, pachytene; Diplo, diplotene.

have led to the elaboration of a molecular scenario occurring in male germ cells during fetal life that provides lifelong protection of the genetic material against retrotransposons. This involves a relay between RNA interference mechanisms centered on piRNAs, and transcriptional control mediated by DNA methylation.

Primordial germ cells (PGCs), the progenitors of the germline, are set aside very early from the other cell lineages of the developing embryo, around 6–7 days post coitum (dpc) in mice (Figure 6.3). These specified germ cells undergo massive reprogramming of their epigenetic repertoire, which culminates at the time they colonize the fetal gonads, around 10 dpc. At 12.5 dpc, expression of the Y-linked *Sry* gene determines a male fate in XY individuals, while its absence leads to female differentiation in XX animals. In males, this phase is immediately followed at 13.5 dpc by mitotic arrest of germ cells, which take the name of prospermatogonia. Germline epigenetic reprogramming implies a genome-wide loss of DNA methylation, which leaves the retrotransposons unleashed [21]. However, fetal prospermatogonia are equipped with a specialized slicing machinery, the PIWI proteins, which can cleave the retrotransposon transcripts into 25–30 nucleotide piRNAs. In mouse, retrotransposon-derived piRNAs depend on MILI

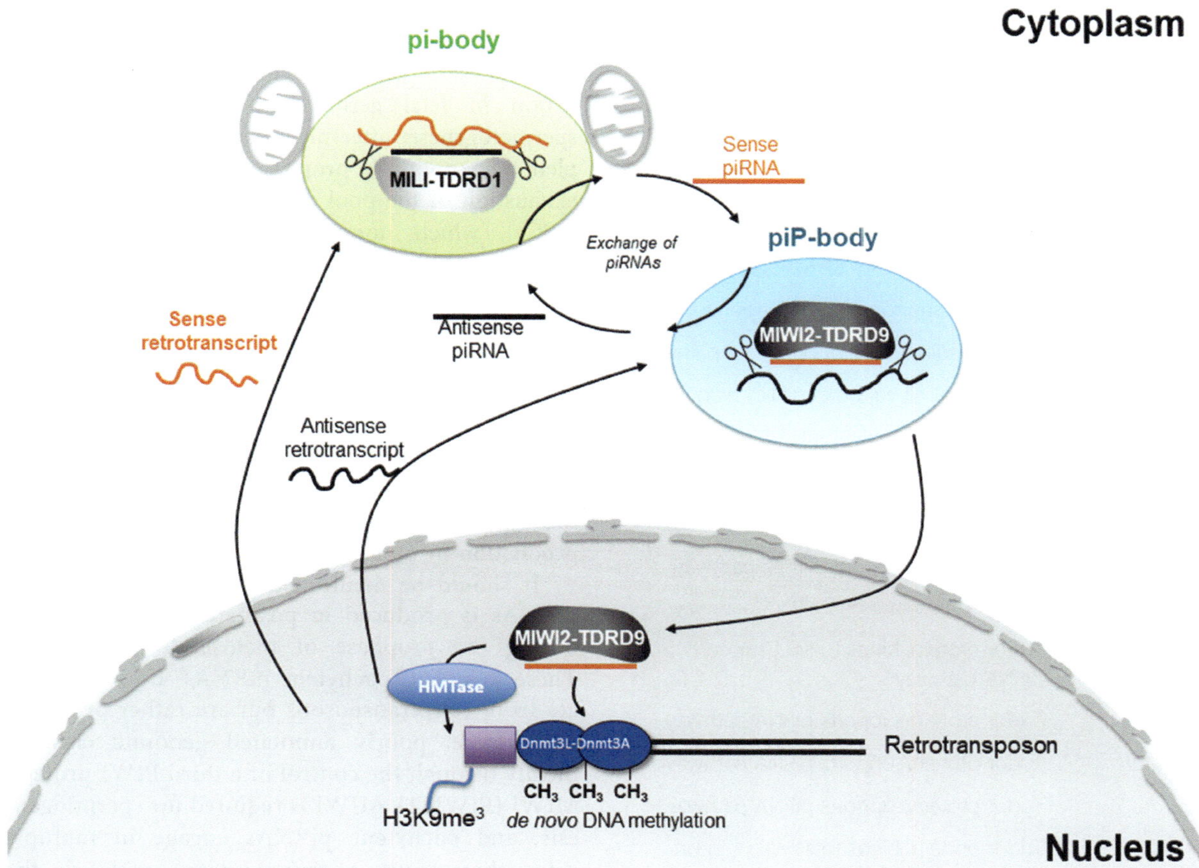

Figure 6.4 piRNA/DNA methylation pathway acting in mammalian germ cells for the repression of retrotransposons. This pathway is active in fetal male germ cells, the prospermatogonia, present from 13.5 dpc to birth in mouse. Sense (gray line) and antisense (black line) mRNAs (retrotranscripts) are produced from genomic retrotransposon sequences interspersed within the nucleus. These mRNAs get cleaved into sense and antisense piRNAs, respectively. Two PIWI proteins, MILI and MIWI2, interacting with Tudor domain-containing (TDRD) proteins 1 and 9, respectively, localize to different cytoplasmic granules (pi- or piP-bodies). Sense and antisense piRNAs associate with MILI-TDRD1 and MIWI2-TDRD9 complexes, respectively, where they will further lead to the recognition and cleavage of retrotransposon mRNAs in opposite orientation, and promote amplification for post-transcriptional degradation of retrotranscripts through the so-called "ping-pong" loop. The relay between post-transcriptional and transcriptional control is achieved by the translocation of the MIWI2–piRNA complex to the nucleus, where it facilitates *de novo* DNA methylation (by the recruitment of the DNMT3L/DNMT3A machinery) to stably silence retrotransposons. Once established, DNA methylation patterns are then stably maintained throughout the end of spermatogenesis.

(PIWIL2) and MIWI2 (PIWIL4) proteins and feed what has been referred to as the "ping-pong" mechanism (Figure 6.4), which relies on the hypothesis that both sense and antisense retrotransposon mRNAs are transcribed at the same time and can serve as substrates for sense and antisense piRNAs, respectively. In this model, the MILI protein loaded with sense piRNAs recognizes and degrades antisense retrotranscripts, generating antisense piRNAs that are loaded into MIWI2. The MIWI2-piRNAs can then target sense retrotransposon mRNAs, generating new sense piRNAs that will bind to MILI, thereby forming a positive amplification loop. This chain of events occurs in specialized cytoplasmic compartments, the pi- and piP-bodies, which gather general components of RNA biogenesis and processing [18]. Following this post-transcriptional degradation of retrotransposons, MIWI2 loaded with antisense piRNAs is able to translocate to the nucleus, where it triggers *de novo* methylation of genomic copies of retrotransposons by recruiting, in an unknown manner, the DNA

Table 6.1. Germline repressors of retrotransposon activity in mouse

Gene	Biochemical function/role
DNMT3A	DNA methyltransferase/*de novo* methylation
DNMT3L	Cofactor of DNMT3A/*de novo* methylation
Fkbp6	FK-506 binding protein 6/supports piRNA biogenesis
Gasz	Unknown/supports piRNA biogenesis
Gtsf1	Unknown/likely to support piRNA pathway
MitoPLD	Lipid signaling/supports piRNA biogenesis
Miwi2	RNaseH (PIWI family)/piRNA biogenesis
Mili	RNaseH (PIWI family)/piRNA biogenesis
Maelstrom	Unknown/supports piRNA pathway
Mov10L1	RNA helicase/supports piRNA pathway
Mvh	RNA helicase/supports piRNA pathway
Tdrd1	Tudor protein, binds to MILI/supports piRNA pathway
Tdrd5	Tudor protein/supports piRNA pathway
Tdrd6	Tudor protein/supports piRNA pathway
Tdrd7	Tudor protein/supports piRNA pathway
Tdrd9	Tudor protein, binds to MIWI2/supports piRNA pathway
Tex19.1	Unknown/supports piRNA pathway

methyltransferase DNMT3A and its cofactor DNMT3L (Figure 6.4). DNA methylation is completed before birth, and then maintained in postnatal germ cells as they undergo spermatogonial proliferation, meiosis, and maturation into spermatids and finally spermatozoa (Figure 6.3), allowing the stable repression of retrotransposons for the rest of the reproductive lifetime of the male mouse.

As an illustration of the importance of this pathway for spermatogenesis, some 17 genes have been discovered to be involved in the relay between piRNA and DNA methylation in mouse, for the purpose of stable repression of retrotransposons in the male germline (Table 6.1). As an example, various Tudor proteins (TDRD) are responsible for organizing the pi- and piP-bodies, through the recognition of arginine modifications carried by MILI and MIWI2 proteins [18]. Various RNA helicases, such as the Mouse Vasa Homolog (Mvh) and Mov10L1, are also required. Single inactivation of any of these 17 genes leads to retrotransposon reactivation in fetal germ cells, profoundly affecting spermatogenesis after birth, in the form of a complete interruption in prophase of meiosis I and rapid exhaustion of the pool of spermatogonial stem cells (SSCs), which normally sustain spermatogenesis throughout life [3]. Retrotransposon reactivation seems therefore deleterious for the process of homologous recombination and germline stem cell renewal. As a result, mutant males do not produce any spermatozoa and exhibit a Sertoli-only testicular phenotype after a few weeks (Figure 6.5). Interestingly, the female germline seems more tolerant to alteration of the piRNA pathway, as none of the aforementioned genes has an effect on oocyte production in mouse mutant models.

It should be mentioned that a second class of piRNAs is produced in postnatal spermatogenesis during the prophase of meiosis I (Figure 6.3). These so-called “pachytene piRNAs” do not originate from retrotransposons but are rather produced from large, poorly annotated genomic clusters, mostly through the control of a third PIWI protein, MIWI (PIWIL1). MIWI is required for spermiogenesis, and pachytene piRNAs engage in multiple roles throughout spermatogenesis, such as the post-transcriptional silencing of trophectoderm transcripts [36].

6.5 Transposons and Fertility in Humans: What We Suppose and What We Really Know

The number of genes involved in the piRNA/DNA methylation pathway of retrotransposon control and the dramatic consequences of their inactivation provides a strong basis for considering retrotransposon repressors as major guardians of fertility in males. Moreover, the piRNA pathway has been shown to be conserved in the germline of all animals: PIWI homologs have been identified and genetically analyzed in *Drosophila*, *Caenorhabditis elegans*, and zebrafish, and they are systematically required for fertility through retrotransposon silencing. What is the evidence so far that the piRNA/DNA methylation pathway may play a similar role in human reproduction?

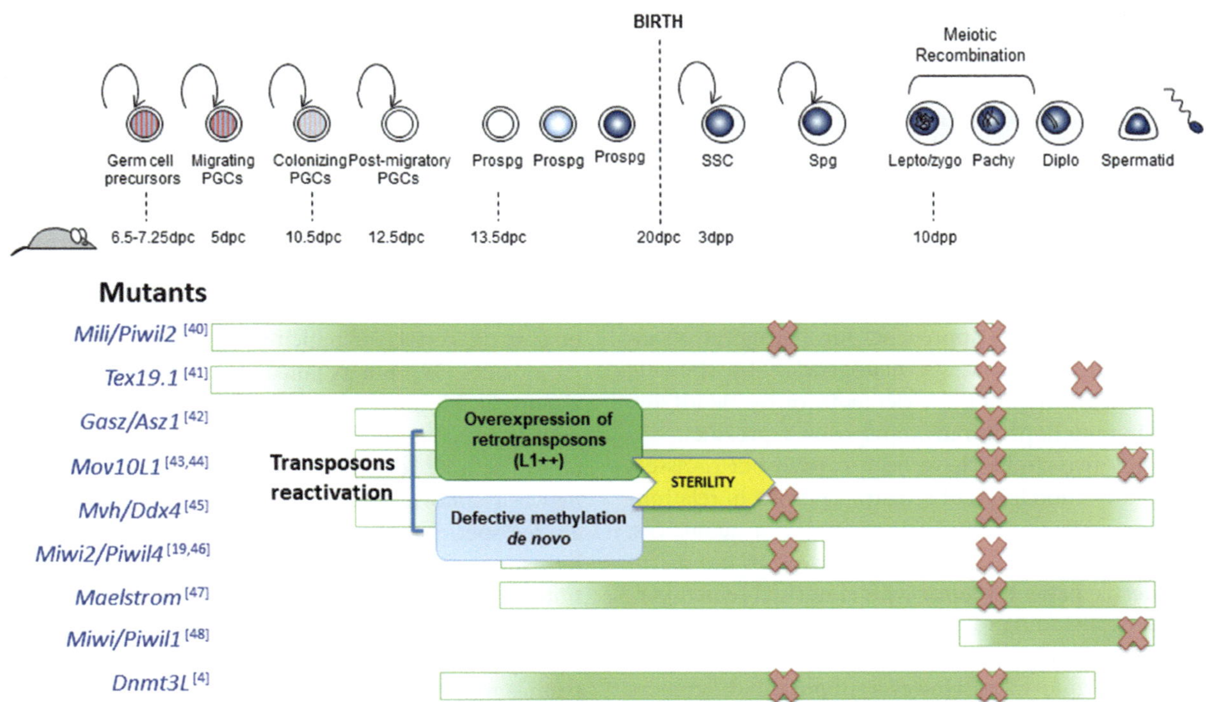

Figure 6.5 Mouse mutants of germline repressors of retrotransposons. Inactivation of the genes involved in the repression of retrotransposons in the male germline (i.e. in the ping-pong cycle such as *Mili/Piwil2* [27], *Tex19.1* [28], *Gasz/Asz1* [29], *Mov10L1* [30,31], *Mvh/Ddx4* [32], *Miwi2/Piwil4* [19,33], *Maelstrom* [34], *Miwi/Piwil1* [35], *DNMT3L* [4]) leads to retrotransposon reactivation (green) in fetal germ cells, which profoundly affects spermatogenesis after birth and thus induces failure to produce spermatozoa in males.

6.5.1 Developmental Kinetics of Human Spermatogenesis

As a first step toward involving retrotransposons as important determinants of male fertility, it is important to anticipate at which stage the piRNA/DNA methylation pathway may be active during human development. The main actors of this pathway are all conserved between mouse and human, from the PIWI proteins to the DNMT3A–DNMT3L *de novo* methylation complex. There are even four PIWI genes in humans, while the mouse genome contains three of them. Although high-throughput sequencing analyses have confirmed the existence of pachytene piRNAs in testis samples of adult men, retrotransposon-derived piRNAs have not been observed in humans so far. This does not imply that they do not exist. It may rather reflect their restricted access linked to their fetal origin, estimated to span the lifetime of prospermatogonia in the mouse, from their emergence at 13.5 dpc to their conversion into dividing SSCs at birth (Figure 6.4). Developmental equivalences would therefore suggest that the piRNA machinery may be active starting around 20 weeks after conception in human male fetuses, at the time when PGCs initiate their differentiation into prospermatogonia. Fundamental differences exist though: in mouse, PGC to prospermatogonia conversion occurs in a synchronous manner and prospermatogonia then become SSCs at birth. In human, prospermatogonia appear throughout the duration of fetal gonad development and are maintained until puberty. Retrotransposon piRNAs may therefore continue being produced after birth in humans.

Cycles of DNA methylation erasure and establishment occur in the developing germline of all mammalian species examined so far. Not only retrotransposons undergo this process, but also notably genes subject to genomic imprinting (also known as imprinted genes, as described in Chapter 5). Although the dynamics of retrotransposon methylation have not been specifically studied in human spermatogenesis, analyses of paternally imprinted loci have discovered that the completion of DNA methylation patterns sometimes occurs postnatally, before puberty. While a large number of studies have tried to make a correlation between methylation anomalies at imprinted genes and fertility in humans, very few have focused on the relationship between retrotransposon control and sperm criteria.

6.5.2 Transposon Control and Sperm Criteria

There are two main approaches for linking retrotransposons to abnormal spermatogenesis: by showing altered repression of these elements, or by identifying genetic mutations in retrotransposon repressors.

Only five studies have been carried out to assess the methylation levels of retrotransposons in relationship to sperm quality. Concerning *Alu* sequences, lower methylation levels were observed in the sperm of infertile patients and/or with abnormal semen parameters, including oligozoospermia, asthenozoospermia, and teratozoospermia, compared with normospermic men. However, reported differences are very subtle, with 23.1% and 24.2% of *Alu* methylation measured in altered versus normal sperm, respectively [37] (Table 6.2). For patients with abnormal spermatogenesis, this tendency was nonetheless associated with decreased pregnancy success and live-birth rate after assisted reproductive technologies (24.1% of methylation in sperm samples leading to birth as compared to 22.7% in those who did not [37]), whereas no significant differences were reported in unexplained infertility cases [38]. Concerning L1 methylation, while a tendency to lower average methylation was reported in one study of a small cohort of patients with abnormal sperm parameters [39], others failed to provide evidence for abnormal L1 methylation in the sperm of infertile men [37,40,41]. Further studies are clearly needed to allow definite conclusions about the relationship between abnormal retrotransposon methylation and altered sperm criteria. Moreover, conclusions should be tempered, as abnormal retrotransposon methylation in infertile patients may be equally a cause, or a consequence of altered spermatogenesis. Additionally, gamete manipulation inherent to human sperm studies may be a cause of abnormal retrotransposon control. Of note, a study has concluded that spermatozoa cryopreservation, one of the most common

Table 6.2. Retrotransposon methylation and sperm parameters in humans

Reference	Studied population (N)	Analyses	Transposable elements	Results
Urdinguio et al. (2015) [38]	NZ men (17) Infertile patients without major sperm defects (27)	Bisulfite pyrosequencing	L1, *Alu* Yb8, NBL2, D4Z4	*Alu* methylation was lower in sperm samples from infertile patients (unexplained) than in fertile individuals No significant differences with respect to ICSI outcome
El Hajj et al. (2011) [43]	NZ men (28) OA and/or T men (106)	Bisulfite pyrosequencing	*Alu* and L1 elements	*Alu* methylation was lower in sperm samples from OA±T vs. NZ men and higher in sperm samples leading to live birth
Boissonnas et al. (2010) [39]	NZ men (17) OAT men (22)	Bisulfite pyrosequencing	L1 elements	OAT patients displayed a tendency to lower methylation levels (46 ± 3.8% in NZ vs. 44 ± 4.1% in OAT)
Marques et al. (2008) [41]	NZ men (1) OAT men (4)	Bisulfite sequencing	L1 elements	Ratio of the methylation showed no significant differences
Kobayashi et al. (2009) [40]	NZ men (14) OAT men (10)	Bisulfite–PCR restriction	L1 and *Alu* elements	Ratio of the methylation showed no significant differences

NZ, normal semen parameters (normozoospermia).
OAT, abnormal sperm parameters (O, oligozoospermia; A, asthenozoospermia; T, teratozoospermia).

sperm manipulations, is safe in terms of *Alu* and L1 methylation [42].

Although functional studies in mice have led to a list of 17 proteins involved in safeguarding fertility (Table 6.1), this pathway has so far drawn very little attention in the field of human reproductive genetics. In mice, complete inactivation of these repressors invariably results in azoospermia. Two main spermatogenetic processes are affected: an interruption at meiosis I, and a rapid exhaustion of SSCs that eventually leads to a complete lack of germ cells, termed Sertoli-cell-only syndrome (SCOS). By analogy, mutations in retrotransposon repressors are expected to be associated with nonobstructive azoospermia (NOA) with a block of spermatogenesis at meiosis I, and SCOS in men. However, causal genetic mutations are more often hypomorphic than nullimorphic in humans, so a wider range of infertility phenotypes could be expected. Recently, genetic mutations were investigated and revealed germline mutations in human PIWI in patients with azoospermia due to aberrant histone retention in late-stage germline cells [44]. In this regard, a single nucleotide polymorphism (SNP) located in the 3′ UTR of the *PIWIL2* gene and a nonsynonymous SNP in the *PIWIL3* gene have been associated with a reduced risk of oligozoospermia in a Chinese population of patients recruited in an infertility clinic [45]. However, the biological significance of these polymorphic variants on retrotransposon control is unknown.

Another study has identified lower expression of genes encoding germline retrotransposon repressors in the testes of cryptorchid boys, a condition leading to a high risk of infertility [46]. These included *PIWIL2*, *PIWIL4*, *MAELSTROM*, the Tudor protein *TDRD9*, and the RNA helicases *VASA* and *MOV10L1*. Although an interesting result, the lower expression of these germline genes may just be a consequence of germ cell depletion in the testes of these cryptorchid boys. Similarly, a study reported a gain of methylation at the promoters of *PIWIL2* and *TDRD1*, along with transcript-level downregulation, in adult infertile patients suffering various degrees of secretory spermatogenic failure [47]. Once again, as the analysis was performed on whole-testis biopsies, this may rather reflect a more somatic-like pattern due to germ cell depletion rather than direct alteration of these *PIWI*-related genes. However, interestingly, a slight defect in LINE1 methylation was specifically observed in some cases of early spermatogenetic interruption, at the spermatogonia or spermatocyte stages. This type of investigation may set the stage for further analysis of retrotransposon control in altered spermatogenesis, performed on cellular fractions or in a more genome-wide manner.

6.5.3 Situations at Risk: Endocrine Disruptors and Synthetic Gametes

Environmental factors could also alter retrotransposon control and lead to altered gamete integrity. Exposure to endocrine disruptors has been linked to abnormal gonadal development and gametogenesis, in both animal models and human epidemiologic studies. These molecules are ever more widely represented in our industrial environments. To cite a few, bisphenol A, phthalates, and perfluorinated substances are found in a variety of household consumables such as plastics, cosmetics, pesticides, and cleaning products. These chemicals have the ability to interfere with the metabolism of endogenous hormones, and to impact on their downstream targets, which include the epigenetic setting of the germline. The fetal window of development has been highlighted as particularly susceptible to these compounds, in agreement with the major phases of sexual determination and germline development occurring before birth. Notably, DNA methylation anomalies have been sporadically reported to occur in response to endocrine disruptor exposure during fetal male germ cell development, mostly in rodent models [48]. However, no study has so far been dedicated to the analysis of retrotransposon control following exposure to these factors. Considering the short- and long-term consequences that retrotransposon reactivation can have on gamete production and development of the next generation, this appears as an area of utmost importance to investigate in order to document the origin of transgenerational effects induced by endocrine disruptor exposure.

The recent development of induced pluripotent stem cells (iPSCs), which consists in the forced reprogramming of a somatic cell (such as a fibroblast) into an embryonic-like state of pluripotency, holds great promise for regenerative medicine, including regenerative reproductive medicine. Consequently, attempts have been made toward producing "synthetic" gametes from the differentiation of these iPSCs in a Petri dish, as an unlimited source of reproductive cells for infertile patients [49]. Protocols are currently being optimized, converging toward an obligate transplantation step into a gonad to promote full gametogenesis. However, conversely to the mouse model, for which fertile offspring can be obtained even from sterile sex chromosome trisomic mice [50], a recent study has raised awareness of the potential risks of this method of iPSC-derived gametes by showing that human iPSC derivation induces a dramatic relaxation of L1 control. These cells exhibit a loss of DNA methylation, increased expression and, even more worryingly, a high rate of retrotransposition of L1 elements when compared with parental fibroblasts [51]. Considering the ability of L1-encoded proteins to also promote the mobilization of other retrotransposon classes, the number of L1-mediated insertional mutations has probably been underestimated. This observation seriously questions the safety of using iPSCs as a source of gametes, in particular in human, as their altered genetic integrity induced by L1 reactivation is likely to promote abnormal phenotypes in children conceived by this extreme technique of medically assisted reproduction.

6.6 Concluding Remarks

Studies dedicated to the biology of retrotransposons are of utmost importance for understanding the life cycle of retrotransposons in the germline, the defense routes preventing their activity, and the impact they have on the production of gametes and the quality of the genetic information that will pass to subsequent generations. Retrotransposon repressors appear to represent major determinants of mammalian reproduction and human fertility. Further research exploring the world of retrotransposons in human spermatogenesis, in both normal and pathologic contexts, may be helpful in increasing our understanding of the etiology of male infertility and, eventually, in considering new therapeutic approaches in human.

Acknowledgments

We would like to thank the members of our teams for useful discussion and especially Natasha Zamudio. Patricia Fauque's research is supported by Agence Nationale pour la Recherche (ANR-17-CE12-0014).

References

1. Ostertag EM, Kazazian HH, Jr. Twin priming: a proposed mechanism for the creation of inversions in L1 retrotransposition. *Genome Res* 2001;11:2059–65.
2. Cordaux R, Batzer MA. The impact of retrotransposons on human genome evolution. *Nature Rev Genet* 2009;10:691–703.
3. Zamudio N, Bourc'his D. Transposable elements in the mammalian germline: a comfortable niche or a deadly trap? *Heredity (Edinb)* 2010;105:92–104.
4. Bourc'his D, Bestor TH. Meiotic catastrophe and retrotransposon reactivation in male germ cells lacking Dnmt3L. *Nature* 2004;431:96–9.
5. Feschotte C, Gilbert C. Endogenous viruses: insights into viral evolution and impact on host biology. *Nature Rev Genet* 2012;13:283–96.
6. Burns KH, Boeke JD. Human transposon tectonics. *Cell* 2012;149:740–52.
7. Stoye JP. Studies of endogenous retroviruses reveal a continuing evolutionary saga. *Nature Rev Microbiol* 2012;10:395–406.
8. Esnault C, Priet S, Ribet D, et al. A placenta-specific receptor for the fusogenic, endogenous retrovirus-derived, human syncytin-2. *Proc Natl Acad Sci USA* 2008;105:17532–7.
9. Bolze PA, Mommert M, Mallet F. Contribution of syncytins and other endogenous retroviral envelopes to human placenta pathologies. *Prog Mol Biol Transl Sci* 2017;145:111–62.
10. Lander ES, Linton LM, Birren B, et al. Initial sequencing and analysis of the human genome. *Nature* 2001;409:860–921.
11. Shi X, Seluanov A, Gorbunova V. Cell divisions are required for L1 retrotransposition. *Mol Cell Biol* 2007;27:1264–70.
12. Yoder JA, Walsh CP, Bestor TH. Cytosine methylation and the ecology of intragenomic parasites. *Trends Genet* 1997;13:335–40.
13. Matsui T, Leung D, Miyashita H, et al. Proviral silencing in embryonic stem cells requires the histone methyltransferase ESET. *Nature* 2010;464:927–31.
14. Bishop KN, Holmes RK, Sheehy AM, et al. Cytidine deamination of retroviral DNA by diverse APOBEC proteins. *Curr Biol* 2004;14:1392–6.
15. Beauregard A, Curcio MJ, Belfort M. The take and give between retrotransposable elements and their hosts. *Annu Rev Genet* 2008;42:587–617.
16. Obbard DJ, Gordon KH, Buck AH, Jiggins FM. The evolution of RNAi as a defence against viruses and transposable elements. *Philos Trans R Soc Lond B Biol Sci* 2009;364:99–115.
17. Kaneko H, Dridi S, Tarallo V, et al. DICER1 deficit induces Alu

RNA toxicity in age-related macular degeneration. *Nature* 2011;471:325–30.

18. Aravin AA, van der Heijden GW, Castaneda J, et al. Cytoplasmic compartmentalization of the fetal piRNA pathway in mice. *PLoS Genet* 2009;5(12):e1000764.
19. Carmell MA, Girard A, van de Kant HJ, et al. MIWI2 is essential for spermatogenesis and repression of transposons in the mouse male germline. *Dev Cell* 2007;12:503–14.
20. Gasior SL, Wakeman TP, Xu B, Deininger PL. The human LINE-1 retrotransposon creates DNA double-strand breaks. *J Mol Biol* 2006;357:1383–93.
21. Popp C, Dean W, Feng S, et al. Genome-wide erasure of DNA methylation in mouse primordial germ cells is affected by AID deficiency. *Nature* 2010;463:1101–5.
22. Hajkova P, Erhardt S, Lane N, et al. Epigenetic reprogramming in mouse primordial germ cells. *Mech Dev* 2002;117:15–23.
23. Lane N, Dean W, Erhardt S, et al. Resistance of IAPs to methylation reprogramming may provide a mechanism for epigenetic inheritance in the mouse. *Genesis* 2003;35:88–93.
24. Lerner-Geva L, Boyko V, Ehrlich S, et al. Possible risk for cancer among children born following assisted reproductive technology in Israel. *Pediatr Blood Cancer* 2017;64(4):e26292.
25. Guo F, Yan L, Guo H, et al. The transcriptome and DNA methylome landscapes of human primordial germ cells. *Cell* 2015;161:1437–52.
26. Tang WW, Dietmann S, Irie N, et al. A unique gene regulatory network resets the human germline epigenome for development. *Cell* 2015;161:1453–67.
27. Kuramochi-Miyagawa S, Kimura T, Ijiri TW, et al. Mili, a mammalian member of piwi family gene, is essential for spermatogenesis. *Development* 2004;131:839–49.
28. Ollinger R, Childs AJ, Burgess HM, et al. Deletion of the pluripotency-associated Tex19.1 gene causes activation of endogenous retroviruses and defective spermatogenesis in mice. *PLoS Genet* 2008;4(9):e1000199.
29. Ma L, Buchold GM, Greenbaum MP, et al. GASZ is essential for male meiosis and suppression of retrotransposon expression in the male germline. *PLoS Genet* 2009;5 (9):e1000635.
30. Frost RJ, Hamra FK, Richardson JA, et al. MOV10L1 is necessary for protection of spermatocytes against retrotransposons by Piwi-interacting RNAs. *Proc Natl Acad Sci USA* 2010;107:11847–52.
31. Zheng K, Xiol J, Reuter M, et al. Mouse MOV10L1 associates with Piwi proteins and is an essential component of the Piwi-interacting RNA (piRNA) pathway. *Proc Natl Acad Sci USA* 2010;107:11841–6.
32. Kuramochi-Miyagawa S, Watanabe T, Gotoh K, et al. MVH in piRNA processing and gene silencing of retrotransposons. *Genes Dev* 2010;24:887–92.
33. Shoji M, Tanaka T, Hosokawa M, et al. The TDRD9–MIWI2 complex is essential for piRNA-mediated retrotransposon silencing in the mouse male germline. *Dev Cell* 2009;17:775–87.
34. Soper SF, van der Heijden GW, Hardiman TC, et al. Mouse maelstrom, a component of nuage, is essential for spermatogenesis and transposon repression in meiosis. *Dev Cell* 2008;15:285–97.
35. Reuter M, Berninger P, Chuma S, et al. Miwi catalysis is required for piRNA amplification-independent LINE1 transposon silencing. *Nature* 2011;480:264–7.
36. Ernst C, Odom DT, Kutter C. The emergence of piRNAs against transposon invasion to preserve mammalian genome integrity. *Nat Commun* 2017;8:1411.
37. Mackay DJ, Temple IK. Transient neonatal diabetes mellitus type 1. *Am J Med Genet C Semin Med Genet* 2010;154C:335–42.
38. Urdinguio RG, Bayon GF, Dmitrijeva M, et al. Aberrant DNA methylation patterns of spermatozoa in men with unexplained infertility. *Hum Reprod* 2015;30:1014–28.
39. Boissonnas CC, Abdalaoui HE, Haelewyn V, et al. Specific epigenetic alterations of IGF2-H19 locus in spermatozoa from infertile men. *Eur J Hum Genet* 2010;18:73–80.
40. Kobayashi H, Hiura H, John RM, et al. DNA methylation errors at imprinted loci after assisted conception originate in the parental sperm. *Eur J Hum Genet* 2009;17:1582–91.
41. Marques CJ, Costa P, Vaz B, et al. Abnormal methylation of imprinted genes in human sperm is associated with oligozoospermia. *Mol Hum Reprod* 2008;14:67–74.
42. Klaver R, Bleiziffer A, Redmann K, et al. Routine cryopreservation of spermatozoa is safe: evidence from the DNA methylation pattern of nine spermatozoa genes. *J Assist Reprod Genet* 2012;29:943–50.
43. El Hajj N, Zechner U, Schneider E, et al. Methylation status of imprinted genes and repetitive elements in sperm DNA from infertile males. *Sex Dev* 2011;5 (2):60–9.
44. Gou LT, Kang JY, Dai P, et al. Ubiquitination-deficient mutations in human Piwi cause

male infertility by impairing histone-to-protamine exchange during spermiogenesis. *Cell* 2017;169:1090–104 .e13.

45. Gu A, Ji G, Shi X, et al. Genetic variants in Piwi-interacting RNA pathway genes confer susceptibility to spermatogenic failure in a Chinese population. *Human Reprod* 2010;25: 2955–61.

46. Hadziselimovic F, Hadziselimovic NO, Demougin P, Krey G, Oakeley EJ. Deficient expression of genes involved in the endogenous defense system against transposons in cryptorchid boys with impaired mini-puberty. *Sex Dev* 2011;5:287–93.

47. Heyn H, Ferreira HJ, Bassas L, et al. Epigenetic disruption of the PIWI pathway in human spermatogenic disorders. *PLoS One* 2012;7(10):e47892.

48. Bromer JG, Zhou Y, Taylor MB, Doherty L, Taylor HS. Bisphenol-A exposure in utero leads to epigenetic alterations in the developmental programming of uterine estrogen response. *FASEB J* 2010;24:2273–80.

49. Yang S, Bo J, Hu H, et al. Derivation of male germ cells from induced pluripotent stem cells in vitro and in reconstituted seminiferous tubules. *Cell Prolif* 2012;45:91–100.

50. Hirota T, Ohta H, Powell BE, et al. Fertile offspring from sterile sex chromosome trisomic mice. *Science* 2017;357:932–5.

51. Wissing S, Munoz-Lopez M, Macia A, et al. Reprogramming somatic cells into iPS cells activates LINE-1 retroelement mobility. *Hum Mol Genet* 2012;21:208–18.

Chromosomal Causes of Infertility

Svetlana A. Yatsenko and Aleksandar Rajkovic

7.1 Chromosomal Causes of Male Infertility

The genetic causes of male infertility are highly heterogeneous, and a large portion of these causes remains unexplained. More than 2300 testes-specific genes may contribute to male infertility [1]. Primary testicular disorders affecting spermatogenesis are commonly associated with abnormal semen parameters, including sperm concentration (oligozoospermia or azoospermia), morphology, motility, and vitality. Studies in infertile men have demonstrated that up to 20% carry constitutional chromosome aberrations [2–5]. Genomic aberrations found in these patients include numerical abnormalities, such as Klinefelter syndrome and its variants; XYY karyotype; testicular disorders of sex development, such as XX males; structural chromosome rearrangements, including Robertsonian translocations, balanced reciprocal translocations and inversions; as well as submicroscopic DNA copy number alterations (microdeletions and microduplications) encompassing genes associated with spermatogenesis or gonadal development.

7.1.1 Numerical Sex Chromosome Abnormalities

7.1.1.1 Klinefelter Syndrome

Klinefelter syndrome (KS) is the most common chromosomal aberration among infertile men, accounting for 14% of azoospermia patients [2–4]. Klinefelter syndrome is characterized by the presence of one or more extra X chromosomes in a normal male karyotype (Figure 7.1). The most common variant, the 47,XXY karyotype (Figure 7.1b), is seen in about 90% of KS men [6]. Klinefelter syndrome variants such as 48,XXXY; 48,XXYY, or 49,XXXXY are much less frequent. The extra X chromosome in KS usually arises from meiotic nondisjunction, when the X chromosome fails to separate during the first or second meiotic division in male or female gametogenesis [3]. About 10% of males with KS are mosaic (47,XXY/46,XY). In such cases, the extra X chromosome in a portion of cells is due to the mitotic nondisjunction in the developing zygote [4].

However, high-resolution techniques such as fluorescence in situ hybridization (FISH) or chromosomal microarray analyses are essential for better characterization.

Because of the variability of the phenotype, KS is underdiagnosed, with only 10% of KS patients recognized prepubertally and an additional 15% identified after puberty [2,5]. In childhood, KS boys may present with language delay and learning and behavioral problems. During the prepubertal period, boys with KS have a normal number and morphology of Sertoli and Leydig cells; a reduced number of spermatogonia; and normal serum levels of testosterone, follicle-stimulating hormone (FSH), luteinizing hormone (LH), and inhibin B [4,5]. The degenerative process in testes is accelerated with a decline in testosterone and a gradual increase in FSH and LH concentrations after the onset of puberty. Infertility and small testes are the most prevalent characteristics in adult KS patients. The testes in adult KS males are characterized by extensive fibrosis, hyalinization of the seminiferous tubules, and impaired spermatogenesis with azoospermia or severe oligozoospermia [2–5]. Although most KS patients are infertile, testicular spermatozoa can be identified and recovered from at least 50% of men with the nonmosaic 47,XXY karyotype [5]. In mosaic and rare nonmosaic KS cases, mature spermatozoa can be found in ejaculates [5]. Testicular sperm extraction (TESE) combined with intracytoplasmic sperm injection (ICSI) allows over 50% of patients with KS to father their own biological children [5,6]. Sperm from KS men usually have a

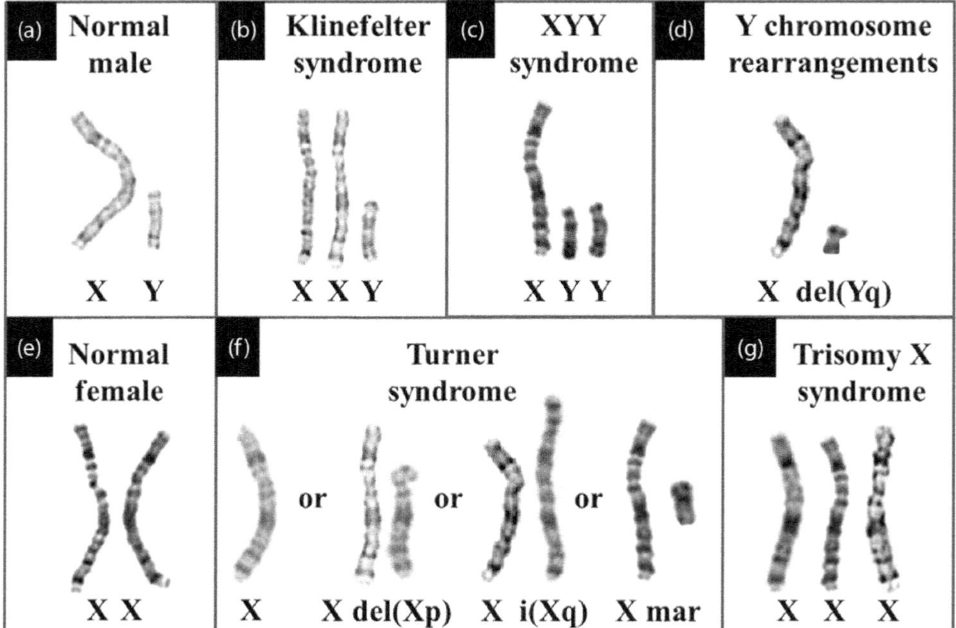

Figure 7.1 Sex chromosome numerical and structural abnormalities associated with human infertility. Normal male (a) and normal female (e) chromosomal complement is shown. The common sex chromosome aneuploidies detected by classical karyotype analyses are shown for (b) Klinefelter syndrome (47,XXY), (c) XYY men, (f) Turner syndrome (45,X), and (g) trisomy X women (47,XXX). (f) Gross structural X and Y chromosome rearrangements include deletions of the short arm of chromosome X (del(Xp)), isochromosome composed of the two long arms of chromosome X (i(Xq)), small marker (mar) chromosome containing X-specific or Y-specific DNA. The X- and Y-chromosome structural rearrangements can be observed by chromosome analysis, such as deletions of the long arm (del(Yq)) (d).

normal 23,X or 23,Y haploid chromosome complement. Despite this, an increased frequency for both autosomal and sex chromosome aneuploidy has been reported in their offspring [6,7].

7.1.1.2 47,XYY Karyotype

An extra copy of the Y chromosome is present in 47, XYY males (Figure 7.1c). This chromosomal aneuploidy occurs in 1 in 1000 live male births in the general population and is seen more frequently in the infertile population [2,4]. Men with the 47,XYY karyotype have a normal phenotype. With regard to fertility, semen analyses may show oligozoospermia or azoospermia in some patients, while the majority of 47,XYY males are fertile with normal semen parameters, and produce normal haploid spermatozoa [6]. The extra Y chromosome is eliminated via apoptosis of the cells carrying two Y chromosomes during the premeiotic stage of spermatogenesis, although the frequency of aneuploidy for the sex chromosomes is slightly increased in mature sperm [4]. Testes biopsies demonstrate that perturbation during meiotic pairing may contribute to sperm apoptosis and subsequent oligozoospermia and infertility in men with the 47, XYY karyotype [2,6].

7.1.2 Y Chromosome Microdeletions

The human Y chromosome contains many genes that are essential for male sex determination and spermatogenesis [1,3,8]. Based on observation of cytogenetically visible deletions, the azoospermia factor (AZF) region has been established within the Yq and extensively studied during the last decade. Microdeletions involving the long arm of chromosome Y (Yq) are one of the most significant pathogenic defects in infertile males, found in about 10% of men with oligozoospermia and in up to 15% of azoospermic patients [2,4]. The AZF region consists of three genetic domains in the long arm of the human Y chromosome: *AZFa*, which is located within Yq11.21, *AZFb*, and *AZFc*. *AZFb* and *AZFc* overlap and map within bands Yq11.22 and Yq11.23, respectively (Figure 7.2). Overall, these three regions contain gene families for 27 distinct proteins [8,9]. Microdeletions involving *AZFa*, *AZFb*, and *AZFc* result in disruption of spermatogenesis at three different stages. *AZFa* deletions cause Sertoli-cell-only

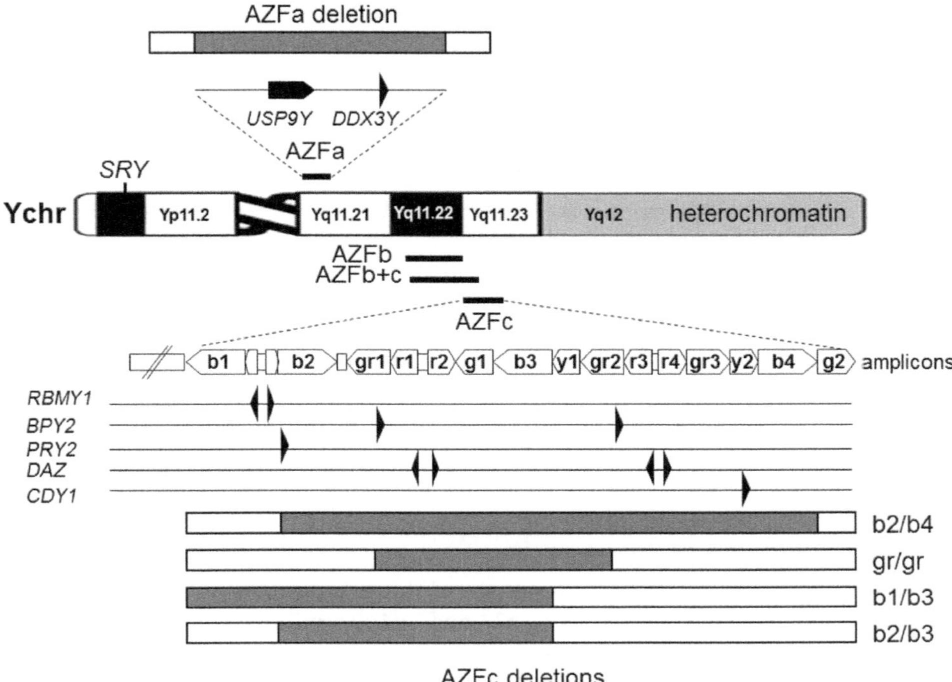

Figure 7.2 The AZF regions of the Y chromosome. Y chromosome (Ychr) schema details the location of the *SRY* gene at Yp11.31, as well as *AZFa*, *AZFb*, and *AZFc* deletion regions. The three AZF regions map to the Yq11.21, Yq.11.22, and Yq11.23 bands. Above the Ychr, the *AZFa* chromosome region is magnified to show the *USP9Y* and *DDX3Y* genes. Below the Y chromosome, complex genomic organization within the *AZFc* region is shown. *AZFb* and *AZFc* regions are formed by amplicons (nearly identical stretches of DNA). Amplicons consist of five sequence families: b, gr, r, g, and y, originally named as blue, green, red, gray, and yellow, respectively. Nearly identical segments within the same sequence family are differentiated by numbers (b1, b2, b3, etc.). Sequence orientation is shown by b ack arrowheads. Functional genes and their 5′–3′ orientation are indicated by black-filled triangles; pseudogenes are not shown. *AZFc* deletions are shown as gray bars.

syndrome with a complete absence of germ cells. Arrest at the spermatocyte stage (normal population of spermatogonia and primary spermatocytes, but no postmeiotic germ cells) was observed in the testes of all patients with the *AZFb* region deletion. *AZFc* deletions affect the postmeiotic spermatid maturation process, resulting in a decreased number of mature germ cells [8,9]. Microdeletions encompassing other genes located on the Y chromosome have been proposed to influence spermatogenesis, although their role remains to be elucidated.

Gross structural abnormalities of the Yq chromosome (Figure 7.1d), such as whole long arm deletions (del(Y)(q11.2)), isochromosome Yp (i(Yp)), and dicentric Yp (dic(Yp)), occur less frequently (up to 10%) than microdeletions and result in complete absence of germ cells [2,6].

7.1.2.1 *AZFa*

The *AZFa* region, a segment of 792 kb in size, is located at the proximal Yq (chromosome position [GRCh37]: chrY:12,900,000–13,700,000) (Figure 7.2) [3,4,9]. This region is flanked by two repetitive DNA elements that are about 10 kb each in size and share 94% sequence identity. Nonallelic homologous recombination between these repeats results in a complete *AZFa* deletion (OMIM #400042), a fairly rare but recurrent event. This deletion removes two genes that are located in the *AZFa* region, *USP9Y* (OMIM #400005) and *DBY* (OMIM #400010) (also called *DDX3Y*), and is associated with Sertoli-cell-only syndrome, a condition characterized by the presence of Sertoli cells in the testes but a lack of spermatozoa in the ejaculate [9]. A testicular biopsy shows degeneration of germ cells within tubules due to failure of differentiation and maturation of spermatocytes and spermatids [2,9]. Partial *AZFa* deletions encompassing either *USP9Y* or *DBY* genes are rarely identified. Deletions involving only the *USP9Y* gene have been found among infertile men with azoospermia, oligozoospermia, or oligoasthenozoospermia, as well as in fertile men, suggesting that the gene might not be critical for sperm production [2]. Small, 98-kb deletions encompassing only the *DBY* gene have been reported in multiple infertile patients; however, this was in a single population study and requires future replications.

7.1.2.2 *AZFb*

Deletions involving the *AZFb* region account for about 30% of all AZF deletions [6,9]. The *AZFb* region spans a 6.2-Mb segment (genomic coordinates [GRCh37] chrY:18,100,000–24,300,000). *AZFb* has a complex genomic structure (Figure 7.2) and contains multiple highly identical sequences organized in opposite orientation to each other (palindromic amplicons) [2]. Similar amplicons are also present distally to *AZFb*, within the *AZFc* region [2,8]. The outcome of nonallelic homologous recombination between these amplicons results in deletion of a 6.2-Mb or 7.7-Mb segment. Therefore, deletions spanning the *AZFb* region are variable in size, have different proximal and distal breakpoints, and may include both *AZFb* and *AZFc* regions (Figure 7.2). There are a number of testis-specific genes located within the *AZFb* region (Figure 7.2). Four main gene families (*CDY*, *HSFY*, *RBMY*, and *PRY*) are present in more than one copy in the *AZFb* deletion interval. A combined *AZFb+c* deletion additionally removes a 1.5-Mb part of the *AZFc* region, and is therefore 7.7-Mb in size. Patients with *AZFb* or *AZFb+c* deletions present with azoospermia due to maturation arrest at the primary spermatocyte stage [9].

7.1.2.3 *AZFc*

The *AZFc* region is located at the distal long arm of the Y chromosome at band Yq11.23 and encompasses a segment of about 3.5 Mb in size (Figure 7.2). Approximately 60% of the Y chromosome microdeletions affect the *AZFc* locus. The region where recurrent *AZFc* deletions occur consists of five large DNA sequences (amplicons) ranging from 115 to 678 kb that are repeated several times and are arranged in either direct or inverted orientation to each other [8]. Such complex genomic structure makes *AZFc* susceptible to genomic rearrangements, including deletions, duplications, and inversions. Complete and partial *AZFc* deletions that affect the dosage of genes implicated in spermatogenesis can cause spermatogenic impairment and infertility [2,8,9]. The *AZFc* region contains at least four protein-coding germline-specific gene families: *BPY2*, *PRY2*, *DAZ*, and *CDY1* [8]. There are four functional copies of the *DAZ* gene, three copies of *BPY2*, two copies of *CDY1*, and a single copy of the *PRY2* gene, as well as their inactive copies (pseudogenes) in the reference human genome (Figure 7.2).

The *DAZ* (deleted in azoospermia, OMIM #400003) genes (*DAZ1–DAZ4*) encode four RNA-binding proteins that are expressed exclusively in the adult testis at all stages of germ cell development. *DAZ* genes regulate translation, are involved in control of meiosis and maintenance of the primordial germ cell population, and are therefore thought to be critical genes in the *AZFc* region [8]. Despite sequence homology, the four DAZ proteins vary in the number of functional domains, and may have overlapping, but not identical, roles in spermatogenesis.

A complete 3.5-Mb *AZFc* deletion, also known as a b2/b4 deletion (Figure 7.2), is the product of intrachromosomal homologous recombination between the b2 and b4 amplicons, and is the most frequently deleted region among infertile men with Y chromosome microdeletions. Men with complete *AZFc* deletions have reduced spermatogenesis, ranging from azoospermia to severe oligozoospermia. *AZFc* deletions cause approximately 12% of nonobstructive azoospermia and 6% of severe oligozoospermia [8,9].

Additionally, four recurrent rearrangements involving a part of the *AZFc* region have been described: b1/b3, b2/b3, gr/gr partial deletions, and gr/gr duplication (Figure 7.2). Partial deletions remove 1.6–1.8 Mb of *AZFc* and reduce the copy number of several testis-specific *AZFc* genes; however, only the gr/gr deletions are associated with increased risk of spermatogenic failure. Men with a gr/gr deletion show significant phenotypic variability ranging from normozoospermia to azoospermia, likely due to a complex interaction of many factors including the influence of ethnic and environmental backgrounds. Deletions b1/b3 and b2/b3, and complete or partial *AZFc* duplications, do not seem to have an effect on semen parameters [4,8].

The vast majority of complete *AZFc* microdeletions occur de novo (i.e. are not present in the patient's father); however, rare cases of natural transmission have been reported. Partial gr/gr deletions can be passed from father to son [4,9]. It has also been shown that the presence of a partial *AZFc* deletion in fathers can increase the risk of a complete *AZFc* deletion in their sons.

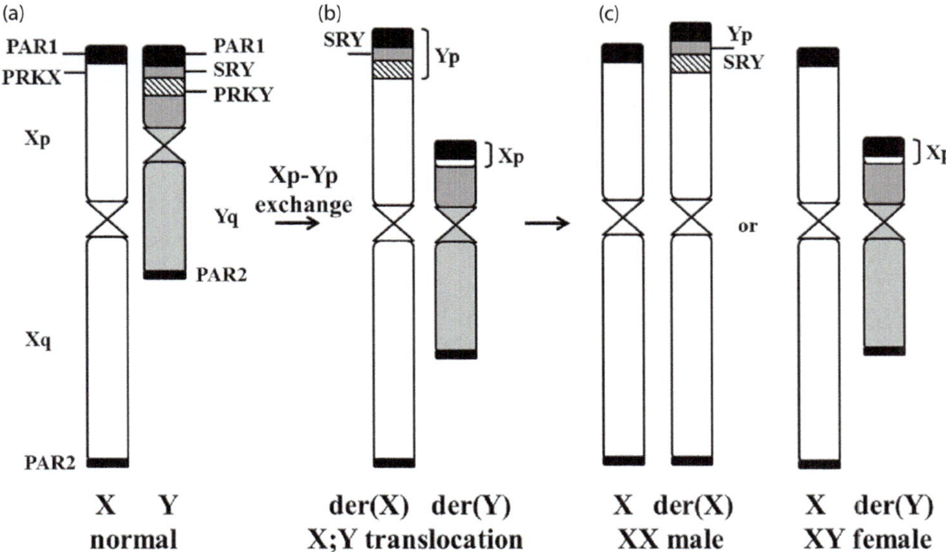

Figure 7.3 X;Y chromosomes translocation. During male meiosis (a), X and Y chromosomes normally pair and recombine within the two homologous regions (pseudoautosomal regions, PAR1 and PAR2) located at the distal short and long arms of sex chromosomes. Aberrant recombination involving highly homologous DNA sequences such as the *PRKX* and *PRKY* genes results in exchange of X-specific and Y-specific DNA segments. The X;Y translocation causes transposition of the *SRY* gene from the Y to the X chromosome resulting in derivative chromosome X (derX), comprising *SRY*, and derivative chromosome Y (derY), deleted for the *SRY* gene (b). Fertilization by a sperm containing the derivative X chromosome will conceive an *SRY*-positive XX male, whereas sperm carrying derivative Y chromosome, deleted for *SRY*, will conceive an XY female (c).

7.1.3 46,XX Male Syndrome or Testicular Disorder of Sex Development

The XX male syndrome, or testicular disorder of sex development (testicular DSD) is a rare condition with a frequency of 1 in 25 000 male newborns, and is characterized by the presence of a 46,XX karyotype and male genitalia [10]. Approximately 20% of boys with testicular DSD have ambiguous genitalia at birth, whereas the remaining 80% present with steroidogenic and spermatogenic dysfunction after puberty.

The majority of males with the 46,XX karyotype are *SRY* positive by FISH or polymerase chain reaction (PCR) amplification [4,10]. The *SRY* gene, also known as the testis-determining factor (located on the Y chromosome at Yp11.31; Figure 7.3), encodes for the sex-determining Y protein, which activates a cascade of male-specific transcription factors essential for male development. In most instances, 46,XX male syndrome is caused by an exchange of segments between the short arms of the X and Y chromosomes (X;Y translocation) during paternal meiosis (Figure 7.3a), resulting in the derivative

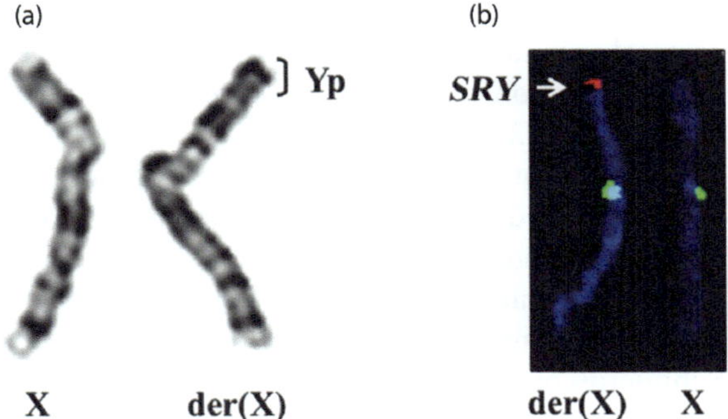

Figure 7.4 Partial karyotype and fluorescence in situ hybridization (FISH) analysis in the XX male. (a) Chromosome analysis detected a derivative chromosome X containing the Yp segment at the distal short arm of X chromosome. (b) FISH analysis with *SRY*-specific probe (red signal, arrow) shows that the *SRY* gene is present in the derivative X chromosome. The centromere of the X chromosome is colored green.

chromosome X containing the *SRY* gene (Figure 7.3b,c), and the derivative chromosome Y deleted for *SRY*. Fertilization with an abnormal gamete, containing either the *SRY*-positive X chromosome (Figure 7.4) or the Y chromosome with the *SRY* gene deletion, will result in a sterile XX

Table 7.1. Genomic imbalances associated with complete gonadal dysgenesis (CGD) in 46,XX males

Locus	Genomic abnormality	Gene	Molecular mechanism	Phenotype	OMIM # [PMID]
1p34.3	Disruption	*RSPO1*	Loss of function	CGD, palmoplantar hyperkeratosis	610644
17q24	Duplication	Regulatory region *SOX9*	Gain of function	CGD	613080
Xp21	Deletion	Regulatory region *MAGEB* and *NR0B1*	Gain of function	CGD	[28483799]
Xq26	Duplication, deletion	*SOX3*	Gain of function	CGD	300833
Yp11	Presence	*SRY*	X;Y or Yp–autosome translocation	CGD	400044

male or XY female, respectively. Translocations between the Y and autosomal chromosomes can also give rise to *SRY*-positive XX males, where the *SRY* gene is located on the autosome [10]. *SRY*-positive males with 46,XX testicular DSD and normal male genitalia have small testes with severe atrophy and absent spermatogenesis, azoospermia, and hypergonadotropic hypogonadism [10].

About 10% of 46,XX men are *SRY* negative and can present with ambiguous genitalia at birth. *SRY*-negative, 46,XX infertile men can also have normal external genitalia [10,11]. *SRY* normally triggers testes formation by activating expression of *SOX9*, located at 17q24.3. Like *SRY*, *SOX9* is necessary for testis differentiation, and its overexpression can lead to male development in the absence of *SRY*. *SOX9* expression is regulated by testis-specific transcriptional enhancer elements mapped within a 1-Mb segment upstream of the *SOX9* gene. Submicroscopic chromosome 17q24.3 duplications and triplications detected by array comparative genomic hybridization (aCGH) analysis, as well as balanced translocations upstream of *SOX9*, have been identified in some infertile XX males that are *SRY* negative [10,11]. In addition to *SRY*, genes encoding steroidogenic factor 1 (*NR5A1*, 9q33.3) and *SOX3* (Xq27.1) can upregulate expression of *SOX9* [11]. Genomic rearrangements that cause *SOX3* gain of function have been identified among *SRY*-negative XX males (Table 7.1). Disruption of the gene encoding R-spondin 1 (*RSPO1*, 1p34.3) is also a rare cause of the XX male phenotype. Genetic etiology in many other XX male *SRY*-negative individuals remains unknown.

7.1.4 Balanced Chromosome Rearrangements

Structural chromosomal abnormalities are frequent in infertile men, with an overall incidence of about 5%, a percentage that is 10-fold higher than the 0.5% prevalence in the general population [4,12]. Chromosome rearrangements are found in approximately 14% of azoospermic and 4.5% of oligozoospermic patients. Autosomal aberrations (3%) are more commonly associated with oligozoospermia, whereas sex chromosome defects (12.6%) predominate among azoospermic men [12–14]. Structural chromosome rearrangements may cause impaired spermatogenesis by adversely affecting chromosome synapsis during meiosis [2,13]. In rare cases, chromosome breaks may result in disruption/inactivation of a single dosage-sensitive gene involved in spermatogenesis, thus resulting in the arrest or abnormal male germ cell development [12,13]. Balanced chromosome rearrangements may be classified as reciprocal translocations, inversions, complex chromosome rearrangements, or Robertsonian translocations (see Chapter 1 for detailed information) (Figure 7.5).

7.1.4.1 Reciprocal Translocations, Inversions

Reciprocal translocations occur when segments are exchanged between two chromosomes with no apparent loss or gain of genetic material at the breakpoint sites. Carriers of balanced chromosome rearrangements usually have a normal phenotype and are often diagnosed during evaluation of their infertility problem or recurrent losses of pregnancy, or following the birth of a child with an unbalanced chromosome

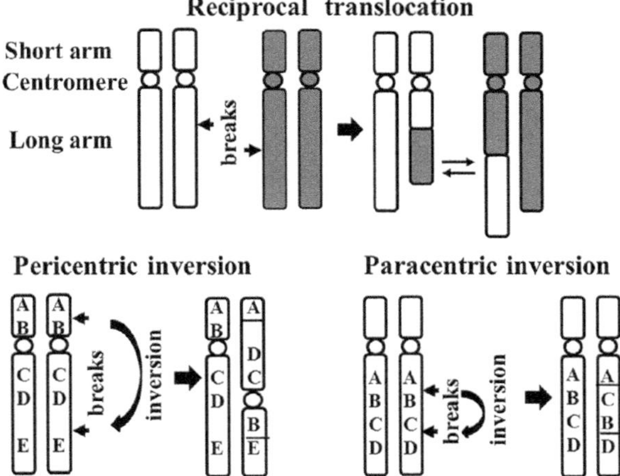

Figure 7.5 Balanced structural chromosome rearrangements. Reciprocal translocations can occur between nonhomologous chromosomes. Breaks in the DNA (arrowheads) at two different chromosomes, followed by exchange of chromosomal segments distal to the break, will lead to two derivative chromosomes. Paracentric and pericentric inversions are associated with two breakpoints on a single chromosome (arrowheads). Balanced rearrangements may disrupt gene functions at the translocation breakpoints and cause chromosomally abnormal offspring and recurrent miscarriages.

complement [2,12,13]. In balanced translocation carriers, meiotic pairing results in the formation of a quadrivalent structure between translocated chromosomes and their normal homologs. Chromosome segregation analyses demonstrate a highly variable proportion (up to 80%) of chromosomally unbalanced spermatozoa among carriers of reciprocal translocations [12,13]. Because segregation of translocated chromosomes depends on many cytogenetic factors, such as chromosomes involved, size of the translocated segments, gene content, chromosome type, and other characteristics, it is hard to predict the proportion of normal spermatozoa. Fertility problems in male carriers can be attributed to disturbance of the meiotic process and various degrees of sperm defects. However, the presence of a balanced chromosome rearrangement is not necessarily associated with spermatogenic failure and infertility [2,13]. Fertilization by an unbalanced gamete does occur, but many resulting embryos do not survive. Therefore, individuals carrying balanced rearrangements benefit from preimplantation genetic testing for structural chromosome rearrangements (PGT-SR) to identify and implant embryos with normal or balanced chromosome complement (see Chapter 13). Individuals who carry chromosome inversions (see Chapter 1) are healthy in general, but infertility, recurrent pregnancy losses, and chromosomally abnormal offspring have been reported [2,13). During meiosis, pairing of inverted segments is achieved by the formation of an inversion loop. The size of the inverted segment and position of the crossover determine the outcome of meiotic pairing. In carriers of paracentric inversions, unbalanced chromosomal complements have been reported in about 1% of spermatozoa; however, such studies have been performed on a limited number of individuals [13]. In contrast, carriers of pericentric inversions may have a high proportion (up to 54%) of spermatozoa with unbalanced recombinant chromosomes [2,6]. In general, large pericentric inversions (encompassing more than half of the chromosome length) are more likely to produce unbalanced chromosomes and are therefore more frequently observed among infertile men [6,13].

7.1.4.2 Complex Chromosome Rearrangements

Complex chromosome rearrangements (CCRs) are rare structural aberrations with at least three breakpoints and an exchange of genetic material between two or more chromosomes and occur in approximately 0.5% of newborns [14]. Unbalanced CCRs are often associated with intellectual disability and congenital abnormalities. Balanced CCRs are seen in phenotypically normal individuals with a history of recurrent abortions and/or infertility. Molecular studies using high-resolution aCGH and analysis of breakpoint sequences demonstrate that many genomic rearrangements are complex events; however, neither the complexity nor the number of breaks can be used to predict infertility or spermatozoa chromosome rearrangements.

General risk of spontaneous abortion for CCR carriers is estimated to be approximately 50% and about 20% for affected offspring. Each CCR is unique and reproductive risks will depend on multiple factors such as chromosome origin, location of breakpoints, number of chromosomes involved, and overall breaks, genome content, rearrangement type, and complexity. There are 64 possible combinations of chromosomes in spermatozoa of a carrier for CCR with three breaks involving three chromosomes. The number of combinations increases with the involvement of additional chromosomes and/or breakpoints. Because of the low proportion of balanced sperm available (~10–20%), ICSI is not recommended in male CCR carriers [13,14].

7.1.4.3 Robertsonian Translocations

Robertsonian translocations are the most common structural chromosomal rearrangement in humans, resulting in a derivative chromosome composed of the long arms of two acrocentric chromosomes (13, 14, 15, 21, and 22) (Figure 7.6; for more information, see Chapter 1). The most frequent Robertsonian translocations are der(13;14) and der(14;21) with incidences of about 1 in 1000 and 1 in 5000, respectively [15,16]. Carriers of Robertsonian translocations have an increased risk for infertility, spontaneous abortions and chromosomally unbalanced offspring, but are otherwise healthy. Studies involving male carriers of der(13;14) showed that in about 80% of cases the partners had spontaneous pregnancies, while in 20% of cases the male carriers were infertile [16]. About 1.6% of infertile male patients carry Robertsonian translocations. The cause of infertility in these individuals has been attributed to meiotic disturbances of rearranged chromosomes with subsequent meiotic arrest, resulting in oligozoospermia or azoospermia [15,16]. During meiosis, pairing of the translocated chromosomes gives rise to a trivalent configuration. Most of the mature spermatozoa (75–90%) are normal or balanced as a result of alternate segregation; however, certain variability exists among patients [15]. It is remarkable that in the mature spermatozoa a much higher proportion of nullisomies versus disomies has been found for chromosomes 13, 14, 15, and 22. These findings correlate with a higher incidence of monosomic embryos detected by PGT-SR [15,16]. Monosomic embryos are usually lost during early pregnancy. In rare cases, these embryos may survive due to the maternal uniparental disomy (UPD). In contrast, male carriers of a Robertsonian translocation involving chromosome 21 are more likely to produce disomic as opposed to nullisomic gametes. Such gametes are more likely to produce trisomy 21 offspring rather than monosomic embryos. However, male carriers of Robertsonian translocations are subfertile due to low sperm counts, and the overall chance of such a male producing trisomy 21 offspring is very low. In addition, some carriers demonstrate an increased risk for aneuploidy of the sex chromosomes in the spermatozoa produced, suggesting an interchromosomal effect [7,17].

Infertile men are found to have significantly increased levels of chromosome abnormalities in their sperm, despite the fact that the majority of them have a normal constitutional karyotype [6]. Increased aneuploidy frequencies have been reported for all chromosomes; however, chromosomes 21, 22, X, and Y are more prone to nondisjunction [6,7]. High aneuploidy levels have been associated with abnormal semen profiles including azoospermia, oligozoospermia, asthenozoospermia, and teratozoospermia [18].

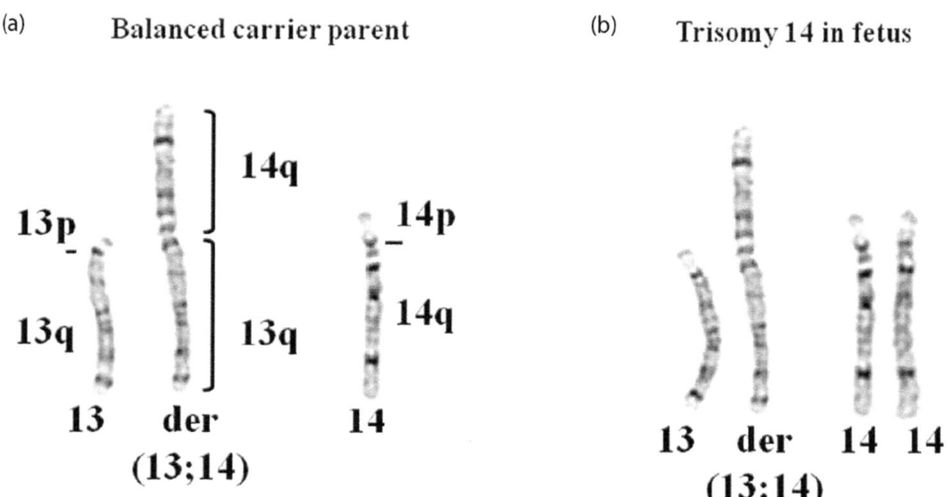

Figure 7.6 Robertsonian translocation between chromosomes 13 and 14. (a) Balanced carrier of the Robertsonian translocation has a derivative chromosome der(13;14) composed of the long arms of chromosomes 13 and 14, as well as one normal homolog of chromosomes 13 and 14. The derivative (13;14) is formed by fusion of two acrocentric chromosomes at the centromere, with accompanying loss of the short arms 13p and 14p. (b) Trisomy 14 in the fetus is due to the inherited Robertsonian translocation. Fertilization with gametes containing both the der(13;14) and normal chromosome 13 or 14 can lead to either trisomy 13 (Patau) syndrome or trisomy 14.

To date, chromosomal causes have been identified in only about 20% of infertile men. However, it is clear that disturbances in male meiosis, particularly in chromosome pairing, synapses, and meiotic recombination, are associated with male infertility, and a number of genes involved in this process remain to be discovered.

7.2 Chromosomal Causes of Female Infertility

Female infertility is often attributed to an impairment of ovarian function that can result from several different genetic mechanisms: numerical X chromosome abnormalities, including Turner and triple X karyotype; balanced structural chromosomal rearrangements; genomic imbalances involving the X chromosome and autosomes; XY gonadal dysgenesis and single gene alterations leading to ovarian dysgenesis; premature ovarian failure; and reproductive dysfunction. X chromosome-linked aberrations play a major role among currently known genetic defects. Here, we review chromosomal and genomic alterations that result in ovarian insufficiency and present in syndromic and nonsyndromic forms.

7.2.1 Sex Chromosome Abnormalities

7.2.1.1 Turner Syndrome

Turner syndrome is a common genetic disorder that results from loss of a sex chromosome (45,X or monosomy X) in a phenotypic female (Figure 7.1f). Turner syndrome occurs in approximately 1 in 2000–3000 female live births as a result of chromosome nondisjunction during meiosis [3]. Monosomy X is a common abnormality among spontaneous abortions and only 1–3% of fetuses survive to term. Clinical manifestations are variable in affected females and include short stature, skeletal abnormalities, congenital heart and kidney anomalies, characteristic physical features (such as a wide and webbed neck, a low hairline at the back of the neck, flat chest), lymphedema in infancy, and gonadal dysgenesis with primary amenorrhea [19,20]. Most women with Turner syndrome have normal intelligence, although cognitive deficits, developmental delays, nonverbal learning disabilities, and behavioral problems are possible.

Approximately half of females diagnosed with Turner syndrome have 45,X chromosome complement, whereas the other half are mosaic. The non-45,X cell lines in mosaics have variable chromosomal complement including cells containing two normal X chromosomes, three normal X chromosomes, or one normal X and a structurally abnormal X or Y chromosome. The X chromosome abnormalities associated with Turner syndrome also include deletions of the short arm (46,X,del(Xp)), isochromosome Xq (46,X,i(Xq)), ring X chromosome (46,X,r(X)), derivative X chromosome (46,X,der(X)), and small marker chromosomes, which usually contain an X chromosome centromere and pericentromeric DNA sequences (Figure 7.1f). Mosaicism for multiple abnormal cell lines can also be found. The chromosome constitution and level of mosaicism influence the resulting phenotype in Turner syndrome patients. Deletions in the distal short arm of chromosome X involving the *SHOX* gene are associated with short stature and skeletal anomalies, whereas variable deletions in the proximal Xp and deletions in the long arm of X chromosome have been observed in patients with gonadal dysgenesis and primary ovarian insufficiency (POI) [19–21]. An abnormal X chromosome is preferentially inactivated, unless rearrangement involves the X inactivation center (*XIST* gene) [3]. Lack of the X inactivation center on the aberrant X chromosome causes a much more severe phenotype. In 80% of females with the 45,X karyotype, the X chromosome is maternal in origin and the paternal sex chromosome has been lost. In 20% of 45,X cases, the paternally derived X is present; however, there is no difference in the phenotype based on the X chromosome parent of origin [3].

Patients with Turner syndrome do not undergo puberty, have infantile internal and external genitalia, and fail to develop secondary sexual characteristics unless they receive hormone therapy. Early diagnosis is important in order to initiate appropriate therapy with growth hormone and estrogen. Girls with short stature should be karyotyped to rule out Turner syndrome. Also, FISH studies on uncultured blood or other tissue samples may be helpful in identifying mosaic monosomy X in cell types other than the T-cell lymphocyte populations used for karyotyping. In Turner girls, primordial germ cells form but are lost rapidly, and hypoplastic "streak" gonads, composed of fibrous tissue, are detected at the time of puberty [19]. In normal females, one X chromosome in every somatic cell is usually inactivated; however, gene expression from both X chromosomes in oocytes is necessary for normal ovarian development [3]. Haploinsufficiency for the X-linked genes is likely responsible for gonadal dysgenesis and infertility in 45,X individuals.

A combination of a 45,X (Turner syndrome) cell line and a normal 46,XX chromosome complement is the most common form of mosaicism in Turner syndrome individuals. Many patients with mosaicism for monosomy X are mildly affected, undergo breast development, menstruate, and may present in clinic only due to infertility or diagnosis of POI [19–21]. 45, X/47,XXX mosaicism occurs less frequently, but clinical manifestations are similar to those in 45,X/46,XX. The ovaries of teenage girls who have Turner syndrome with X chromosome mosaicism contain follicles that secrete estrogen but their ovarian function rapidly declines [19,20]. The degree of 45,X mosaicism that associates with ovarian dysfunction is unclear. Ovarian mosaicism in mosaic Turner cases is inferred from blood lymphocytes. Small studies suggest that the threshold for ovarian phenotype is when mosaicism reaches 10% of karyotyped lymphocytes, although pregnancies are reported in a small percentage of women with a high grade (>10%) mosaicism for monosomy X [22].

7.2.1.2 47,XXX Karyotype

Trisomy X, triple X, or 47,XXX karyotype (Figure 7.1g) is a fairly common chromosome aneuploidy condition caused by a nondisjunction event of the X chromosome, either during gametogenesis or after conception [3,23]. Trisomy X affects approximately 1 in 1000 girls. It is estimated that only 10% of cases are diagnosed, as a majority of these women are normal. Some women with 47,XXX can be taller than average and may present with learning disabilities, delayed development of motor skills, and speech and language problems. Most females with trisomy X syndrome have normal pubertal onset and sexual development, and are fertile [20,23]. However, some individuals are not able to conceive due to POI or genitourinary malformations. Trisomy X syndrome is found in approximately 3% of females with POI [20,21,23].

A majority of trisomy X is maternal in origin, derived from meiosis I (~60%) or meiosis II (~17%) errors, and results in nonmosaic 47,XXX karyotypes [3]. Mosaicism occurs in about 20% of cases due to X chromosome nondisjunction during the early development of an embryo. Overall phenotype and fertility is affected by the presence of abnormal cells such as 45,X (Turner) or 48,XXXX (tetrasomy X). Polysomy X (48,XXXX or 49,XXXXX) is associated with more severe intellectual disability, developmental delay, and a range of congenital anomalies [23].

The risk of trisomy X, similar to other chromosome trisomies, significantly increases with advanced maternal age. Fertile females with trisomy X produce normal haploid gametes, with no particularly increased risk for a 47,XXX or 47,XXY child [3,23].

7.2.1.3 X-Chromosome Structural Rearrangements

Structural abnormalities of the X chromosome, including deletions, duplications, inversions, complex rearrangements, and balanced and unbalanced X–autosome translocations, are frequently correlated with a normal or mild phenotype in females, but are associated with infertility, repeated miscarriages, and chromosomal imbalances in offspring [20,21,24–26]. In females, X inactivation is established early during embryo development at the late blastocyst stage, and is maintained in all somatic cells and transmitted to the daughter cells during mitosis. Therefore, one X chromosome is randomly inactivated in each cell [3]. Thus, each female has two cell populations; one with the paternal X chromosome and another with the maternal functional X chromosome. When one of the X chromosomes is abnormal, the X inactivation pattern is usually skewed toward the unbalanced clone. Derivative X chromosomes resulting from unbalanced X autosomal translocation, isochromosome, and the X chromosome carrying deletions or duplications, are typically inactive.

In contrast to somatic cells, the inactive X chromosome is reactivated in female germ cells so that mature oocytes have two active X chromosomes; therefore, X chromosome rearrangements are more likely to affect oogenesis [3,20,21]. Based on analysis of partial X chromosome monosomies in women with a Turner syndrome phenotype or isolated ovarian insufficiency, four regions critical for ovarian function have been delineated. Deletions and translocations detected within Xp11–p13.1 (POF4), Xq13.3–q22 (POF2), Xq22–q25, and Xq27–q28 (POF1), are associated with POI. Deletions involving the Xq13 region are associated with primary amenorrhea, lack of breast development, and ovarian insufficiency in the majority of patients [19–21]. Women with an Xq21–q24 deletion have a less severe phenotype than individuals with Xq13 deletions. Primary ovarian

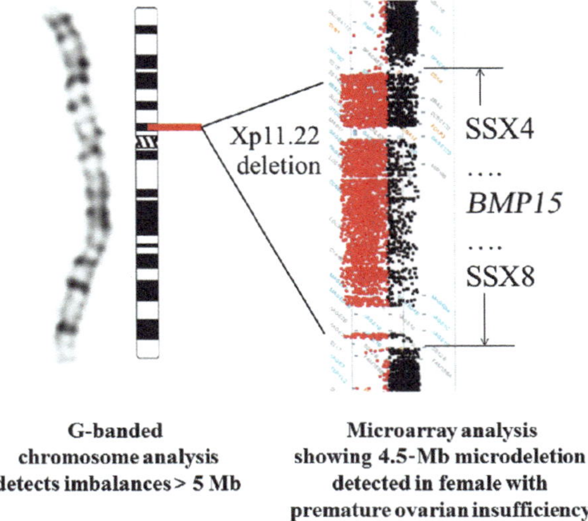

Figure 7.7 rray comparative genomic hybridization can detect submicroscopic Xp deletion. Partial high-resolution G-banded karyotype and ideogram of the X chromosome shows a normal X chromosome in a woman with primary ovarian insufficiency. Array comparative genomic hybridization detected a 4.5-Mb deletion in the Xp11.22 region encompassing the *BMP15* gene.

insufficiency is more commonly associated with Xq25–q28 deletions [19–21]. To date, the number and precise location of genes relevant to X-linked POI are still under investigation. Despite a wealth of evidence implicating X chromosome in ovarian reserves, alterations in few X-linked genes such as *BMP15*, *DIAPH2*, and the premutation *FMR1* alleles have been associated with POI [20,21]. The *BMP15* gene (bone morphogenetic protein 15, OMIM #300247) is a member of the transforming growth factor-β superfamily, and maps to the Xp11.2 region. Submicroscopic deletions (Figure 7.7) and mutations of the *BMP15* gene are observed in women with POI or primary amenorrhea. Disruption of the *DIAPH2* gene (POF2A, Xq21.33, OMIM #300108) has been identified in mother and daughter with POI and a balanced translocation t (X;12). These data demonstrate that haploinsufficiency of ovary-specific genes is one of the molecular mechanisms responsible for X-linked POI [19–21,24]. Premutation *FMR1* alleles account for 2–3% of sporadic POI cases and 10–15% of familial POI cases. Other X-linked genes such as *USP9X*, *ZFX*, *XPNPEP2*, *XIST*, and *SPANX* have been proposed as candidate genes for POI, but their role remains to be elucidated [21,24,25]. The application of high-resolution aCGH analyses of the X chromosome, as well as whole-exome/genome sequencing technologies, is likely to uncover new X-linked genes involved in POI [21,24,26].

In order to identify novel candidate genes, mapping of the breakpoints has been performed for a significant number of patients with POI and balanced X-autosome translocations [19,20]. However, in some patients, X chromosome breakpoints were identified within the genomic regions free of transcribed sequences (so-called "gene deserts"), suggesting an alternative molecular mechanism for POI, such as positional effects of the X chromosome on autosomal genes and polygenic models for POI [21]. Gene expression can be greatly influenced by regulatory elements that can be located far from the actual gene. It is also possible that integrity of the X chromosome influences expression of key autosomal genes required for proper oogenesis.

7.2.2 Autosomal Structural Rearrangements

About 1 in 500 people in the general population is a carrier of a balanced chromosome rearrangement. Phenotypically normal carriers of balanced chromosomal rearrangements (translocations and inversions; see Chapter 1) have an increased risk of infertility and cytogenetically abnormal offspring [14,19,20,26]. Beside sex chromosome abnormalities, rearrangements involving autosomes are common in infertile patients. Several mechanisms may account for gonadal insufficiency and infertility in carriers of autosomal structural rearrangements: (1) rarely, rearrangements may disrupt expression of an ovary-specific gene at the breakpoint by reducing dosage or causing abnormal expression; (2) rearrangements may also cause "position effect" by disrupting regulatory elements that influence expression of genes near the breakpoint; (3) rearrangement may cause high predisposition to form chromosomally unbalanced gametes with low survival rate; or (4) a higher aneuploidy rate for other chromosomes that are not involved in the rearrangement, but are mis-segregated due to an interchromosomal effect or meiotic disturbance during gametogenesis. *FOXL2* (OMIM#605597), *NOBOX* (OMIM#610934), *FIGLA* (OMIM#612310), and *NR5A1* (OMIM#612964) genes are a representative subset of autosomal genes required for normal ovarian development, differentiation, and oogenesis [19–21]. Disruption of any of these genes by structural rearrangements may adversely affect ovarian function. Disruption of an ovary-specific gene by a translocation breakpoint is a rare cause of infertility, and can be associated with syndromic or nonsyndromic POI. Women who carry a balanced chromosomal abnormality have a much higher risk for infertility due to early pregnancy loss or miscarriage of an unrecognized pregnancy [26].

The reproduction outcomes greatly depend on the chromosome involved, breakpoint location, gene content, and the complexity of the rearrangement in each individual case. However, on average, about half of the pregnancies in a person carrying a balanced chromosome rearrangement will be lost in miscarriage [12,14,16,26]. Both male and female carriers of constitutional structural chromosomal rearrangements are likely to produce genetically unbalanced gametes, resulting in partial monosomy or trisomy in the embryo [12–16,26]. In males, numerous studies have been performed to determine meiotic segregation patterns in spermatozoa [12–16], whereas the cytogenetic analysis of female gametes and embryos remains difficult to study in humans. In cases of balanced chromosome rearrangements, most imbalances at birth result from a rearrangement carried by the mother [26]. The more stringent meiotic checkpoints in males cause unbalanced segregation of structurally abnormal chromosomes to result in oligozoospermia/azoospermia, male infertility, and therefore significantly reduced likelihood of successful conception and transmission to the offspring [12–16]. On the other hand, female gametogenesis has a less stringent meiotic checkpoint control (see Chapter 3 for more details) and oocyte maturation is less affected by autosomal genomic imbalances leading to unbalanced conceptions. Successful pregnancy in these families can be achieved with IVF and PGT to transfer embryos with euploid chromosome complements (see Chapter 13 for detailed information).

7.2.3 46,XY Female (Swyer Syndrome)

In mammals, the gonads in both sexes have the potential to develop into either ovaries or testes. Normal male sexual differentiation in 46,XY individuals depends on proper function and complex interaction of numerous testis-determining genes, including *SRY*, *SOX9*, *NR5A1/SF1*, *WT1*, *NR0B1*, *AR*, *DHH*, and *CBX2* [10]. Failure in the normal male sex differentiation process can cause a complete or partial 46,XY gonadal dysgenesis. Partial 46,XY gonadal dysgenesis is characterized by impaired testicular development and ambiguous external genitalia, whereas individuals with complete 46,XY gonadal dysgenesis, or Swyer syndrome, have normal female external genitalia and internal organs, but also have bilateral streak gonads [10,19,20]. Swyer syndrome has been estimated to occur in approximately 1 in 30 000 individuals. Affected females are typically tall, lack secondary sexual characteristics, may have mild clitoromegaly, and are infertile. This condition commonly remains undiagnosed until adolescence, when puberty fails to occur. Females with a 46,XY karyotype (Figure 7.3a) have an increased risk of developing gonadoblastoma or dysgerminoma and therefore streak gonads are usually removed shortly after diagnosis.

Complete 46,XY gonadal dysgenesis is a heterogeneous disorder that results from chromosomal abnormalities (deletions, duplications, structural rearrangements) or point mutations of genes implicated in sexual differentiation [10,20,27].

Despite considerable advances in understanding the genetic factors involved in gonadal determination and differentiation, a molecular diagnosis is made in only about 20% of cases with complete 46,XY gonadal dysgenesis. Mutations and deletions of the *SRY* gene are the cause of complete 46,XY gonadal dysgenesis in approximately 10–15% of patients with Swyer syndrome [27]. The *SRY* gene is located on the short arm of the Y chromosome, within a 35-kb sequence proximal to the pseudoautosomal region boundary (Figure 7.2). Structural Y chromosome rearrangements resulting in loss of the *SRY* gene include Yp deletion, dicentric Y isochromosomes composed of the long arm (idic(Yq)), ring Y chromosomes (r(Y)), small marker Y chromosomes, and Y autosome translocations [19,20,27]. Structurally abnormal Y chromosome can be detected by conventional G-band chromosome analysis in some cases; however, molecular cytogenetic studies such as FISH and aCGH analyses are essential for accurate diagnosis. Y chromosome rearrangements are frequently accompanied with mosaicism for multiple Y chromosome-containing abnormal cell lines or 45,X chromosome complement.

X chromosomal rearrangements in 46,XY females have led to the identification of a dosage-sensitive sex locus at the Xp21 region containing the *NR0B1*

Table 7.2. Genomic imbalances associated with complete gonadal dysgenesis in 46,XY females

Locus	Genomic abnormality	Gene	Molecular mechanism	OMIM # [PMID]
1p36.12	Duplication	*WNT4*	Gain of function	603490
5q11.2	Deletion	*MAP3K1*	Loss of function	613762
8p23.1	Deletion downstream	*GATA4*	Regulatory region	600576
9p24.3	Deletion	*DMRT1, DMRT2*	Loss of function	154230
9q33.3	Deletion	*NR5A1/SF1*	Loss of function	612965
10q25	Deletion	Unknown	Unknown	[19558528]
11p13	Deletion	*WT1*	Loss of function	607102
12q13.1	Homozygous deletion	*DHH*	Loss of function	233420
Xp21	Duplication, deletion upstream	*NR0B1/DAX1*	Gain of function, regulatory region	300018
Yp11.31	Deletion	*SRY*	Loss of function	400044

(*DAX1*) gene. Patients with cytogenetically visible Xp21 duplications, containing multiple genes in addition to *NR0B1*, have a complex phenotype with congenital anomalies, dysmorphic features, intellectual disability, and gonadal dysgenesis (Table 7.2). Isolated 46,XY gonadal dysgenesis has been reported in two siblings carrying an Xp21.2 duplication of 637 kb in size that encompasses *DAX1* as well as four *MAGEB* genes. A submicroscopic 257-kb deletion upstream of *DAX1* has been described in a 46,XY female with primary amenorrhea and gonadal dysgenesis. This deletion likely affects regulatory sequences leading to altered *DAX1* expression and 46,XY gonadal dysgenesis.

Many autosomal genes are implicated in disorders of sexual development in humans. Cytogenetically visible chromosome abnormalities including deletions of 9p22, 9q33, 10q25, 11p13, 13q32–q34, and 17q24; duplication of 1p34; and balanced translocations involving 17q24 have been identified in patients with XY gonadal dysgenesis (Table 7.2). These rearrangements comprise multiple genes and are usually associated with multiple congenital anomalies and intellectual disabilities (syndromic XY gonadal dysgenesis) [10,19,20,25].

Isolated or nonsyndromic forms of XY gonadal dysgenesis are most probably due to a single gene defect and are unlikely to be detected by classical karyotype. Detection of small deletions and duplications, encompassing a single gene, are beyond the resolution of classical G-band chromosome and FISH analyses, and require application of a high-resolution genome analysis technique such as aCGH [25]. Table 7.2 summarizes several genes that are known to cause nonsyndromic XY gonadal dysgenesis, including those with microdeletions or microduplications affecting the gene or its regulatory regions. Using high-resolution whole-genome and sex chromosome aCGH analyses, the cause of 46,XY gonadal dysgenesis can be elucidated in up to 30% of affected individuals, while the remaining patients will benefit from next generation sequencing (NGS) or exome sequencing tests.

7.2.4 Submicroscopic DNA Copy Number Alterations

Cytogenetically visible numerical and structural chromosomal rearrangements are already known to cause a substantial number of human diseases. These abnormalities are commonly identified by conventional karyotype analysis (see Chapter 1), a low-resolution technique that has limited ability to detect genomic imbalances less than 4–10 Mb in size (Figure 7.7). With the development of microarray technology, aCGH and SNP-array platforms can be used to detect genomic deletions and duplications (losses and gains) as small as 1 kb (1000 bp) in size [21,24,25]. Submicroscopic chromosomal imbalances or DNA copy number variations (CNVs) are present in the genome of each individual and can be classified as benign, pathogenic, or variants of unknown clinical significance. It is now established that CNVs cause genetic syndromes, isolated congenital defects, disease susceptibility, miscarriages, or reproductive failure [21,24,25]. In addition, CNVs can also result in the unmasking of a recessive mutation, functional polymorphism, or epigenetic or environmental interactions. Many CNVs are benign and exist in healthy individuals. Databases of genomic variants can help distinguish normal variation from pathogenic. However, the reproductive history of individuals in these databases is rarely known.

Application of microarray technology in basic research and clinical diagnosis of male and female infertility has led to the identification of multiple autosomal loci implicated in POI, male infertility, abnormal fetal development, and placental dysfunction [21,24,25]. Remarkably, such genomic imbalances contain many genes involved in meiosis, DNA repair, and ovarian folliculogenesis [21,24].

The advent of high-resolution genome analyses such as aCGH and high-throughput sequencing will enable identification of many genes implicated in human sex determination, differentiation, and reproduction. These new powerful tools will revolutionize clinical diagnosis and personalized reproductive management in the future.

7.3 Practical Clinical Approach

Identification of genetic causes underlying male and female infertility is an essential part of the clinical evaluation, genetic counseling, and successful treatment of the infertile couple. Accurate genetic diagnosis presents the couple with an opportunity to guide treatment options, to achieve natural conception with their own gametes, and provides important information regarding the health and reproductive potential of an affected individual. Moreover, genetic counseling of the couple is essential to provide information about the risk of transmitting a genetic abnormality to the offspring. IVF, in its current form and genomic technologies, represents the best option for maximizing genetic diagnosis and preventing transmission of many, but not all genetic disorders.

Genetic evaluation is indicated for couples that fail to achieve pregnancy after 12 months of regular unprotected intercourse, as well as for patients with a clinical diagnosis or medical history of a chromosomal or genetic disorder. Careful family history is necessary to determine if there is a clear familial pattern of infertility, miscarriages, skewed gender ratios (e.g. complete androgen insensitivity syndrome), rapid aging, and/or syndromic causes associated with infertility (Fanconi anemia, Bloom syndrome, ataxia telangiectasia). However, negative family history does not rule out any genetic contribution as de novo genetic events likely account for a substantial number of sporadic cases. Examples include most chromosomal abnormalities such as Klinefelter syndrome.

Karyotype analysis of peripheral blood samples should be performed as an initial component of evaluation for male or female infertility to identify sex chromosome aneuploidy and gross structural chromosome rearrangements (Figure 7.8). Despite gonadal failure in most affected patients with Klinefelter or Turner syndrome, 5–20% of individuals with sex chromosome aneuploidy may have a limited number of mature germ cells, enabling the live birth of biological children. The number of available germ cells significantly decreases with age. Cryopreservation of ovarian follicles or retrieved testicular spermatozoa as an infertility treatment option may be feasible for some patients with Turner or Klinefelter syndrome, respectively [3,5]. Women with Swyer syndrome cannot produce eggs, but successful pregnancies have been achieved in some patients using donated eggs or embryos [20,27].

Spontaneous pregnancy in Turner syndrome is reported to be possible in about 2% of patients with mosaic monosomy X chromosome constitution. Pregnancy and delivery, either spontaneous or more commonly from donor oocytes, are associated with a 2% risk of death in nonmosaic Turner syndrome due to dissection and rupture of the aorta. Pregnancy is a relative contraindication in women with Turner syndrome who have a negative preconception cardiac evaluation but an absolute contraindication in those with documented cardiac anomaly [28].

Structural chromosome abnormalities detected by karyotype analysis confer an increased risk for spermatogenic failure, miscarriages, stillbirth, and liveborn children with congenital defects and chromosomal aberrations. In some patients, balanced chromosome abnormalities are present but they are below the resolution of detection by conventional cytogenetic analysis. Microarray analysis should be considered on DNA from spontaneous abortions and stillbirth, if available. For couples undergoing IVF, PGT-A and PGT-SR should be performed to

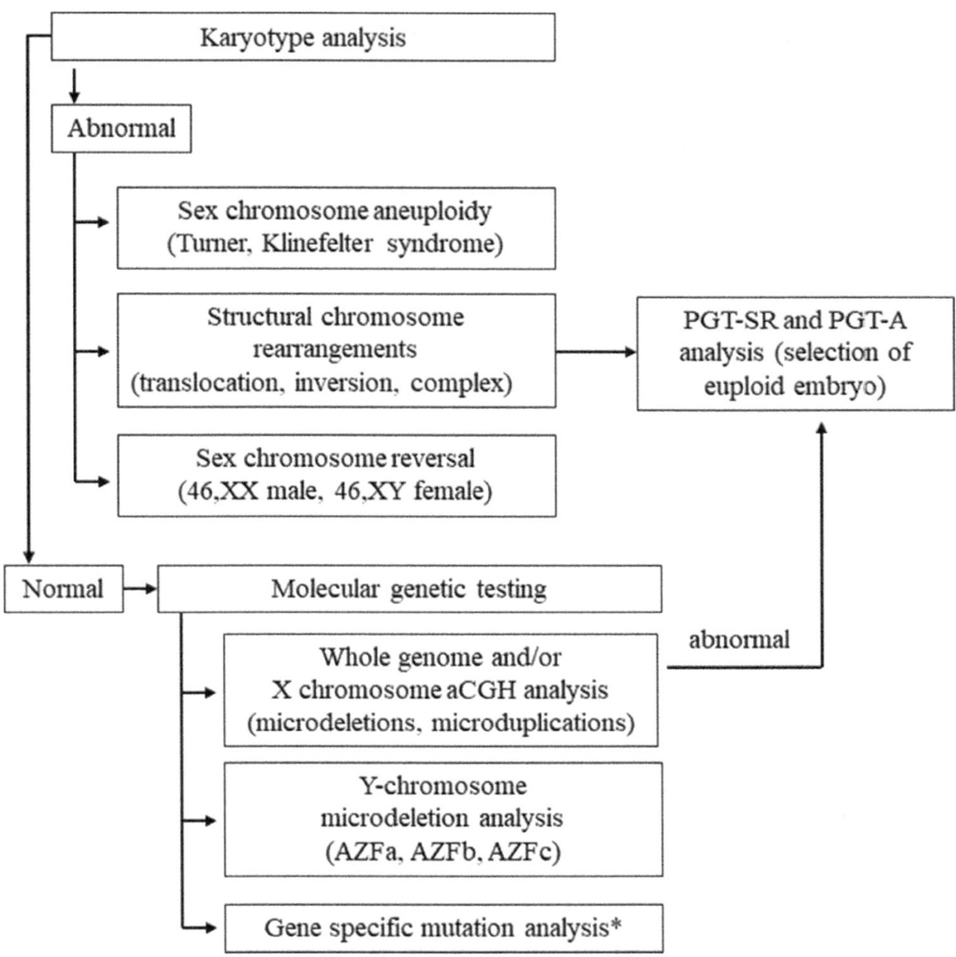

Figure 7.8 Genetic evaluation in patients with infertility. PGT-SR, preimplantation genetic testing for structural rearrangements; PGT-A, preimplantation genetic testing for aneuploidy. See Chapter 8 for implicated genes.

discover suitable euploid embryos for transfer (Figure 7.8).

In the case of a normal karyotype, expanded genetic testing should include microarray analysis to detect submicroscopic chromosome abnormalities, Y chromosome microdeletions and, depending on gathered clinical information, a possible individual gene mutation analysis (Figure 7.8; for more information see Chapters 8 and 9). It is critical that men with nonobstructive azoospermia undergo molecular and cytogenetics analyses for Y chromosome rearrangements and microdeletions in order to receive an accurate diagnosis and proper genetic counseling regarding utility of assisted reproduction. In about 60% of males with *AZFc* deletions, sperm can be found in the testes and retrieved using testicular sperm extraction (TESE) or testicular sperm aspiration (TESA) techniques. In men with partial *AZFb* (either alone or associated with *AZFc*), TESE is productive in about 7%, while in patients with *AZFa* or complete *AZFb* deletions TESE is invariably unsuccessful [29]. The Y chromosome microdeletions are not associated with health problems; however, male offspring will inherit the AZF microdeletion from infertile fathers in pregnancies achieved by assisted reproduction.

High-resolution whole-genome and X chromosome microarray analyses are recommended as part of the genetic evaluation of infertility in patients with normal karyotype (Figure 7.8). Gonadal failure and recurrent pregnancy losses of male fetuses have been associated with submicroscopic X chromosome deletions and duplications [3,21,24,25]. It is important to note that negative genetic testing does not exclude genetic pathology, as there are other, presently unknown, genes that are implicated in normal gametogenesis. To date, mutations have been found in approximately 300 genes, of which 70 syndromes are known to be associated with reproductive disorders, and this list continues to grow. Multigene panels are widely used for detection of disease-causing variants among clinically relevant genes. Although gene panel testing provides approximately 100% coverage in targeted genes, whole-exome and whole-genome approaches are now a logical extension, providing advances in analysis of a much broader gene list, including newly discovered disease loci and alterations in noncoding gene regulatory regions.

Genetic counseling should be ideally provided before a genetic test is offered, so that the couple understands the pros and cons of genetic testing. Post-test genetic counseling is necessary for the couple to understand the significance of each possible outcome: normal results, pathologic test findings, and findings of unknown clinical significance. The results of several studies have indicated an increased risk of birth defects in pregnancies and children conceived by infertile couples, a substantial portion of which is attributed to the undiagnosed parental chromosomal rearrangements, higher risk of chromosomal aneuploidies, epigenetic disturbances, and complications related to multiple gestations [30]. It is essential that fertility specialists be engaged with clinical genetic experts, including those in the genomic laboratories, in order to provide the most optimal and appropriate testing for their patients.

References

1. Schultz N, Hamra FK, Garbers DL. A multitude of genes expressed solely in meiotic or postmeiotic spermatogenic cells offers a myriad of contraceptive targets. *Proc Natl Acad Sci USA* 2003;100:12201–6.
2. Krausz C, Riera-Escamilla A. Genetics of male infertility. *Nat Rev Urol* 2018;15:369–84.
3. Heard E, Turner J. Function of the sex chromosomes in mammalian fertility. *Cold Spring Harb Perspect Biol* 2011:3(10):a002675.
4. Walsh TJ, Pera RR, Turek PJ. The genetics of male infertility. *Semin Reprod Med* 2009;27:124–36.
5. Wikström AM, Dunkel L. Testicular function in Klinefelter syndrome. *Horm Res* 2008;69:317–26.
6. Martin RH. Cytogenetic determinants of male fertility. *Hum Reprod Update* 2008;14:379–90.
7. Hennebicq S, Pelletier R, Bergues U, et al. Risk of trisomy 21 in offspring of patients with Klinefelter's syndrome. *Lancet* 2001;357:2104–5.
8. Kuroda-Kawaguchi T, Skaletsky H, Brown LG, et al. The AZFc region of the Y chromosome features massive palindromes and uniform recurrent deletions in infertile men. *Nat Genet* 2001;29:279–86.
9. Krausz C, Casamonti E. Spermatogenic failure and the Y chromosome. *Hum Genet* 2017;136:637–55.
10. Kousta E, Papathanasiou A, Skordis N. Sex determination and disorders of sex development according to the revised nomenclature and classification in 46,XX individuals. *Hormones (Athens)* 2010;9:218–31.
11. Grinspon RP, Rey RA. Disorders of sex development with testicular differentiation in SRY-negative 46,XX individuals: clinical and genetic aspects. *Sex Dev* 2016;10:1–11.
12. Ferguson KA, Wong EC, Chow V, Nigro M, Ma S. Abnormal meiotic recombination in infertile men and its association with sperm aneuploidy. *Hum Mol Genet* 2007;16:2870–9.
13. Marchetti F, Wyrobek AJ. Mechanisms and consequences of paternally-transmitted chromosomal abnormalities. *Birth Defects Res C Embryo Today* 2005;75:112–29.
14. Madan K. Balanced complex chromosome rearrangements: reproductive aspects. A review. *Am J Med Genet A* 2012;158A:947–63.
15. Roux C, Tripogney C, Morel F, et al. Segregation of chromosomes in sperm of Robertsonian translocation carriers. *Cytogenet Genome Res* 2005;111:291–6.
16. Engels H, Eggermann T, Caliebe A, et al. Genetic counseling in Robertsonian translocations der (13;14): frequencies of reproductive outcomes and infertility in 101 pedigrees. *Am J Med Genet A* 2008;146A:2611–16.
17. Machev N, Gosset P, Warter S, et al. Fluorescence in situ hybridization sperm analysis of six translocation carriers provides evidences of an interchromosomal effect. *Fertil Steril* 2005;84:365–73.
18. Viville S, Mollard R, Bach M-L, et al. Do morphological anomalies reflect chromosomal aneuploidies? *Hum Reprod* 2000;15:2563–6.
19. Simpson JL, Rajkovic A. Ovarian differentiation and gonadal failure. *Am J Med Genet* 1999;89:186–200.

20. Yatsenko SA, Rajkovic A. Genetics of human female infertility. *Biol Reprod* 2019;101:549–66.

21. Yatsenko SA, Wood-Trageser M, Chu T, et al. A high-resolution X chromosome copy-number variation map in fertile females and women with primary ovarian insufficiency. *Genet Med* 2019;21: 2275–84.

22. Doğer E, Çakıroğlu Y, Ceylan Y, et al. Reproductive and obstetric outcomes in mosaic Turner's syndrome: a cross-sectional study and review of the literature. *Reprod Biol Endocrinol* 2015;13:59.

23. Tartaglia NR, Howell S, Sutherland A, Wilson R, Wilson L. A review of trisomy X (47, XXX). *Orphanet J Rare Dis* 2010;5:8.

24. McGuire MM, Bowden W, Engel NJ, et al. Genomic analysis using high-resolution single-nucleotide polymorphism arrays reveals novel microdeletions associated with premature ovarian failure. *Fertil Steril* 2011;95:1595–600.

25. Rajcan-Separovic E. Chromosome microarrays in human reproduction. *Hum Reprod Update* 2012;18:555–67.

26. Desjardins MK, Stephenson MD. "Information-rich" reproductive outcomes in carriers of a structural chromosome rearrangement ascertained on the basis of recurrent pregnancy loss. *Fertil Steril* 2012;97:894–903.

27. Jorgensen PB, Kjartansdóttir KR, Fedder J. Care of women with XY karyotype: a clinical practice guideline. *Fertil Steril* 2010;94:105–13.

28. Gravholt CH, Andersen NH, Conway GS, et al. International Turner Syndrome Consensus Group. Clinical practice guidelines for the care of girls and women with Turner syndrome: proceedings from the 2016 Cincinnati International Turner Syndrome Meeting. *Eur J Endocrinol* 2017;177: G1–G70.

29. Stahl PJ, Masson P, Mielnik A, et al. A decade of experience emphasizes that testing for Y microdeletions is essential in American men with azoospermia and severe oligozoospermia. *Fertil Steril* 2010;94:1753–6.

30. Berntsen S, Söderström-Anttila V, Wennerholm UB, et al. The health of children conceived by ART: "the chicken or the egg?" *Hum Reprod Update* 2019;25:137–58.

Genetics of Human Male Infertility: The Quest for Diagnosis and Treatment

Stéphane Viville and Özlem Okutman

8.1 Introduction

The World Health Organization (WHO) defines infertility as the inability to conceive within 12 months despite regular unprotected intercourse, a condition that concerns about 10–15% of couples globally. Infertility is considered as primary or secondary depending on whether a couple has experienced a prior pregnancy or not.

The causes of infertility can be highly diverse and include hormonal disorders, inflammatory diseases, genital tract obstruction, abnormal gametogenesis, implantation failures, and sexual dysfunctions, such as those involving erection or ejaculation. Infertility can even have a physiologic or a psychiatric origin. In addition, lifestyle and environmental factors, which may be tightly linked, can also affect reproductive ability.

The underlying origin of infertility can be female (~50%), male (20–30%) or mixed (20–30%). Because of the complexity of the reproductive process, standard fertility examinations fail to identify an etiology in 15–30% of infertile couples and such cases are then classified as "idiopathic." It is believed that about half of the idiopathic cases could be explained by a genetic defect [1]. This chapter is devoted to the genetics of male nonsyndromic infertility. Female infertility and its genetic causes are discussed in Chapter 9.

Diagnosis of male infertility is generally based on standard semen analysis. Abnormal semen parameters are classified according to different criteria, including sperm concentration, morphology, motility, and vitality. Measurements are made on semen samples obtained following 2–7 days of abstinence and compared with reference values defined by WHO.

Sperm concentration represents the number of spermatozoa per unit volume. When a patient presents a complete absence of spermatozoa in the ejaculate, he is characterized as azoospermic and diagnosed as suffering from azoospermia. A patient with less than 15 million spermatozoa per milliliter is oligozoospermic. When spermatozoa are only observed in a centrifuged pellet, the patient is characterized as cryptozoospermic.

Abnormalities if morphology and mobility are termed teratozoospermia and asthenozoospermia, respectively. Patients with less than 4% morphologically normal spermatozoa are considered teratozoospermic, while asthenozoospermic patients have less than 32% progressively motile spermatozoa.

Defects in concentration, motility, and morphologic parameters simultaneously are defined as oligoasthenoteratozoospermia (OAT). Sperm vitality, the percentage of living sperm in the semen, is especially important for samples with less than about 40% progressively motile spermatozoa. Necrozoospermia refers to the low percentage of live and high percentage of immotile spermatozoa in the ejaculate.

Syndromic disorders, characterized by their pathology, may also be associated with infertility. This group is defined as syndromic infertility. In most instances, infertility is a minor feature of the clinical condition and not at all a concern for the patient or even goes unrecognized. However, some syndromic pathologies may have variable penetrance and mild forms may only affect the fertility of the patient. This is the case for instance in cystic fibrosis (CF), an autosomal recessive disease caused by mutations in the cystic fibrosis transmembrane regulator (*CFTR*) gene. Affected patients may present with a severe form of CF and in such cases they usually do not consult for infertility. However, progress achieved in the care of these patients has drastically improved their life expectancy, to the extent that more of these patients may have a reproductive project. The most minor form of CF affecting adult males is characterized by an isolated congenital bilateral absence of vas deferens (CBAVD) which leads to an obstructive azoospermia in 80–90% of cases. Another example that needs to be differentially diagnosed from CF is primary ciliary dyskinesia, also known as immotile cilia syndrome, a genetically heterogeneous autosomal recessive disorder resulting from loss of normal ciliary function and progressive damage to the respiratory system. About half of the male patients are infertile due to asthenozoospermia.

When infertility is the only symptom, it is considered as nonsyndromic. Research into the monogenic forms of infertility, both male and female, has been very successful since many gene mutations responsible for absent or abnormal spermatogenesis, without any other symptoms, have recently been described.

The genetic causes of infertility can be classified into three groups: chromosomal abnormalities (discussed in Chapter 7), mitochondrial DNA mutations (discussed in Chapter 12), and monogenic mutations.

Any loss, gain, or abnormal arrangement of genetic material at the chromosomal level is defined as a chromosomal abnormality and it has long been recognized as a possible cause of infertility because of "abnormal" spermatogenesis [2]. The next most frequent abnormality found in these patients is microdeletion of the Y chromosome, found respectively in 15% and 10% of azoospermic and oligozoospermic patients (see Chapter 7). Mitochondrial DNA mutations are associated with poor motility and diminished fertility of human sperm (discussed in Chapter 12).

Male infertility due to genetic defects may be, as in females, attributable to abnormalities in sexual development, deficiencies of the hypothalamic–pituitary–gonadal axis, impaired spermatogenesis [3], or other anomalies leading to problems with the delivery of sperm as well as mixed phenotypes involving defects in sperm count (oligozoospermia or azoospermia), morphology (teratozoospermia), and motility (asthenozoospermia). Considering the concerted action of more than 2000 genes in mouse spermatogenesis, a similar number of genes no doubt play a role in the human, so theoretically, a mutation in any of these genes may be responsible for a spermatic defect [4].

The identification of genes involved in fertility not only benefits our basic knowledge of the molecular mechanisms underlying human gametogenesis but also enables us to offer better care to infertile couples, at the least by being able to establish the genetic etiology of male infertility, but also by improving assisted reproductive technology (ART) and allowing genetic counseling.

This chapter is dedicated to monogenic causes of nonsyndromic male infertility. In the first section, the methods allowing the identification of new genes involved in male infertility are described and illustrated with some examples. Monogenic mutations causing infertility in the male as discovered over the last decade are then described and discussed. In the final section, a workflow of who may benefit from genetic diagnosis of male infertility is overviewed.

8.2 Identification of New "Infertility" Genes

Study of the "genetics of infertility" began less than two decades ago with the identification of the first gene mutations that cause infertility and was the result of the combination of two events, a human one and a technological one. Indeed, the establishment of preimplantation diagnosis in the 1990s led geneticists to become interested in human reproduction and to ask what genes may be involved in infertility. In addition, the field has benefited from the recent technological improvements in molecular biology and more precisely in molecular genetics. In the last two decades, the rapid evolution of analytical tools has revolutionized the field of genetics, increasing the speed of identification of genes and mutations involved in human pathologies including infertility. Indeed, since the beginning of the twenty-first century, the technology has moved from the analysis of single genes to whole genomes. The improvements established by high-throughput sequencing (HTS) technologies and the reduction in their cost have made the whole-exome sequencing strategy (described below) the preferred one in the field of genetics, leaving on the sidelines other whole-genome analyses such as single nucleotide polymorphism (SNP)-array or array comparative genomic hybridization (aCGH) which are less and less used for identifying gene mutations.

The parallel sequencing of millions of DNA molecules by high-throughput methods has considerably speeded up the discovery of the genetic causes of diseases. Such methods are often termed "next generation sequencing "(NGS), which is not the most appropriate term since a variety of technologies evolve quickly into the "next" generation; "high-throughput sequencing" should be the preferred term

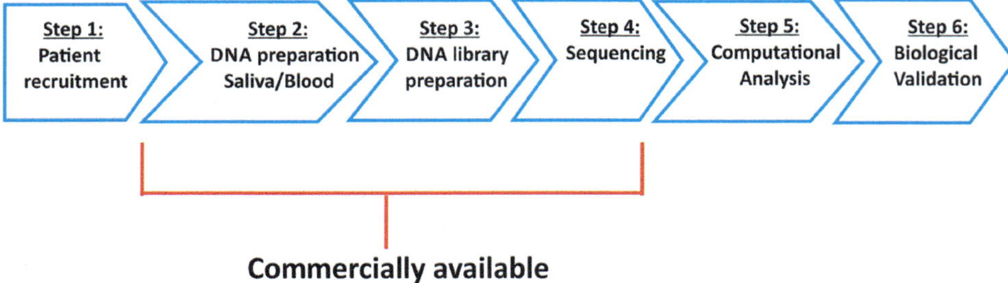

Figure 8.1 Workflow for exome sequencing in the quest for causative genes in male infertility.

because it is independent of the sequencing technology used. Using these methods, different strategies are available involving either the whole genome without distinction between coding and noncoding sequences or only enriched parts of the genome. The type of enrichment chosen will depend on the purpose of a sequencing project. For diagnostic purposes it can be limited to the genes known to be involved in the pathology studied, known as gene panel sequencing, while for research purposes with the goal of identifying new genes, enrichment should include all the known coding parts of the genome. Indeed, a popular variant of HTS is exome sequencing which limits sequencing only to exons, about 1% of the genome. Since it is estimated that 85% of the disease-causing mutations are located in coding and functional regions of the genome [5], that the complexity of the bioinformatics analysis increases with the size of the part of the genome analyzed, and that exome sequencing is still considerably cheaper than whole-genome sequencing, most studies are carried out by sequencing the exome and not the whole genome.

Whole-exome sequencing (WES) has emerged as a powerful and efficient strategy for identifying the genes responsible for Mendelian disorders and complex diseases and has already led to the identification of dozens of genes involved in infertility within families or patient cohorts. Overall, WES includes six major steps: recruitment of patients, sample preparation, library preparation, sequencing, data processing, analysis, and validation as outlined in Figure 8.1.

Recruitment of patients, computational analysis (also called bioinformatics analysis), and biological validation (respectively steps 1, 5 and 6 in Figure 8.1) are the most difficult steps and require a high level of expertise. On the contrary, DNA preparation (which can be done on different types of samples, but most commonly blood or saliva samples), DNA library preparation and sequencing (respectively steps 2, 3 and 4 in Figure 8.1) are technical and available through different commercial companies, which on most occasions will not, or only partially, perform the bioinformatics analysis. In the following paragraphs we focus on steps 1, 5 and 6.

8.2.1 Recruitment of Patients (Step 1)

As for any other study in the quest for new genes or gene defects, material from affected patients is needed as well as samples from well-selected control individuals. Needless to say, an accurate clinical diagnosis is of utmost importance because any incorrect diagnosis will introduce errors in recruitment and render the genetic analysis inaccurate. This underlines the importance of the quality of the collaboration between clinicians diagnosing and selecting patients and geneticists carrying out the research. Before embarking on a project, together they should carefully elaborate the strategy by defining the selection criteria and the procedure for recruitment. The minimum requirement is to follow the institutional directives already established for the diagnosis of the pathology being studied, to which additional criteria can be added according to the scientific hypotheses linked to the study. In the field of male genetic infertility, WHO instructions need to be followed and at least two detailed spermiograms performed at an interval of at least 3 months in order to define the defect in terms of sperm number (azoospermia or oligozoospermia) and the motility of the sperm cells present (asthenozoospermia). If the study is focused on morphologic defects (teratozoospermia), both a spermiogram and a spermocytogram are required. The study can involve a combination of these defects, some of them (oligoasthenozoospermia, asthenoteratozoospermia, oligoteratozoospermia), or all (oligoasthenoteratozoospermia). In the case of nonobstructive azoospermia, it can be worthwhile, but not required, to have information on testis histology. Indeed, such anatomopathologic reports will provide specific information on the stage of developmental arrest of sperm cells. Such reports allow the best classification of the cases studied.

Genetic analysis can be performed on different cohorts of patients, including: (1) large families with some degree of consanguinity where at least two brothers are affected and at least one other is not; (2) a number of small families, at least three, possibly with some degree of consanguinity, in which patients display the same phenotype; and (3) large groups of patients showing the same spermiogram defect. These can be either individuals presenting the same phenotype without any relationship or individuals originating from a restricted area or living in a limited sociocultural enclave, which could correspond to a founder effect. This implies that historically the mutation occurred in a single person from whom it was inherited from generation to generation with a recessive mode of transmission and eventually, because of lack of diversity, was carried by both partners of a couple who then transmitted it as a homozygous mutation to their sons.

The ability to identify a causative mutation depends on the number of patients and their familial relationship. As a rule, large consanguineous families are more readily studied than smaller ones, whereas the latter are more favorable for genetic investigations than cohorts of patients or groups of individual patients (Figure 8.2). Large consanguineous families are preferred because the probability that all affected brothers display the same causative mutation is very high, whereas uncertainty remains in other instances due to the poor accuracy of the diagnosis.

One of the difficulties in recruiting patients, mostly familial cases, is due to the taboo associated with male infertility. Infertile male patients, because of the cultural link between fertility and virility, do not speak about their incapacity to conceive, even within their own family, and therefore even if

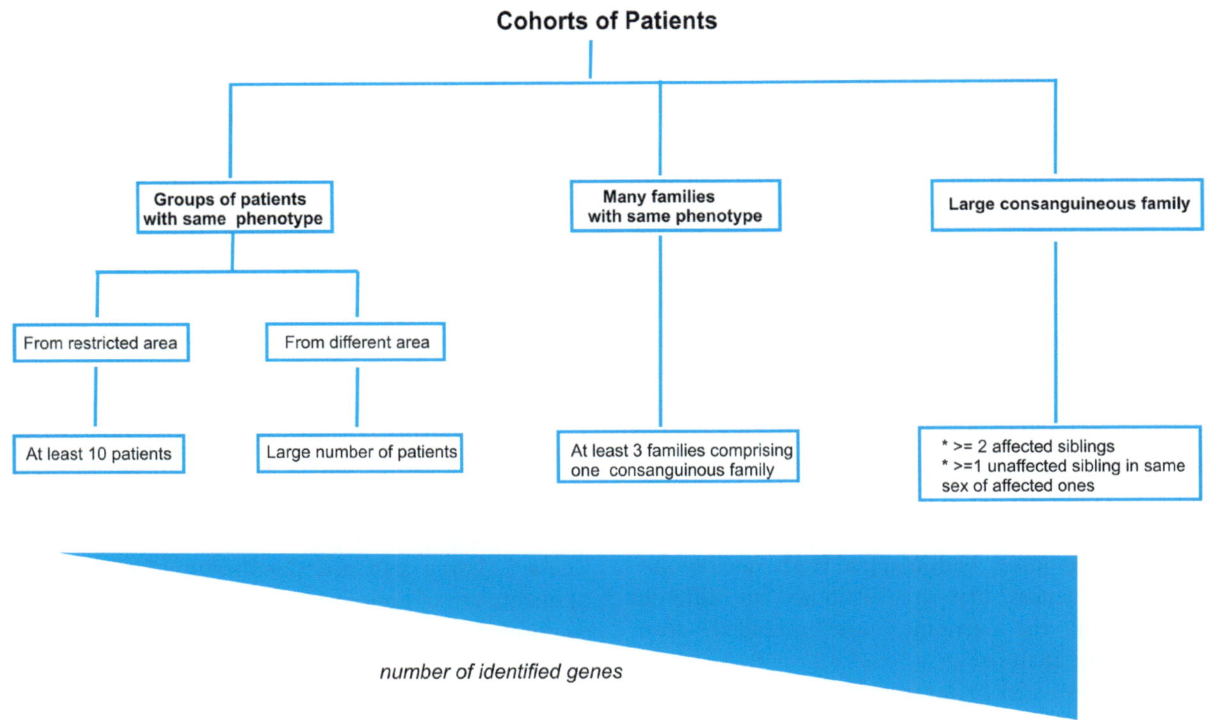

Figure 8.2 The different cohorts of patients that can be analyzed in the quest for infertility genes, the success in identifying genes depending on the cohort studied and the optimal number of patients to be analyzed. Cohorts can consist of groups of patients presenting the same phenotype, which can be possibly divided into two subgroups from a restricted area (where a founder effect is expected) or from different geographic locations. Family-based studies provide an alternative approach to identifying genes involved in infertility. The studies can be based on either numerous small families or one large family. Source: adapted from Okutman et al. [17] .

several brothers are affected it may remain unacknowledged.

The choice of which samples will be sequenced depends on the type of recruitment. For familial cases, most of the time all affected brothers and possibly one unaffected brother with parents if available are sequenced. For individual cases, without a familial link, all of them must be sequenced. Therefore, recruitment has a significant impact on the cost of a study.

8.2.2 Computational Analysis: Data Processing and Bioinformatics Analysis (Step 5)

In the quest for causative rare variants, the quality of the processing of samples to be sequenced is critical. If samples have to be sent to a company to be sequenced, it is important to ensure that the data provided are of the highest standard.

Of note, use of the terms "variant" and "mutation" are still a matter of debate. In our case, we use "variant" for any change in the sequence when compared with the reference sequence and "mutation" for any variant identified as causative of disease.

Briefly, two main criteria are used to evaluate the quality of a sequencing run; coverage and read depth. Coverage is a measure of what proportion of the genome intended to be sequenced is represented in the dataset. The read depth, on the other hand, describes the number of times that a given nucleotide in the genome has been read in an experiment. Good-quality HTS guarantees that more than 95% of the region of interest is covered by at least 30 reads [6].

Dedicated pipelines can be used to process data (Figure 8.3). Many pipelines are available and each possesses its own specificity. It is important to have a minimum of experience with the pipeline used. All have in common the ability to trim off bad-quality parts of reads and to map the good-quality reads onto a reference genome. Multimapped reads, corresponding to sequences that are attributed to different regions of the genome, are excluded from downstream analysis.

Using powerful bioinformatics tools, the variants detected are scored and ranked from the potentially most pathogenic to the least pathogenic in the area sequenced. Variant filtering for frequency will select rare variants, which are expected for rare to very rare pathologies. For this purpose, publicly available sequence databases, such as Genome Aggregation Database (gnomAD) (http://gnomad.broadinstitute.org/) and 1000 Genomes project data (www.internationalgenome.org/), which comprise several hundred thousand individual sequences, are the most helpful tools for filtering allele frequencies. However, data for certain populations are limited or not available. In such cases, the pathogenicity of the mutation may be challenged by checking for the absence of the variation in at least 100 sequenced healthy fertile controls matching the same geographic region and/or same ethnicity of analyzed patients.

Two more criteria can help to categorize variants: RNA or protein expression, which can be checked through online expression databases (i.e. Amazonia, BioGPS, EMBL-EBI expression atlas, Human Protein Atlas), and a literature scan for a possible role in the phenotype studied. For example, a high level of expression in the testes supports the role of a variant identified in a male infertility phenotype. In contrast, an absence of testes expression will argue against considering the variant. If the protein has been studied in a model system such as the mouse, which can be diverse depending on its conservation during evolution and displays a function compatible with the phenotype studied, it can also be an indication of the pertinence of the selected variant.

A classification system based on a five-class scheme for clinical reporting of genetic variants was proposed by the International Agency for Research on Cancer (IARC) in 2008 [7]. Nonpathogenic variants are classified as Class 1, likely not pathogenic variants as Class 2. When there is too little information to make any recommendation and it is not known whether the variant has any effect on gene function which might cause the phenotype being studied, the variant is reported as of uncertain significance (VUS) or Class 3. Variants which have significant but disputable evidence for pathogenicity are classified as Class 4. Finally, pathogenic variants are classified as Class 5. Using standard classification facilitates the transmission of appropriate information to clinicians and improves clinical use of the information.

The selected variation(s), identified by HTS, must be confirmed via Sanger sequencing. All affected patients within the same family should carry the variant with the same zygosity, homozygous or compound heterozygous for recessive inheritance and heterozygous for dominant inheritance. Unaffected patients should not carry the variant for dominant

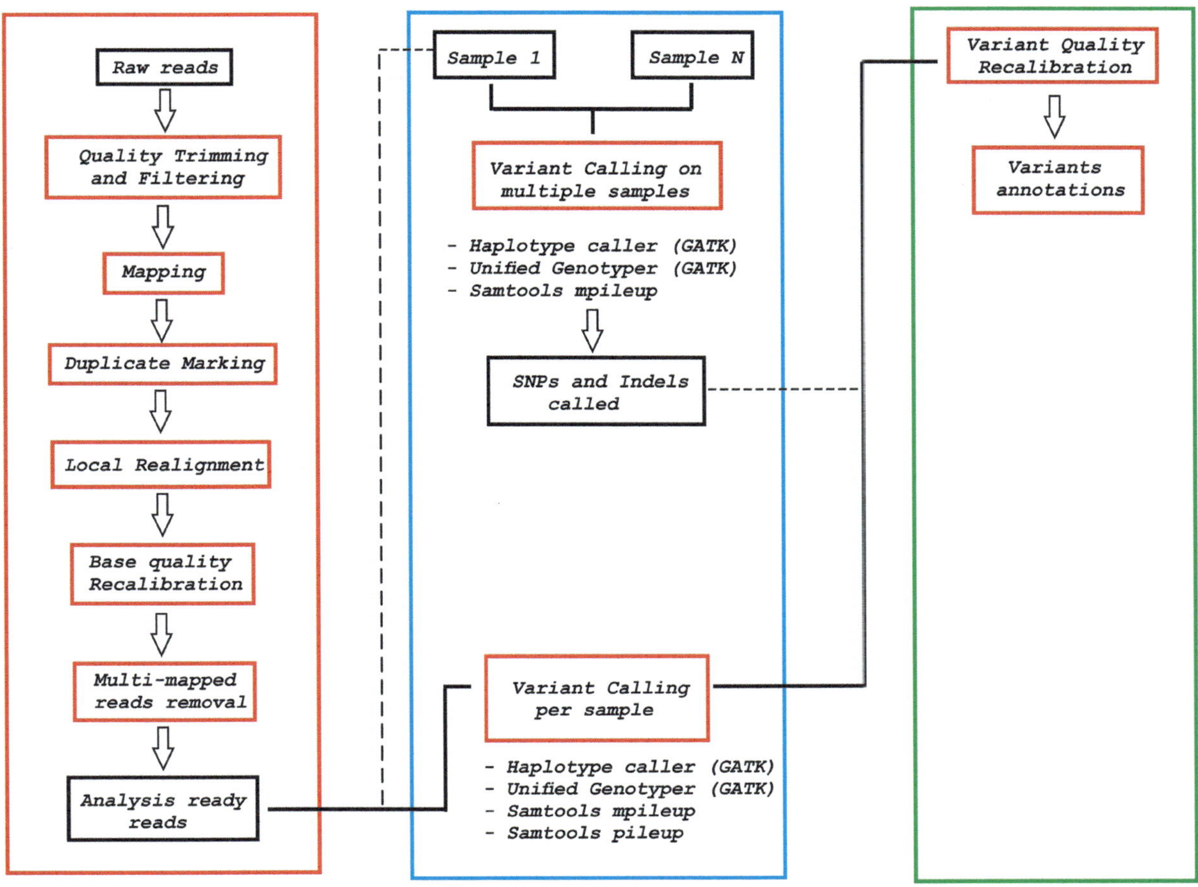

Figure 8.3 Schematic representation of a bioinformatics pipeline for data processing (https://gatk.broadinstitute.org). Briefly, the raw reads obtained from HTS are subjected to data processing for quality assessment, with refinement of the sequence reads, which involves removal of the adapter sequences and the trimming and discarding of reads based on quality. High-quality clean reads are then further processed by the alignment of the sequence reads to a reference genome. Variant detection is based on sequence differences between the reference and sample reads under investigation. The variants are then annotated to elucidate their functional and biological relevance. Source: copyright (c) 2009–2020, Broad Institute, Inc. All rights reserved.

inheritance whereas they can be heterozygous for the variant or homozygous for the wild-type sequence for recessive inheritance. Once the mutation is confirmed, the largest possible cohort of patients needs to be screened in order to determine the prevalence and type of mutations possibly affecting the gene identified. Even if the loss of function of the gene has been studied in an animal model, this analysis is paramount to the confirmation of the identified gene and its pathologic character in humans.

8.2.3 Biological Validation (Step 6)

Unfortunately, nothing is black and white in biology and genetics does not escape this rule. The ultimate effect of a genetic variation can be drastically different depending on its precise type and its position within a gene. There may be either no consequence at all or complete loss of function, with all the intermediate situations where a gene conserves a partial function. In addition, as seen in dominant pathologies, a mutated gene can encode a protein that acquires a dominant negative function precluding the function of the normal protein or even a new detrimental function. Different types of genetic variations with relative molecular changes and possible phenotypic consequences are shown in Figure 8.4.

Sequence variations can be synonymous, where the amino acid encoded is the same but translated from a different codon. In such instances, the

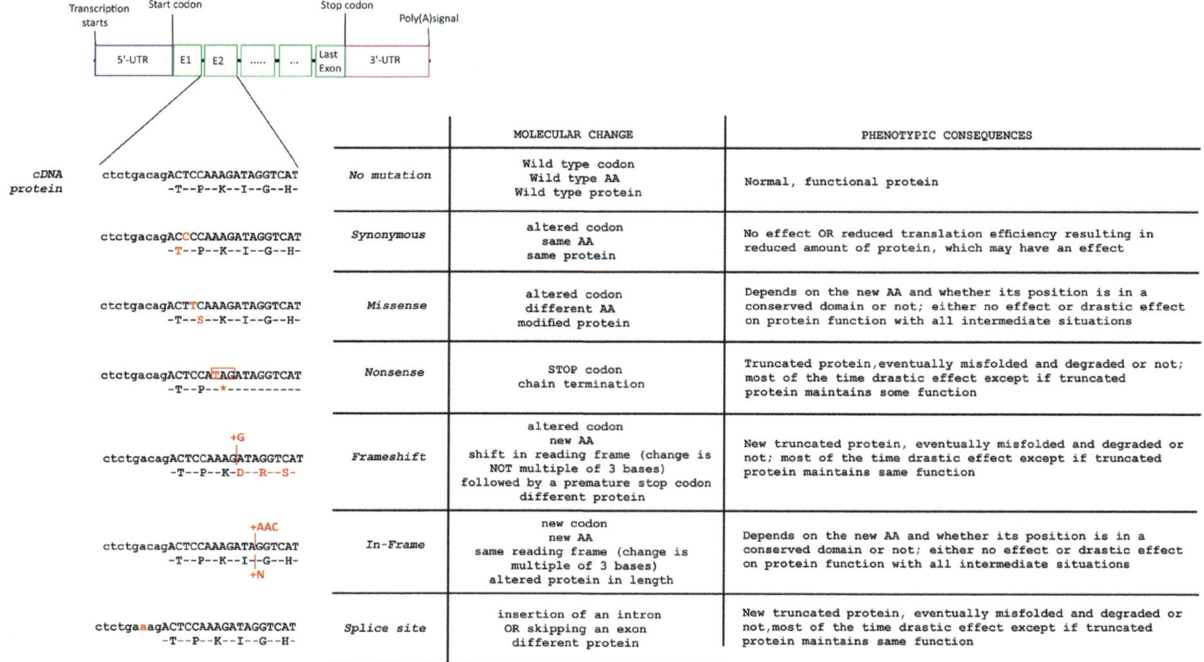

Sequence (cDNA / protein)		MOLECULAR CHANGE	PHENOTYPIC CONSEQUENCES
ctctgacagACTCCAAAGATAGGTCAT -T--P--K--I--G--H-	*No mutation*	Wild type codon Wild type AA Wild type protein	Normal, functional protein
ctctgacagACCCCAAAGATAGGTCAT -T--P--K--I--G--H-	*Synonymous*	altered codon same AA same protein	No effect OR reduced translation efficiency resulting in reduced amount of protein, which may have an effect
ctctgacagACTTCAAAGATAGGTCAT -T--S--K--I--G--H-	*Missense*	altered codon different AA modified protein	Depends on the new AA and whether its position is in a conserved domain or not; either no effect or drastic effect on protein function with all intermediate situations
ctctgacagACTCCATAGATAGGTCAT -T--P--*----------	*Nonsense*	STOP codon chain termination	Truncated protein,eventually misfolded and degraded or not; most of the time drastic effect except if truncated protein maintains some function
+G ctctgacagACTCCAAAGATAGGTCAT -T--P--K-D--R--S-	*Frameshift*	altered codon new AA shift in reading frame (change is NOT multiple of 3 bases) followed by a premature stop codon different protein	New truncated protein, eventually misfolded and degraded or not; most of the time drastic effect except if truncated protein maintains same function
+AAC ctctgacagACTCCAAAGATAGGTCAT -T--P--K--I-+-G--H- +N	*In-Frame*	new codon new AA same reading frame (change is multiple of 3 bases) altered protein in length	Depends on the new AA and whether its position is in a conserved domain or not; either no effect or drastic effect on protein function with all intermediate situations
ctctgaaagACTCCAAAGATAGGTCAT -T--P--K--I--G--H-	*Splice site*	insertion of an intron OR skipping an exon different protein	New truncated protein, eventually misfolded and degraded or not,most of the time drastic effect except if truncated protein maintains same function

Figure 8.4 Schematic representation of pre-mRNA, sequence variations with respective molecular change, and possible protein consequences. For frameshift and in-frame mutations only insertion of base(s) is given as an example; indeed, deletion of base(s) entails the same consequences. AA, amino acid; STOP codon, a nucleotide triplet within mRNA that signals a termination of translation into proteins. There are three different stop codons in the standard genetic code.

variation frequently does not affect the protein's function, but it may eventually reduce the efficiency of mRNA translation leading to a reduced amount of protein, which may have a phenotypic effect.

Missense variations substitute one amino acid for another. Again, such changes can have either no effect at all or a very drastic effect. This will depend on the change, since some amino acids are similar and others quite distinct in term of structure but also their chemical and physical properties, while the position within the protein sequence can be critical. Some positions can tolerate changes while others cannot. The comparison of proteins from a large variety of species, if possible, can allow one to determine if a particular amino acid is maintained throughout evolution and part of a conserved domain and, if so, this can be an indication, but only an indication, that it is important for the function of the protein.

Insertions or deletions (Indel) of several bases but not multiples of three will alter the reading frame of the mRNA, corresponding to a frameshift mutation. In such cases, most of the time, this introduces a premature stop codon that will affect the size of the protein encoded. When the number of inserted or deleted base pairs is a multiple of three, the reading frame of the gene is not disrupted. This results in an addition or deletion of amino acids. If it does not involve a conserved domain, it may not affect the function of the protein.

Nonsense mutations change a normally functional amino acid into a stop codon, which results in premature termination of the protein chain. As for Indels, at best they allow the production of a truncated protein, but most of the time produce misfolded proteins that will be quickly degraded. In both cases, with rare exceptions regarding truncated protein production, they will affect protein function.

Mutations outside the coding sequence can also impact protein function. Indeed, a splice site mutation, which usually refers to the insertion, deletion or change of nucleotides in the consensus splicing sequences, will interfere with proper constitutional intron splicing during the process of mRNA maturation, thus generating an aberrant transcript including an intron or skipping an exon. Here again, this mostly results in the introduction of a premature stop codon, leading to the production of a truncated protein or its absence. Eventually, if the two newly joined exons are in-frame, it can lead to a truncated protein missing the amino acid sequence coded by the missing exon.

Recent high-throughput analyses have demonstrated that numerous alternative splicing events can affect 95–100% of multi-exon human genes and it has been suggested that this occurs in a tissue-specific manner [8]. In mammalian exons, there are *cis*-acting splicing regulatory elements that consist of exonic splicing enhancers and silencers (ESE and ESS); these enhance or inhibit splicing of the pre-mRNA and contribute to alternative splicing. Any change in these sequences may disrupt splicing. Indeed, a synonymous c.2667C>T mutation in exon 8 of the androgen receptor was the first reported exonic splicing mutation affecting human genital development [9]. The mutation creates a new ESS site and leads to aberrant splicing in a patient with partial androgen insensitivity syndrome.

Although point mutations may disturb an ESE or ESS thereby resulting in splice defects, the mechanisms that regulate alternative RNA splicing are only partially understood. There are different methods for estimating alternative splicing events at present; for instance, the intronic and exonic sequences can be easily analysed by Human Splicing Finder (HSF) which is available for noncommercial users (www.umd.be/HSF/index.html).

Initiation of transcription is controlled by short sequence elements called promoters, immediately upstream (5′) of genes; depending on the location and the nature of changes, a promoter mutation can decrease or increase the level of mRNA and the encoded protein and thereby affect its function. The downstream untranslated region (3′-UTR) is the sequence that immediately follows the stop codon and which contains regulatory regions that post-transcriptionally influence gene expression. The 3′-UTR plays a crucial role in gene expression, since any changes in the sequence may influence the localization, stability, export, and translation efficiency of an mRNA.

Because of this complexity, it is critical to carry out functional studies. The choice of a useful model system and strategy depends on the type of gene and mutation(s) identified, its profile of expression, and its evolutionary conservation. It would be simpler if human gametogenesis could be carried out in vitro. Unfortunately, so far there are no human cells capable of recapitulating gamete differentiation in vitro. One supplementary layer of difficulty lies in the fact that the expression profile of most genes involved in male fertility is restricted to the testes. Thus, it is not a simple task to develop functional tests to demonstrate the pathogenicity of a mutation that has been identified.

In some cases, very little will be known about the gene identified and before starting any functional analyses it will be necessary to establish its profile of expression. Some data can be gathered from public databases. For instance, it can be useful to check if the expression profile is conserved between species, for example between mouse and human. An absence of expression in the testes strongly suggests that the variant identified plays no role in fertility. However, there is no reason to limit an investigation to genes whose expression is strictly limited to the testes. A mutation in any gene expressed in the testes may have a specific testes effect. This can be explained either because a broadly expressed gene may interact with testes-specific proteins, or by a redundancy phenomenon where other family members of the same gene family, which are not expressed in the testis, compensate for the absence of the mutated gene. A discrepancy between mouse and human expression would need confirmation of the expression profile, if possible, on human tissues. RNA of different species and organs are commercially available to carry out reverse transcription polymerase chain reaction (RT-PCR) experiments.

If testes expression is confirmed, functional analyses can be considered. We briefly describe, without being comprehensive, different models that can be used from the most complicated but the most pertinent, to the easiest but not necessarily the most convincing.

When a mutation affects the production of a protein, the best option is to check for its presence or level of expression directly in patients' tissues. However, obtaining tissue biopsies can be difficult, and even more so for testes. For obvious ethical reasons, it is impossible to ask the patient for a testicular biopsy for research purposes only.

In the absence of a human model, the most appropriate one is the mouse model [10]. It has the main advantage of being mammalian, but the production of such models is expensive. If a mutation leads to a frameshift or premature stop codon, the resulting effect can be predicted as a complete absence of protein or the production of a truncated one. This can also be the case with splice site mutations. In such situations, knockout of the gene in the mouse is the best approach. A survey of the literature may show that a mouse model has already been established. If the phenotype is similar or identical to the one observed in humans, proof of its role in male infertility is very strong. Otherwise, a mouse model can be created using the CRISPR/Cas9 technology.

Even if several tools for predicting the effect of missense mutations are available online, the discrepancy of the predictions observed between them underlines the limitations of these software programs. For missense mutations, a knock-in strategy, based on CRISPR/Cas9 technology allowing the insertion of the exact sequence variation, can be used. In the case of a substitution, it is feasible to introduce the same amino acid change in the mouse protein. Such a strategy is justified only if the amino acid is well conserved throughout evolution.

Other models such as zebrafish, *Drosophila*, or yeast can be used. They are less relevant, but if the protein is well conserved they can be pertinent. All these models allow meiosis to be studied. Their genomes can be manipulated, but not necessarily more easily than the mouse genome. However, their cost will be more affordable, especially for the yeast model, which also has the advantage of fast growth and can thus speed up results. This may be done in collaboration, although finding laboratories using one of these models, having the competence in the field of reproduction, and willing to work on the protein can also be a challenge. Finally, if the gene is ubiquitously expressed or if it is a splice mutation, a cellular model can be used. Many human cell lines are available that can be transfected with the gene under study to analyze its function. For example, a splice site mutation can be easily tested by constructing a minigene, thereby allowing analysis of the splicing of the target intron after transfection into human cells such as HEK-293 [11].

8.3 Genes Involved in Male Infertility

Using SNP or CGH arrays, and more recently HTS, the list of genes involved in male infertility has grown considerably. The identification of such genes enhances the range of diagnostic tools for patients. Interestingly, Oud et al. [12] have established a list of all published genes thought to be involved in male infertility and ranked them according to the extent the evidence supports their biological role and thus the trust they can be granted. In total the authors identified 23 526 publications describing 521 genes that they placed in six categories of evidence: definitive, strong, moderate, limited, no evidence, or not evaluated. Of the 521 genes, only 92 were classified in the first three categories; including the fourth category increased the list to 185 (Table 8.1), less than 36% of the total number of genes claimed to be involved in human male infertility [12]. Considering that mouse spermatogenesis is under the concerted action of more than 2000 genes [4] and that a similar number most probably controls human male gametogenesis, many genes remain to be identified.

This study illustrates the current state of the genetics of male infertility, leaving the reproductive biologist and the geneticist with some uncertainty and explaining the difficulty involved in deciding whether a particular gene should be included when developing a panel of genes to be sequenced for diagnosis. Therefore, the choice of gene panels can still be governed by the level of subjectivity, and some genes are included or not according to the subjective judgment of the geneticist in charge. The set of infertility genes identified so far, for which the data are strong enough to consider as responsible for nonsyndromic male infertility when mutated and which therefore should be included in a diagnostic panel, are shown in Table 8.2. Genes are categorized according to phenotype: teratozoospermia [with as subcategories,

Table 8.1. Number of genes eligible for clinical validity assessment in male infertility

Level of evidence	Definitive	Strong	Moderate	Limited	No evidence	Not evaluated	Total number of genes
Number of genes	38	22	32	93	160	176	**521**

Source: Oud et al. [12].

Table 8.2. The genes identified for nonsyndromic male infertility

Phenotype		Genes
Teratozoospermia	Globozoospermia	*SPATA16* (609856)
		DPY19L2 (613893)
	Macrocephaly	*AURKC* (603495)
	Acephalic sperm	*BRDT* (602144)
		PMFBP1 (618085)
		CEP112 (618980)
		SUN5 (613942)
		TSGA10 (607166)
	Multiple morphologic abnormalities of the flagella (MMAF)	*DNAH8* (603337)
		DZIP1 (608671)
		CFAP58 (619029)
		DNAH1 (603332)
		CFAP43 (617558)
		CFAP44 (617559)
		CFAP69 (617949)
		FSIP2 (615796)
		WDR66 (618146)
		AK7 (615364)
		SPEF2 (610172)
		DNAH2 (603333)
		TTC29 (618735)
		CEP135 (611423)
		ARMC2 (618424)
		QRICH2 (618304)
Asthenozoospermia		*CATSPER1* (606389)
		GALNTL5 (615133)*
		SLC26A8 (608480)
		DNAH17 (610063)
		EIF4G1 (600495)*
		SPAG17 (616554)*
		AKAP4 (300185)*
Total fertilization failure		*PLCZ1* (608075)
Spermatogenic failure		*SYCP3* (604759)
		DAX1 (300473)
		TEX11 (300311)
		TEX15 (605795)
		MAGEB4 (300153)*
		TAF4B (601689)
		ZMYND15 (614312)
		HSF2 (140581)
		KLHL10 (608778)
		MEIOB (617670)
		TEX14 (605792)
		DNAH6 (603336)*
		M1AP (619098)
		XRCC2 (619145)
		SPINK2 (605753)
		SOHLH1 (610224)
		NPAS2 (603347)*
		SYCE1 (616950)
		TDRD9 (617963)
		NANOS1 (608226)
		NR5A1 (184757)
		FANCM (609644)
		MEI1 (608797)
		SYCP2 (604105)
		STAG3 (608489)*
		Wt1 (607102)
		DMC1 (602721)*
Mixed phenotype		*SEPT12* (611562)
		CFAP65 (614270)
		TTC21A (611430)
		PMFBP1 (618085)
		CFAP70 (618661)

Candidate genes with limited evidence that needs further confirmation in human are labeled with asterisks (*); others are listed as causative for male infertility in OMIM or validated at least as moderately linked to nonsyndromic male infertility phenotype. Source: Oud et al. [12].

globozoospermia, macrocephaly, microcephaly, acephaly, defective sperm head, and multiple morphologic abnormalities of the flagella (MMAF)], asthenozoospermia, complete lack of fertilization, spermatogenic failure (which includes azoospermia or oligozoospermia), and mixed phenotypes including sperm count, motility and/or morphology. This list presents a picture of the current situation and is based on our personal analysis. Indeed, the discovery of new genes is progressing so rapidly that the list of genes must be updated at least once a year.

Recently, HTS-based gene panel tests to examine selected genes for specific infertility phenotypes have been developed and proposed as diagnostic tools, with the limitations we have described above. These panels have several advantages in clinical applications: (1) they allow a precise diagnosis and identify the etiology, which is important for the psychological well-being of the patient; (2) they may help to choose the best treatment and therefore improve care and advice not only for the patient but also for his spouse and family; and (3) they are cheaper than whole-genome or -exome sequencing.

Targeted gene sequencing panels allow the analysis of multiple genes simultaneously in a single investigation. In contrast to whole-genome testing, targeted DNA studies focus the analysis on specific areas of interest. The main idea is to build specific libraries, by amplifying regions of interest using PCR with specific sets of pooled primers. If done in a traditional manner, designing such primers and optimizing PCR conditions is time-consuming; however, available web-based applications allow investigators to create and order DNA panels overlapping genes of interest. Regions of interest are defined based on target genes, genomic coordinates, or selection parameters. The main advantage of targeted sequencing lies in high-depth sequencing of key genes or regions of interest, which increase trust in the identification of rare variants. In addition, because the level of complexity is lower than when performing whole-genome or whole-exome sequencing, it simplifies the interpretation of data by limiting the number of genes and reducing the number of variants of unknown significance.

Gene panels can be purchased with preselected content or custom designed to include genomic regions of interest. The gene composition of custom panels is based on public information and in-house research; it may vary from one laboratory to another. There are as yet no fixed criteria for the validation of an HTS-based assay in molecular diagnostics. Although guidelines for diagnostic validation have been published by several groups, available reports are often generalized or focused on specific topics of the HTS pipeline [13]. In any case, a diagnostic laboratory should prove its ability to accurately detect and report the different types of mutations. With a complex process such as HTS, design and optimization are exceedingly important. Each panel must be validated by assessing its analytical sensitivity, a feature that measures reproducibility, and specificity, which relies on "true negative" detection. A second technology, usually Sanger sequencing, is employed for the latter.

Confirmation of the identity of a sample is mandatory in order to ensure the quality of the data and the scientific validity of results drawn from the data. Single nucleotide polymorphisms, or SNPs, are the most common type of sequence variation in the human genome. Each SNP represents a difference in a single nucleotide. Apart from identical twins, each individual has a unique combination of SNPs that can be used to identify DNA samples. SNP profiles can be determined by real-time PCR analysis of the purified DNA using a selection of SNPs (control SNPs) and the profiles obtained can then be compared with profiles from panel results to confirm the identity of the sample [14]. Control SNPs are selected based on allele frequency ($\sim$0.5) and chromosomal location (i.e. all SNPs should be situated on different chromosomes). In addition, the SNPs should not reside within a coding region or the regulatory sequence of a gene. There are commercially available SNP genotyping products that are validated by analyzing several DNA samples from different ethnic origins.

During the development of tests, one should carefully consider the selection of genes as well as the total number of genes since the size of the panel may affect sequencing reagent costs, depth of sequencing, laboratory productivity, and complexity of analytical and clinical interpretation [15].

8.4 Analysis of Gene Panels as a New Diagnostic Tool in Assisted Reproductive Technology

Semen analysis is initially the routine test for assessment of male infertility. It should be prescribed as soon as possible after couples consult for infertility. The reference values for semen parameters were first established in the 2010 WHO manual [16] in which updated, standardized, evidence-based procedures and recommendations are provided. Nevertheless, such analysis is of limited quantitative (concentration) and poor qualitative (mobility, morphology) value. Furthermore, it does not allow the identification of the etiology nor measure the fertilization potential of spermatozoa, except in extreme cases such as azoospermia, total loss of mobility, or total necrospermia.

The increasing number of "infertility" genes identified offers a new diagnostic tool for medical doctors in charge of infertile couples. For a large proportion of patients, it should establish the cause and, at least in some cases, provide indications for the best treatment. Therefore a genetic test should be requested in order to decipher the etiology and to advise the patient and his family.

We describe a workflow to be followed for the diagnosis of male infertility (Figure 8.5).The results of semen analysis may prompt an examination of the karyotype, a search for Y chromosome microdeletions and/or gene panel tests. The workflow proposed

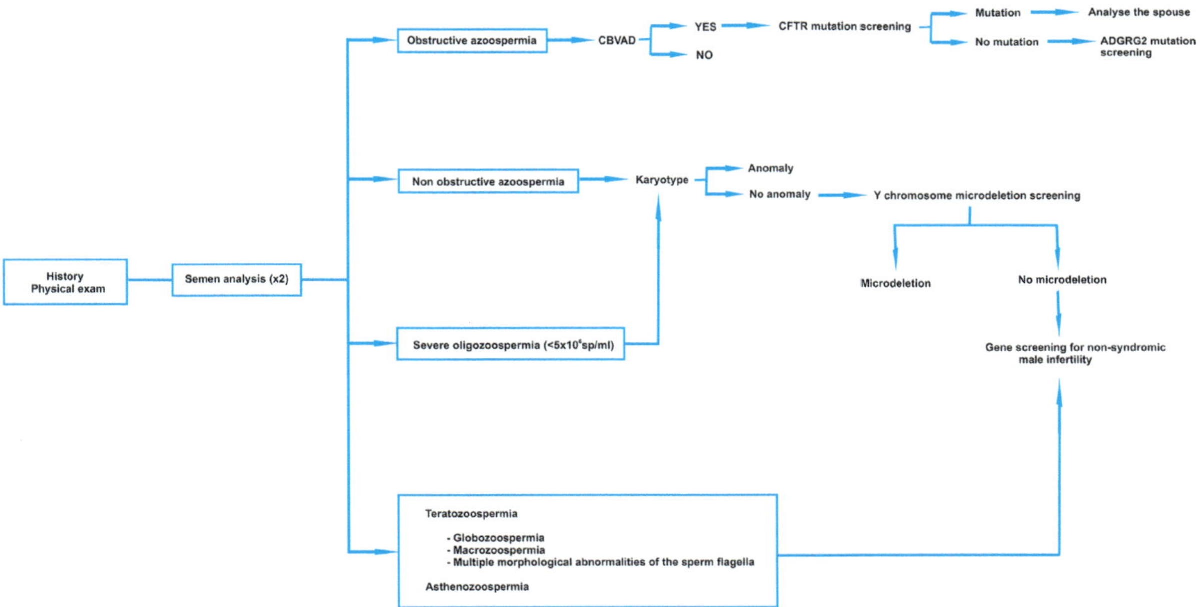

Figure 8.5 Workflow for genetic tests in male infertility. According to semen analysis, karyotype, Y chromosome microdeletion testing, or specific gene mutation screening can be offered to patients. Source: Okutman et al. [17].

is based on the spermiogram and, for rare cases, on spermocytogram results.

8.4.1 Obstructive Azoospermia

In obstructive azoospermia, despite normal spermatogenesis no spermatozoa are found in the ejaculate because of the obstruction of the seminal tracts. The most common cause is congenital bilateral absence of the vas deferens (CBAVD). In about 80–90% of males with CBAVD, the cause is correlated with mutations in the cystic fibrosis transmembrane conductance regulator (*CFTR*, OMIM #602421) gene, the gene defective in the autosomal recessive disease cystic fibrosis (CF) [18].

If *CFTR* mutations are identified in a CBAVD patient, it is essential to screen his partner for *CFTR* mutations at least in European countries and in the USA, where the carrier frequency of CF is high (around 1 in 20–30). If the partner is indeed a carrier, the couple should receive genetic counseling and be informed about the risk of the birth of a child with a severe form of CF, a risk that can reach 25–50%. The couple may then opt for preimplantation genetic testing (PGT-M; see Chapter 13) as they will need IVF and intracytoplasmic sperm injection (ICSI) in order to conceive. Embryos carrying the normal *CFTR* allele of the female partner will be selected for transfer.

Recently, loss-of-function mutations of the adhesion of G protein-coupled receptor G2 (*ADGRG2*, OMIM #300985) gene have been reported by different groups as responsible for CBAVD in patients with no *CFTR* mutation. Although the incidence of *ADGRG2* gene mutations is as yet unknown, results so far reveal the importance of sequencing both *CFTR* and *ADGRG2* among CBAVD patients.

8.4.2 Spermatogenetic Failure (Nonobstructive Azoospermia or Oligozoospermia)

For patients presenting nonobstructive azoospermia or oligozoospermia, a first step to be recommended is determination of the karyotype and a search for microdeletion of the Y chromosome. In the case of no abnormalities, then a mutation screen using a panel of genes should be offered.

8.4.2.1 Karyotype

Chromosomal anomalies are one of the major causes of male infertility (discussed in Chapter 7). Indeed, about 14% of azoospermic men and 4.5% of oligozoospermic men present chromosomal anomalies, rates that are much higher than in the general population (approximately 0.6%) [19]. These anomalies can be numerical or structural.

The most common genetic cause of azoospermia is Klinefelter syndrome (KS) with a 47,XXY karyotype. Concerning the structural anomalies, reciprocal translocations involving the X chromosome lead to azoospermia in most cases, while translocations involving autosomes, whether reciprocal or Robertsonian, have highly variable effects on spermatogenesis. This can range from azoospermia to almost normal semen parameters. In general, Robertsonian translocations induce moderate but variable oligozoospermia. Karyotyping is crucial for genetic counseling. Indeed, the structural defects of chromosomes in these patients, who present a normal phenotype other than their infertility, can be the origin of repeated miscarriages or birth of a child with congenital defects due to abnormal chromosomal segregation during parental gametogenesis.

8.4.2.2 Yq Microdeletions

The AZF locus, which can be divided into three subregions, *AZFa*, *AZFb*, and *AZFc*, on the Yq11 locus of the Y chromosome has been shown, as early as chromosome translocation, to be involved in human spermatogenesis. The locus is a complex region including numerous genes, some of them with multiple copies whose precise role in gametogenesis has not being solved. Microdeletions of Y chromosomes are found in about 10% of men with oligozoospermia and in up to 15% of azoospermic patients (see Chapter 7). Routine screening for Yq microdeletion is recommended for patients with azoospermia and severe oligozoospermia [20].

Deletions of the *AZFc* region are most commonly found (~80%), and this condition causes a heterogeneous phenotype ranging from severe or mild oligozoospermia to azoospermia. In about 70% of these patients, sperm cells are found in either the ejaculate or testes, and these patients can have their own genetically related children following an ICSI procedure. Appropriate genetic counseling should be given to these couples because the transmission of the defective Y chromosome will be associated with impaired spermatogenesis in their sons. There is still debate as to whether oligozoospermia of *AZFc*-deleted patients evolves to azoospermia over time [21]. Hence, sperm cryopreservation may be an option to offer oligozoospermic patients with *AZFc* microdeletion.

AZFa (0.5–4%) and *AZFb* (1–5%) deletions are less common. Most patients with an entire *AZFa* deletion present a Sertoli-cell-only syndrome, while *AZFb* patients present with maturation arrest of spermatogenesis. Complete deletion of multiple regions (*AZFabc* or *AZFbc* region) is accompanied by complete azoospermia. In a few cases, sperm retrieval has been reported but as yet no clinical or chemical pregnancy after ICSI in cases with complete *AZFb*/*AZFbc* microdeletions has been reported [22]. The possibility to enter a testicular sperm extraction (TESE)/ICSI program should be discussed during genetic counseling.

Partial deletions of the *AZFc* region, so-called gr/gr deletions, show an increased frequency in males with fertility problems, but are also found in men with normal sperm parameters. Therefore, these deletions should be considered a risk factor for male infertility and not a causative factor. These deletions are not screened routinely in infertile males and their prognostic value needs to be better defined.

8.4.2.3 Spermatogenic Failure "Gene Panel" Analysis

If no chromosomal defects or Yq microdeletions have been detected, it is worth offering to screen for mutations responsible for a spermatogenic failure. The HTS of selected genes, known as a "gene panel," is the most recent test available in genetics of infertility. Considering the recent interest in the genetics of infertility, the number of genes in the panel will inevitably evolve as new genes are identified.

Among genes identified for spermatogenic failure, there is no predominant one. So far, research on genes for azoospermia has not been able to determine if any of them could be used as a biomarker to predict the presence or absence of spermatozoa after TESE. The main reason for this is the scarcity of patients with mutations, although this information may be available in the future since an increasing number of patients will be tested.

One of the main benefits of a genetic test is to be able to establish a prognosis of the future fertility potential of oligozoospermic patients. It is important to know whether the underlying mutation can be responsible for a progressive decrease in sperm leading patients to become azoospermic. A time frame may exist for the retrieval of sperm from the ejaculate or surgical retrieval of sperm before the onset of azoospermia and complete loss of residual testicular spermatogenesis. In such a situation, it would be wise to offer patients sperm cryopreservation as soon as possible, but also to offer to test patients' siblings and propose a protocol of fertility preservation if it turns out that they are also carriers of the same mutation. So far, *TEX15* and *KLHL10* genes may fall into this category [23,24] even if it still needs to be formally proven. Further research is needed to determine whether other genes behave similarly and which mutations are critical since they can affect the function of the encoded protein to a variable extent.

8.4.3 Teratozoospermia/Asthenozoospermia

Gene panel analysis can also be proposed for teratozoospermic and/or asthenozoospermic patients. Genes have been identified in three well-defined phenotypes, namely globozoospermia, macrocephalia, and multiple morphologic abnormalities of the flagella (MMAF). Mutations in the *DPY19L2* gene, mainly a recurrent deletion of the entire gene, are found in about two-thirds of globozoospermia cases [25,26]. Besides *DPY19L2* mutations, only one other gene has been well documented as involved in globozoospermia, *SPATA16* [27]. The infertility of globozoospermic patients can be overcome by ICSI in combination with artificial oocyte activation. This provides a good option for couples by enabling a satisfactory overall fertilization rate and therefore pregnancy [28].

The *AURKC* gene has been identified as causative for the macrocephalia phenotype. Two recurrent mutations have been identified; c.144delC in the North African population and p.Y248* in the European, suggesting a founder effect. At least for mutation c.144delC, it has been clearly established that the majority of macrocephalic spermatozoa are tetraploid, suggesting a defect in the process of cytokinesis that accounts for the size of spermatozoa and their DNA content. What has been found for this mutation is most probably true for other mutations of *AURKC*. Therefore, finding an *AURKC* mutation should lead to the interruption of any ART procedure. In such a situation, alternative solutions should be offered, ranging from sperm donation to adoption or renouncing a parental project.

A handful of genes have been formally correlated with the MMAF phenotype (see Table 8.2). So far, known MMAF-associated genes account for approximately 60% of human MMAF cases [29] . Although the amount of data is still limited, it seems that ICSI can be proposed safely to patients and leads to similar pregnancy rates as those of routine ICSI cases.

A patient may present with combined phenotypes, meaning that semen analysis could reveal a combination of count, morphology, and/or motility problems together (e.g. oligoasthenoteratospermia). A few genes have already been identified for mixed phenotypes, such as *SEPT12*, *TTC21A*, *PMFBP1*, *CFAP65* and *CFAP70*.

8.5 Conclusion and Future Trends

This chapter has discussed gene defects known to cause male infertility (genetics of female infertility is dealt with in Chapter 9). Besides monogenic defects, numerical or structural chromosomal anomalies are also known to cause infertility. Of note, only a few dominant conditions have so far been described. This is explained by the difficulty of conducting such studies. Indeed, in the case of infertility, a dominant mode of action is mostly a dead end for a mutation that cannot, by definition, be transmitted, which is not the case for recessive conditions. This means that a dominant mutation is either a neomutation or transmitted by one sex when affecting the other one.

It is also interesting to note that only a few "infertility genes" are shared between men and women. This can be explained in at least two ways: first, genes identified in male infertility cases have not been investigated in female cases and vice versa, or second because male and female gametogenesis implies different mechanisms. The latter reason is supported by studies of the mouse model where a large number of gene knockouts have shown infertility in only one sex – most of the time males are fertile while females are infertile. Among these genes, it is worth mentioning the piRNA pathway genes. However, a more in-depth analysis of these genes may show that some are shared between both sexes.

Despite significant progress for many couples, the final diagnosis remains "idiopathic infertility." This means that more research is necessary to identify the unknown causes of infertility, and in many cases a genetic origin may be found. For a variety of reasons, the genetic studies presented in this chapter are of tremendous importance. First, patients will benefit by having a molecular diagnosis followed by appropriate counseling and care. Secondly, the findings will contribute to a better understanding of the physiopathologic processes of human reproduction. This understanding will have clinical and fundamental consequences. Thirdly, in the longer term, a better understanding of the genes involved in spermatogenesis will help to identify extrinsic factors that compromise spermatogenesis and prevent the deterioration of fecundity. Indeed, a general decline in fertility or rather sperm quality has been observed, which is obviously not due to genetic causes but is caused, at least in part, by increased concentrations of environmental endocrine disruptors such as environmental estrogen-like molecules. We believe that a better understanding of spermatogenesis, starting with an in-depth knowledge of all the genes and molecular processes involved, will eventually be instrumental in identifying and understanding the effects of these toxic chemicals and in finding solutions to stop their adverse effects or, better, to promote their removal from the market. Taken together, this should also improve the practice of artificial reproductive technologies, not only for couples who are infertile due to a genetic defect, but for all couples. However, even if a large proportion of infertility can be explained by a genetic defect – a proportion still to be precisely determined – genetics is not everything, and other influences such as environmental factors are also involved.

Ultimately, the genetics of infertility should improve the treatment for couples both directly and indirectly. Directly by providing diagnostic tools, genetic counseling for patients and their families, and biomarkers to help choose the most appropriate treatment and establish a better prognosis. Indirectly, the improvement of our basic knowledge of the underlying processes of gametogenesis will allow the emergence of new ART practices.

New technologies combined with the ever-growing knowledge of genome structure and gene regulation will explain impaired gametogenesis, embryogenesis, and implantation as well as miscarriages and help to find solutions for infertile couples who wish children.

References

1. Krausz C. Male infertility: pathogenesis and clinical diagnosis. *Best Pract Res Clin Endocrinol Metab* 2011;25 (2):271–85.
2. Searle AG, Beechey CV, Evans EP. Meiotic effects in chromosomally-derived male sterility of mice. *Ann Biol Anim Biochim Biophys* 1978;18(2B):391–8.
3. Gordon UD. Assisted conception in the azoospermic male. *Hum Fertil (Camb)* 2002;5(1 Suppl):S9–S14.
4. Matzuk MM, Lamb DJ. The biology of infertility: research advances and clinical challenges. *Nat Med* 2008;14(11):1197–213.
5. Majewski J, Schwartzentruber J, Lalonde E, Montpetit A, Jabado N. What can exome sequencing do for you? *J Med Genet* 2011;48 (9):580–9.
6. Matthijs G, Souche E, Alders M, et al. Guidelines for diagnostic next-generation sequencing. *Eur J Hum Genet* 2016;24(1):2–5.
7. Plon SE, Eccles DM, Easton D, et al. Sequence variant classification and reporting: recommendations for improving the interpretation of cancer susceptibility genetic test results. *Hum Mutat* 2008;29(11): 1282–91.
8. Song H, Wang L, Chen D, Li F. The function of pre-mRNA alternative splicing in mammal spermatogenesis. *Int J Biol Sci* 2020;16(1):38–48.
9. Hellwinkel OJ-C, Holterhus P-M, Struve D, et al. A unique exonic splicing mutation in the human androgen receptor gene indicates a physiologic relevance of regular androgen receptor transcript variants. *J Clin Endocrinol Metab* 2001;86(6):2569–75.
10. Kherraf Z-E, Conne B, Amiri-Yekta A, et al. Creation of knock out and knock in mice by CRISPR/Cas9 to validate candidate genes for human male infertility, interest, difficulties and feasibility. *Mol Cell Endocrinol* 2018;468:70–80.
11. Okutman Ö, Demirel C, Tülek F, et al. Homozygous splice site mutation in ZP1 causes familial oocyte maturation defect. *Genes (Basel)* 2020;11(4):382. DOI 10.3390/genes11040382
12. Oud MS, Volozonoka L, Smits RM, et al. A systematic review and standardized clinical validity assessment of male infertility genes. *Hum Reprod* 2019;34 (5):932–41.
13. Froyen G, Broekmans A, Hillen F, et al. Validation and application of a custom-designed targeted next-generation sequencing panel for the diagnostic mutational profiling of solid tumors. *PLoS One* 2016;11(4):e0154038.
14. Huijsmans R, Damen J, van der Linden H, Hermans M. Single nucleotide polymorphism profiling assay to confirm the

identity of human tissues. *J Mol Diagn* 2007;9(2):205–13.

15. Jennings LJ, Arcila ME, Corless C, et al. Guidelines for validation of next-generation sequencing-based oncology panels: a joint consensus recommendation of the Association for Molecular Pathology and College of American Pathologists. *J Mol Diagn* 2017;19(3):341–65.
16. Cooper TG, Noonan E, von Eckardstein S, et al. World Health Organization reference values for human semen characteristics. *Hum Reprod Update* 2010;16(3):231–45.
17. Okutman O, Rhouma MB, Benkhalifa M, Muller J, Viville S. Genetic evaluation of patients with non-syndromic male infertility. *J Assist Reprod Genet* 2018;35:1939–51.
18. de Souza DAS, Faucz FR, Pereira-Ferrari L, Sotomaior VS, Raskin S. Congenital bilateral absence of the vas deferens as an atypical form of cystic fibrosis: reproductive implications and genetic counseling. *Andrology* 2018;6(1):127–35.
19. Ravel C, Berthaut I, Bresson JL, Siffroi JP, Genetics Commission of the French Federation of CECOS. Prevalence of chromosomal abnormalities in phenotypically normal and fertile adult males: large-scale survey of over 10,000 sperm donor karyotypes. *Hum Reprod* 2006;21(6):1484–9.
20. Krausz C, Hoefsloot L, Simoni M, Tüttelmann F. EAA/EMQN best practice guidelines for molecular diagnosis of Y-chromosomal microdeletions: state-of-the-art 2013. *Andrology* 2014;2(1):5–19.
21. Krausz C, Quintana-Murci L, McElreavey K. Prognostic value of Y deletion analysis: what is the clinical prognostic value of Y chromosome microdeletion analysis? *Hum Reprod* 2000;15(7):1431–4.
22. Soares AR, Costa P, Silva J, et al. AZFb microdeletions and oligozoospermia: which mechanisms? *Fertil Steril* 2012;97(4):858–63.
23. Okutman O, Muller J, Baert Y, et al. Exome sequencing reveals a nonsense mutation in TEX15 causing spermatogenic failure in a Turkish family. *Hum Mol Genet* 2015;24(19):5581–8.
24. Araujo TF, Friedrich C, Grangeiro CHP, et al. Sequence analysis of 37 candidate genes for male infertility: challenges in variant assessment and validating genes. *Andrology* 2020;8(2):434–41.
25. Koscinski I, Elinati E, Fossard C, et al. DPY19L2 deletion as a major cause of globozoospermia. *Am J Hum Genet* 2011;88(3):344–50.
26. Elinati E, Kuentz P, Redin C, et al. Globozoospermia is mainly due to DPY19L2 deletion via non-allelic homologous recombination involving two recombination hotspots. *Hum Mol Genet* 2012;21(16):3695–702.
27. Dam AHDM, Koscinski I, Kremer JAM, et al. Homozygous mutation in SPATA16 is associated with male infertility in human globozoospermia. *Am J Hum Genet* 2007;81(4):813–20.
28. Kuentz P, Vanden Meerschaut F, Elinati E, et al. Assisted oocyte activation overcomes fertilization failure in globozoospermic patients regardless of the DPY19L2 status. *Hum Reprod* 2013;28(4):1054–61.
29. Liu C, He X, Liu W, et al. Bi-allelic mutations in TTC29 cause male subfertility with asthenoteratospermia in humans and mice. *Am J Hum Genet* 2019;105(6):1168–81.

Genetics of Human Female Infertility

Svetlana A. Yatsenko and Aleksandar Rajkovic

9.1 Introduction

Infertility is a genetically heterogeneous condition affecting about 10% of women of reproductive age. Genetic studies on animal models have identified thousands of candidate genes that are essential for gonadal development, germline cell differentiation, complex oocyte–granulosa intercellular signaling, gametogenesis, fertilization, and fetal development. A subset of these candidate genes derived from animal models has been found to cause ovarian dysfunction and infertility in humans.

In many instances the individual wants to know the cause of their infertility. An accurate genetic diagnosis has important clinical significance with direct implications for a woman's reproductive and general health. For example, some genetic causes of infertility can also be associated with increased predisposition to cancer, cardiac and other conditions. For many of the newly discovered genes over the past 10 years, the full phenotypic spectrum beyond infertility is unknown. Accurate genetic diagnosis can minimize the risk of transmitting pathogenic variants through assisted reproductive techniques or natural pregnancy and is critical for developing future successful treatment approaches.

Recent application of the next-generation sequencing technologies has led to significant progress in the discovery of genes and delineation of molecular mechanisms leading to syndromic and nonsyndromic forms of gonadal dysgenesis, primary ovarian insufficiency (POI), and infertility [1]. In this chapter, we focus on nonsyndromic causes and review the known genetic causes of female factor infertility in humans and focus on the current state of knowledge regarding monogenic, polygenic, and genomic alterations associated with ovarian dysgenesis, impaired oogenesis and oocyte maturation, POI, fertilization failure, embryonic pathologies instigated by maternal risk factors, recurrent pregnancy losses, and idiopathic infertility.

9.2 Genes That Disrupt Ovarian Development

A woman's reproductive health and ability to have children directly depend on the development of functional ovaries and mature eggs competent to support the development of embryos. In mammals, the primitive gonads are bipotential, comprising progenitor cells that can differentiate into either testes or ovaries. By 3–5 weeks of gestation, outgrowths of coelomic epithelium give rise to the urogenital ridges. The pool of gonadal somatic progenitor cells must be established before gonadal sex determination takes place. During female fetal development, proper ovarian morphogenesis requires several processes including migration and mitotic proliferation of primordial germ cells, gonadal sex differentiation, follicle assembly, and initiation of meiosis. Germ cells migrate to the genital ridge of the primitive gonad and rapidly proliferate (Figure 9.1), forming clusters or nests of germ cells connected by cytoplasmic bridges to each other. These clusters of germ cells will break down to generate a pool of approximately 7 million primordial follicles in utero. By the time of birth, this pool of primordial follicles will be reduced to about 400 000 in human newborns likely due to apoptosis governed by poorly understood factors, in part related to faulty meiosis. During the reproductive lifespan, the "primordial follicle reserve" is gradually depleted as 30–40 primordial follicles are recruited monthly and the vast majority are lost due to atresia to produce one mature egg [2]. The selection of the dominant follicle is poorly understood, but likely involves an interplay of intraovarian growth factors and extraovarian gonadotropins to ensure survival.

Both human and animal studies have helped elucidate genes and mechanisms of early ovarian development [2–6]. *WT1* and *GATA4* are expressed early during the genital ridge formation and migration of

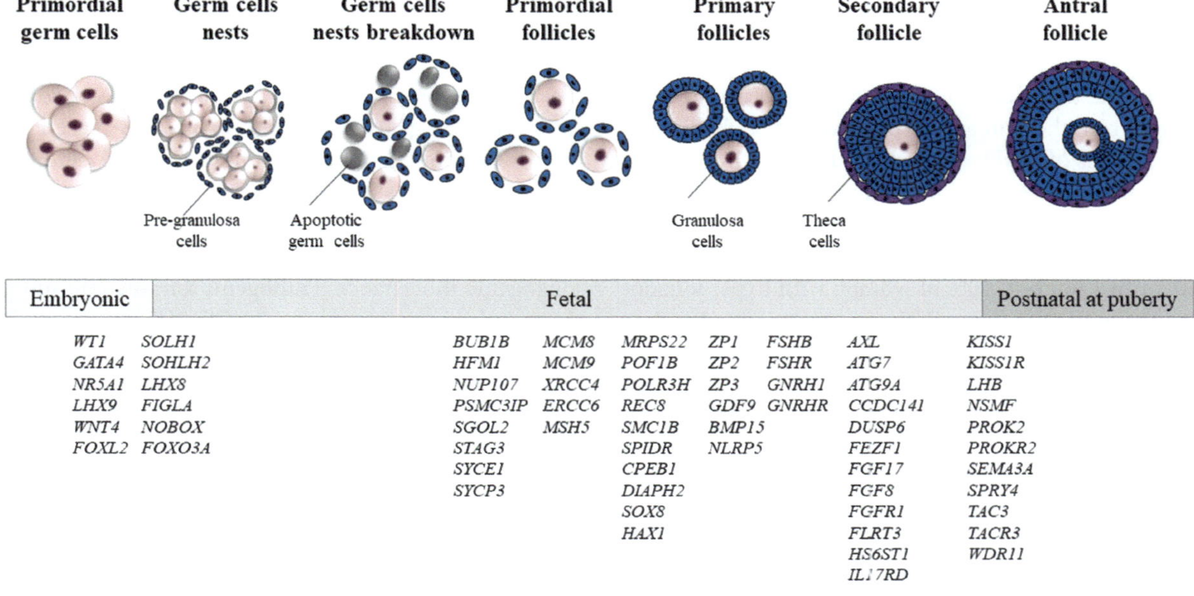

Figure 9.1 Stages of folliculogenesis: the timeline and development of human oocytes from primordial germ cells to antral follicles. The genes involved in the regulation of folliculogenesis and pathogenic variants which are known to result in ovarian dysgenesis, POI and human infertility are outlined at the bottom.

primordial germ cells, followed by *NR5A1* (also known as *SF1*), *LHX9*, *EMX2*, and *CBX2* expression. In embryos with two X chromosomes, the commitment toward the female fate depends on the continuous activation of pro-ovary genes. Several transcription factors, including FIGLA, FOXL2, LHX8, LHX9, NOBOX, KIT, SOHLH1, SOHLH2, and WNT4, are known to regulate ovarian development in humans [7]. Heterozygous and homozygous sequence variants in transcriptional regulators *FIGLA*, *NOBOX*, *LHX8*, *SOHLH1*, and *SOHLH2* have been identified in patients with ovarian dysgenesis and nonsyndromic POI. In mice, these genes cause phenotypes in autosomal recessive fashion, and this is also likely the mode of inheritance in humans. Whether heterozygous variants in these genes can also lead to pathology in humans [6–8] and whether heterozygous variants in different genes within the same pathway cause ovarian dysfunction is unknown. Animal gene knockout models have shown that oocyte-specific transcription factors (*FIGLA*, *NOBOX*, *LHX8*, *SOHLH1*) are required for primordial follicle formation and regulate expression of genes important for folliculogenesis (*GDF9*, *BMP15*, *KIT*) as well as fertilization (*ZP1*, *ZP2*, *ZP3*), and embryonic activation (*MATER*, *ZAR1*, *DNMT1*). Mouse models suggest that early folliculogenesis is therefore essential for proper egg development and early postfertilization embryonic activation.

The size of the primordial follicle pool (ovarian reserve) is one factor that determines reproductive lifespan in human females [9,10]. Perturbation at any stage of primordial follicle formation and maintenance may result in premature loss of oocytes and ovarian dysgenesis. XX complete ovarian dysgenesis presents with primary amenorrhea, lack of secondary sexual characteristics, and gonads that are mainly composed of fibrous tissue (streak gonads). Oocytes are deficient in their ability to produce cholesterol, and depend on surrounding somatic cells for transport and synthesis of certain metabolic substrates. Growing oocytes depend on cumulus cells to provide them with the specific amino acids and products of glycolysis. Folliculogenesis requires complex bidirectional interaction between the oocyte and granulosa, theca and other cells within the follicle [11]. Oocytes are in direct contact with granulosa cells and promote the expression of granulosa genes to coordinate endocrine and paracrine signaling. Mammalian oocytes secrete transforming growth factor (TGF)-β superfamily proteins, including growth differentiation factor 9 (GDF9), bone morphogenetic protein 6 (BMP6), bone morphogenetic protein 15 (BMP15), and fibroblast growth factor 8B (FGF8B). GDF9 and

BMP15 have a synergistic effect on granulosa cells, directing their proliferation, expansion, apoptosis, and steroidogenesis. BMP15 stimulates follicle maturation and, together with GDF9, promotes oocyte transition from the primary to the secondary follicle stage. High levels of GDF9 and BMP15 in the follicle correlate with the presence of developmentally competent oocytes and good embryo morphology. Pathogenic variants in the *GDF9* and *BMP15* genes have been identified in up to 10% of women with hypergonadotropic hypogonadism, POI, primary and secondary amenorrhea [12,13], and polycystic ovary syndrome.

9.3 Hypothalamic–Pituitary Genetic Pathology and Ovarian Function

As the follicle continues to grow, an antrum, a fluid-filled cavity, is formed around the oocyte (Figure 9.1). Antral follicles are responsive to the pituitary gonadotropins follicle-stimulating hormone (FSH) and luteinizing hormone (LH) by acting on their receptors – FSHR and gonadotropin-releasing hormone receptor (GnRHR), respectively – and participate actively in the hypothalamic–pituitary–gonadal (HPG) axis. Unless rescued by FSH, the granulosa cells of most early antral follicles undergo widespread apoptosis and ultimately follicular death. It has been observed that more than 85% of germ cells degenerate during folliculogenesis. In dominant follicles, LH stimulates thecal cell androgen production, while FSH stimulates mural granulosa cell proliferation, aromatization of androgens to estrogens, and LH receptor expression. Follicle-derived estrogens exert positive feedback on both the hypothalamus and pituitary to trigger the mid-cycle gonadotropin surge of LH that precedes ovulation. Pathogenic variants in the *FSHR* gene cause hypergonadotropic hypogonadism, characterized by elevated gonadotropin levels due to lack of negative feedback from ovary to pituitary. The c.566C>T variant in *FSHR* is common among Finnish women with POI, but infrequent in other ethnic groups. Defects in hypothalamic and pituitary hormones cause hypogonadotropic hypogonadism (HH; OMIM #147950). Diagnosis of HH is suspected in women with low serum concentrations of the gonadotropins LH and FSH in the presence of low circulating concentrations of sex steroids. Women with nonsyndromic HH have a 46,XX karyotype, normal physical characteristics, and phenotype limited to delayed or failed puberty, and primary or secondary amenorrhea due to gonadal dysfunction. In patients with pathogenic variants in the *FSHB* (FSH beta subunit) and *FSHR* genes, maturation of follicles beyond the preantral stages is interrupted. To date, more than 50 genes have been reported to cause HH, of which about 50% are inherited in three different ways: X-linked, autosomal recessive, and autosomal dominant; in 10–20% of patients, pathogenic variants are present in more than one gene, suggestive of oligogenic inheritance. Pathogenic variants are more frequently encountered in *ANOS1* (*KAL1*), *CHD7*, *FGFR1*, *GNRHR*, *IL17RD*, *PROKR2*, *SOX10*, and *TACR3* genes, while pathogenic variants in *AXL*, *DMXL2*, *FEZF1*, *FGF17*, *FGF8*, *GNRH1*, *HESX1*, *HS6ST1*, *KISS1*, *KISS1R*, *LEP*, *LEPR*, *NR0B1*, *NSMF*, *OTUD4*, *PCSK1*, *PNPLA6*, *PROK2*, *RNF216*, *SEMA3A*, *SEMA7A*, *TAC3*, and *WDR11* are less common.

9.4 Meiosis, DNA Damage Repair Genes and Reproductive Lifespan

DNA breakage occurs frequently and must be repaired to maintain genome stability and to ensure accurate DNA replication during mitotic and meiotic divisions. Recombination is one of the fundamental steps in sexual reproduction and a key source of genetic variation between generations. Recombination involves programmed DNA double-strand breaks and crossover events, exchanges of DNA between parental chromosomes during meiosis. Only a fraction of DNA breaks (~10%) will result in a crossover. Broken DNA sites must be quickly and efficiently repaired using templates from homologous DNA regions located on either sister chromatids or homologous chromosomes. In response to unrepaired DNA breaks, DNA damage response pathways are activated, leading to cell cycle arrest, induction of DNA repair, and elimination of damaged cells by apoptosis. Tissue-specific DNA repair pathways, such as those involved in meiosis, are highly conserved across species, and deficiencies in these DNA repair machineries can introduce errors in DNA replication and recombination, leading to subsequent depletion of germ cells and accelerated wastage of the follicle reserve.

In female primordial germ cells, meiosis is initiated during fetal development, at around 20 weeks of human gestation. In germ cells, chromosomes undergo several complex processes, including pairing of homologs, alignment, synapsis, and crossover, after which cells are arrested at the diplotene stage of the first meiotic prophase (MI) (see Chapters 3 and 4 for more information). Meiotic errors frequently result in chromosome segregation defects and aneuploidy that prompt a programmed elimination of abnormal or damaged cells at every stage of oogenesis.

Accelerated apoptosis can lead to premature loss or complete elimination of germ cells, resulting in human POI, dysgenic gonads, and infertility in females with defects in genes involved in regulation of chromosome synapsis and meiotic segregation, such as *BUB1B*, *CPEB1*, *DMC1*, *HFM1*, *NUP107*, *REC8*, *SGOL2*, *SMC1B*, *STAG3*, *SYCE1*, *SYCP3*, and *PSMC3IP* (Table 9.1). Disruption during any of these early stages of development can lead to complete loss or severe reduction in the number of oocytes and gonadal dysgenesis.

Female meiosis is susceptible to chromosome segregation errors, particularly in women of 35 years of age and older. During meiosis, sister chromatids are held together by protein complexes. Maintaining cohesion is crucial for pairwise alignment of chromosomes and proper chromosome segregation, preventing oocyte aneuploidy [14]. Components of meiosis-specific cohesin complexes are encoded by the *SMC1A*, *SMC1B*, *RAD21*, *RAD21L*, *REC8*, *STAG1*, *STAG2*, and *STAG3* genes, mutations in which are the cause of multi-chromosome aneuploidy, resulting in follicle atresia. Pathogenic variants in these genes have been observed in patients with POI, idiopathic infertility, and recurrent miscarriages. The gradual loss of cohesins as a result of aging may lead to improper cleavage of cohesion complexes by cohesin-related proteins such as SGOL2 and has been implicated in meiotic nondisjunction and age-related aneuploidy and infertility [15].

There is growing evidence that defective DNA repair causes gonadal dysgenesis [8,16,17], affects follicle maturation, compromises oocyte quality, and leads to an accelerated loss of the germ cells, declining fertility and reproductive lifespan, and early onset of menopause. Women with homozygous or biallelic defects in DNA repair genes may present with a spectrum of phenotypes ranging from gonadal dysgenesis (*BRCA2*, *SPIDR*, *MCM8*, *MCM*, *MSH5*) to POI. Patients with pathogenic variants in these genes have increased susceptibility to chromosome breakage and genome instability, as well as susceptibility to cancer [16,17]. Fanconi anemia genes (*BRCA1*, *BRCA2*, *BRIP1*, *ERCC4*, *FANCA*, *FANCB*, *FANCC*, *FANCD2*, *FANCE*, *FANCF*, *FANCI*, *MAD2L2*, *PALB2*, *PHF9*, *RAD51*, *RAD51C*, *RFWD3*, *SLX4*, *UBE2T*, *XRCC2*, *XRCC9*), known to be involved in DNA repair in mitosis, are also essential in meiosis. Approximately 30% of adult patients with biallelic variants in Fanconi anemia genes do not exhibit congenital malformations and present with nonsyndromic manifestations such as POI, infertility, short stature, and/or skin pigmentation lesions. In women with gonadal dysfunction, the findings of additional phenotypes such as growth retardation, hyperpigmented/hypopigmented skin lesions, frequent infections, or endocrine dysfunctions (hypothyroidism, diabetes, adrenal insufficiency, metabolic syndrome) should trigger genetic evaluation and testing for genes associated with chromosome breakage conditions. Pathogenic homozygous and compound heterozygous variants in another meiotic gene, *HFM1* (ATP-dependent DNA helicase, homolog of helicase family member 1), have been found among women with POI (Table 9.1).

Autosomal recessive mode of inheritance accounts for most of the cases. The consequences of heterozygous variants in DNA repair genes on gonadal function are unknown. Most of the heterozygous variant carriers appear healthy and fertile, although fibroblast studies on individuals with *MCM8* and *MCM9* pathogenic variants showed an increased number of chromosome breakages in heterozygous carriers when compared with the wild-type family members [16,17]. Multiple studies have identified a higher frequency of heterozygous variants in DNA repair genes such as *FANCA*, *FANCL*, and *MSH5* among women with POI in comparison to control groups. These carriers also had reduced protein expression levels, or abnormal protein localization, raising the possibility of compromised DNA repair in heterozygotes. A nonsynonymous single nucleotide polymorphism (rs16991615) in the *MCM8* gene shows the strongest association with the age of menopause in Caucasian women, and replicates in African American and Hispanic women in genome-wide association studies. These findings indicate that heterozygous variants in DNA repair genes may affect reproductive lifespan and may confer additional risk for diminished ovarian reserves (DOR), infertility, POI, and premature menopause [18]. DOR is a more prevalent condition than POI and likely is part of the spectrum of ovarian insufficiency disorders, caused by less penetrant genetic variants.

Table 9.1. Genes involved in nonsyndromic ovarian dysgenesis, POI, and female infertility

Gene	Condition	Phenotype MIM/[PMID]	Mode of inheritance
ATG7	Primary ovarian insufficiency	[30224786]	AD
ATG9A	Primary ovarian insufficiency	[30224786]	AD
BMP15	Ovarian dysgenesis, primary amenorrhea, ODG2	#300510	X-linked
CPEB1	Primary ovarian insufficiency	[201,202]	AD
DIAPH2	Primary or secondary amenorrhea, POF2A	#300511	X-linked
EIF4ENIF1	Primary ovarian insufficiency	[23902945]	AD
ERCC6	Primary ovarian insufficiency, POF11	#616946	AD
ESR1	Estrogen resistance, primary amenorrhea	#615363	AR
ESR2	Ovarian dysgenesis, ODG8, primary amenorrhea	#618187	AD
FIGLA	Ovarian insufficiency, POF6	#612310	AD, AR
FSHR	Ovarian dysgenesis, primary amenorrhea, ODG1	#233300	AR
GDF9	Primary ovarian insufficiency, POF14	#618014	AR, digenic
HAX1	Primary ovarian insufficiency, short stature	[28681255]	AD
HFM1	Ovarian dysgenesis, ovarian insufficiency, POF9	#615724	AR
LHCGR	Luteinizing hormone resistance, amenorrhea, and infertility	#238320	AR
LHX8	Ovarian dysgenesis, ovarian insufficiency	[27603904]	AD
MCM8	Ovarian dysgenesis, primary amenorrhea, POF10	#236700	AR
MCM9	Ovarian dysgenesis, primary amenorrhea, ODG4	#616185	AR
MRPS22	Ovarian dysgenesis, ODG7, primary amenorrhea	#618117	AR
MSH5	Primary ovarian insufficiency, POF13	#617442	AR
NOBOX	Ovarian dysgenesis, primary ovarian insufficiency, POF5	#611548	AD, AR
NR5A1	Gonadal dysgenesis, secondary ovarian insufficiency, POF7	#612964	AD
NUP107	Ovarian dysgenesis, ODG6, primary amenorrhea	#618078	AR
POF1B	Primary ovarian insufficiency, POF2B	#300604	X-linked
POLR3H	Primary ovarian insufficiency	[30830215]	AR
PRLR	Hyperprolactinemia, amenorrhea	#615555	AR
PSMC3IP	Ovarian dysgenesis, ODG3, primary amenorrhea	#614324	AR
REC8	Ovarian dysgenesis, ovarian insufficiency	[27603904]	Undetermined
SALL4	Primary ovarian insufficiency	[30603774]	AD
SGOL2	Primary ovarian insufficiency	[27629923]	AR
SMC1B	Ovarian dysgenesis, ovarian insufficiency	[27603904]	Undetermined
SOHLH1	Ovarian dysgenesis, ODG5, primary ovarian insufficiency	#617690	AR
SOHLH2	Primary ovarian insufficiency	#616066	AD
SOX8	Ovarian dysgenesis, primary or secondary amenorrhea	[29373757]	AD

Table 9.1. *(cont.)*

Gene	Condition	Phenotype MIM/[PMID]	Mode of inheritance
SPIDR	Ovarian dysgenesis	[27967308]	AR
STAG3	Ovarian dysgenesis, ovarian insufficiency, POF8	#615723	AR
SYCE1	Ovarian dysgenesis, ovarian insufficiency, POF12	#616947	AR

9.5 X Chromosome and Ovarian Development

In women with two X chromosomes, in contrast to males with a single X chromosome, a subset of X-linked genes is subject to unique regulation known as X chromosome inactivation. X-linked genes that escape X chromosome inactivation are expressed at an increased dosage, leading to female-specific sex determination, and regulate ovarian determination pathways. Numerical and structural abnormalities involving the X chromosome are observed in at least 10% of women affected with gonadal dysgenesis or POI. Both the dosage of X-linked genes and structural integrity of the X chromosome are important factors for proper ovarian development and meiosis [12]. Despite the obvious importance of the X chromosome for female fertility, search for an X-linked monogenic etiology of ovarian dysgenesis and POI have yielded few genes, but include *BMP15* (MIM *300247, Xp11.22), *FMR1* (MIM *309550, Xq27.3), *POF1B* (MIM *300604, Xq21.1), and *PGRMC1* (MIM *300435, Xq24). Numerous studies utilizing next-generation sequencing and copy number profiling indicate a polygenic etiology of POI [19].

X chromosome alterations, if transmitted to male conceptuses, may affect male embryonic development or fetal viability, leading to recurrent male fetal losses [20]. Approximately 6% of women carry pathogenic copy number variations (CNVs) or single-nucleotide variants (SNVs) in the X-linked genes that are known to be associated with genetic disorder, mostly affecting males. Pathogenic changes in the *BCOR*, *EBP*, *FLNA*, *GPKOW*, *HCCS*, *IKBKG*, *MECP2*, *OFD1*, *OTC*, *PORCN*, and *REP1* genes are known to cause male-lethal phenotypes [20]. So far, about 70% of X-linked genes are not yet linked to any phenotype in humans. Moreover, the absence of deleterious changes in these genes among men may suggest that they represent novel candidate genes for male-lethal conditions. Losses of female fetuses affected by dominant X-linked disorders may also occur, explained by skewed X chromosome inactivation of the unaffected X chromosome or maternal germline mosaicism.

9.6 Genetic Causes of Fertilization Failure

Each human oocyte is surrounded by a highly specialized extracellular matrix, the zona pellucida, which is primarily composed of four glycoproteins: ZP1, ZP2, ZP3, and ZP4. These glycoproteins form long filaments, providing an interactive oocyte–granulosa cell environment. The zona pellucida is critical for protection of the growing oocyte, oocyte–follicle cell interactions, sperm recognition and binding during fertilization, and blastocyst development [21].

Women with biallelic variants in *ZP1* or *ZP2* genes are unable to form functional zonae pellucida; ovulate few, if any, eggs; and are infertile (Table 9.2). Heterozygous pathogenic variants in the *ZP3* gene were identified among women who underwent unsuccessful ovarian stimulation during in vitro fertilization (IVF); no oocytes were present in the mature follicles, a condition known as empty follicle syndrome (EFS). Despite normal follicular development and estradiol levels, oocytes lacking zonae pellucidae degenerate before ovulation. In some families, *ZP3* pathogenic variants were inherited from the patients' fathers, although the infertile phenotype is limited to females, as *ZP3* is expressed exclusively in oocytes. EFS is also caused by biallelic *ZP1* pathogenic variants and biallelic inactivating *LHCGR* mutations (Table 9.2). Digenic inheritance, whereby heterozygous variants in different genes cause a phenotype, has also been reported in a woman with impaired oocyte maturation and infertility, who carried one heterozygous variant in the *ZP2* gene and one heterozygous variant in the *ZP3* gene.

Table 9.2. Genes associated with fertilization defects, embryonic arrest, and recurrent miscarriage in females with idiopathic infertility

Gene	Condition	Phenotype MIM/[PMID]	Mode of inheritance
ANXA5	Susceptibility to recurrent pregnancy loss, RPRGL3	#614391	AD
AURKB	Susceptibility to recurrent pregnancy loss	[28369513]	AD
AURKC	Susceptibility to recurrent pregnancy loss	[28369513]	AD
BUB1B	Mosaic variegated aneuploidy, MVA1	#257300	AD
C11orf80	Infertility, recurrent hydatidiform mole, HYDM4	#618432	AR
CEP57	Mosaic variegated aneuploidy, MVA2	#614114	AR
F2	Susceptibility to recurrent pregnancy loss, RPRGL2	#614390	AD
F5	Susceptibility to recurrent pregnancy loss, RPRGL1	#614389	AD
KHDC3L	Infertility, recurrent hydatidiform mole, HYDM2	#614293	AR
LHCGR	Primary infertility, oocyte maturation defect	[29912377]	AR
MEI1	Infertility, recurrent hydatidiform mole, HYDM3	#618431	AR
NLRP2	Primary infertility, oocyte maturation defect	[30877238]	AR
NLRP5	Primary infertility, oocyte maturation defect	[30877238]	AR
NLRP7	Infertility, recurrent hydatidiform mole, HYDM1	#231090	AR
PADI6	Preimplantation embryonic lethality, PREMBL2	#617234	AR
PANX1	Primary infertility, oocyte maturation defect, OOMD7	#618550	AD
PATL2	Primary infertility, oocyte maturation defect, OOMD4, early embryonic arrest	#617743	AR
PLCZ1	Sperm-related oocyte activation defects, fertilization failure	[31953539]	AR
REC114	Primary infertility, early embryonic arrest	[31704776]	AR
SYCP3	Susceptibility to recurrent pregnancy loss, RPRGL4	#270960	AD
TLE6	Preimplantation embryonic lethality, PREMBL1	#616814	AR
TRIP13	Mosaic variegated aneuploidy, MVA3	#617598	AR
TUBB8	Primary infertility, oocyte maturation defect, OOMD2	#616780	AD, AR
WEE2	Primary infertility, oocyte maturation defect, OOMD5	#617996	AR
ZP1	Primary infertility, oocyte maturation defect, OOMD1	#615774	AR
ZP2	Primary infertility, oocyte maturation defect, OOMD6	#618353	AR, digenic
ZP3	Primary infertility, oocyte maturation defect, OOMD3	#617712	AD

9.7 Postfertilization Preimplantation Embryonic Failure

It has been estimated that about 30–60% of the human embryos produced in IVF are arrested at early stages of development. The diagnosis of preimplantation embryonic lethality refers to a condition when all of the patient's embryos fail to progress beyond the cleavage stages of embryogenesis. Early embryonic arrest is one of the causes of female infertility. Recent application of whole-exome sequencing to women with recurrent embryonic arrest has led to the discovery of several causative genes. Biallelic maternal alterations in *PADI6*, *PATL2*, *TLE6*,

NLRP2, *NLRP5*, *TRIP13*, *PANX1*, and *WEE2*, and heterozygous and homozygous pathogenic variants in *TUBB8* [22], were found in patients with early embryonic lethality (Table 9.2). Further investigation showed that, beside early embryonic arrest, these genes cause a spectrum of phenotypes, including oocyte maturation arrest and fertilization, and implantation failure.

After fertilization, two gametes are united, creating a zygote. The zygotic genome is not transcriptionally active as it transitions into a totipotent state. The first few cell divisions of the embryo are driven by maternally derived cytoplasmic factors, such as RNA, proteins, and subcellular organelles, accumulated during oocyte maturation. Homozygous and compound heterozygous defects in *BTG4*, *REC114*, and *CDC20* genes are the cause of early embryonic arrest associated with poor embryo morphology due to zygotic cleavage failure and pronuclei formation. The OOEP, NLRP5, TLE6, and KHDC3L proteins are subunits of the subcortical maternal complex present in the cytoplasm of mature oocytes. Protein-damaging variants in the *TLE6* and *NLRP5* genes were reported in patients whose embryos failed to form blastocysts. Pathogenic variants in *KHDC3L* and *NLRP7* are associated with recurrent hydatidiform mole and early embryonic arrest. In women with pathogenic variants in these genes, the maternal genome failed to establish a maternally specific gene expression profile, leading to an abnormal imprinting and a phenotype like androgenetic molar pregnancy. A key role of maternal cytoplasmic factors is establishing embryo developmental competence. It is likely that as more cases are sequenced, additional genes responsible for embryonic failure will be discovered.

9.8 Postimplantation Pathologies Resulting in Infertility and Recurrent Pregnancy Loss

It is estimated that about 60% of human embryos are aneuploid, resulting in the miscarriage of nonviable conceptions. Recurrent pregnancy loss (RPL) refers to a condition in which at least two clinically recognized pregnancies were lost before 20 weeks of gestation. It is estimated that up to 5% of couples experience two consecutive pregnancy losses, while the loss of three or more pregnancies affects about 1% of patients. Around 50% of first- and second-trimester miscarriages are due to meiotic nondisjunction in oocytes, the rate of which increases significantly with advancing maternal age. Young women with RPL have higher incidences of recurrent aneuploidy than expected for their age group, raising the possibility of genetic predisposition to aneuploidy or polyploidy [23]. Recurrent trisomy for the same chromosome is usually explained by parental gonadal mosaicism, when a subpopulation of germ cells within a female or male gonad is aneuploid. Mosaic trisomy for every chromosome has been reported in the literature. Individuals with low-level mosaicism or tissue-limited mosaicism are more likely to be unaffected phenotypically, although aneuploidy in their germ cells and oocytes can significantly contribute to the depletion of ovarian reserves, preimplantation failure, or prenatal death. Meta-analysis of couples with more than one child affected by trisomy 21 suggests that parental germline mosaicism may be a cause of recurrent aneuploidy in up to 10% of such families [24]. Other possible explanations include pathogenic variants in genes that regulate meiosis.

A history of multiple miscarriages and infertility is also associated with recurrent triploid conceptions and familial recurrent hydatidiform moles. Triploid human diandric pregnancy (also known as partial molar pregnancy) is a conception that contains an additional paternal set of chromosomes. A complete hydatidiform mole is a pregnancy derived from two copies of a paternal haploid genome in the absence of a maternal complement. Both partial and complete molar pregnancies are associated with early embryonic arrest or first-trimester miscarriage (Table 9.2). Although rare, recurrent familial cases of partial and complete hydatidiform mole conceptions strongly suggest a genetic predisposition. Analysis of familial cases showed that biallelic pathogenic variants in *MEI1*, *KHDC3L*, *NLRP7*, *REC114*, or *TOP6BL* affect the oocyte's ability to complete meiotic division, extruding all chromosomes with the first polar body and creating an "empty" egg (Table 9.2).

Chromosome nondisjunction or premature separation of sister chromatids during meiosis is the mechanism leading to meiotic aneuploidy in oocytes. The vast majority of aneuploid conceptions fail to implant or are lost early, before pregnancy can be clinically recognized. Identification of human genes that affect chromosome segregation during meiosis is difficult, as patients with such defects may present with either idiopathic infertility or POI due to loss of germ cells.

Pathogenic variants in the *AURKB* and *AURKC* genes have been identified in women with extremely high aneuploidy rates among miscarried fetuses. Heterozygous and biallelic alterations in the *BUB1B*, *CEP57*, and *TRIP13* genes are rare findings among fetuses and liveborn individuals affected by mosaic variegated aneuploidy. Because of impaired spindle checkpoint, aneuploidy for multiple different chromosomes in all or a proportion of the cells might be observed in the products of conception.

In up to 12% of couples with RPL, one of the partners is a carrier of a balanced chromosomal aberration [25]. The carriers of balanced chromosomal rearrangements (translocations, inversions, insertions) are phenotypically normal but are at increased risk of chromosomally unbalanced conception, leading to a nonviable embryo or subsequent miscarriage (see Chapter 7). Parental karyotype analysis may reveal a chromosomal abnormality, although about 40% of chromosomal alterations in couples with RPL are submicroscopic, yielding normal results. In the past, assessment of genetic abnormalities was limited to karyotypes obtained during culturing of placental or fetal tissues. With advances in molecular genetic technologies, subtle chromosomal imbalances can be reliably identified. Microarray analysis of the product of conception after a miscarriage is a powerful tool that can be utilized and is recommended for detection of chromosomal rearrangements leading to RPL and infertility.

Parents may carry recessive pathogenic variants in lethal genes, defects which are incompatible with a normal embryo/fetal development. Over 600 OMIM genes can cause early pregnancy loss of conceptions with normal karyotype and microarray testing. A significant number of genes (~40%) currently included in preconception expanded carrier screening are associated with lethality. This is just a small number of genes, as bioinformatics analyses of human genes and animal models suggest there are over 3000 potentially lethal fetal genes [26]. The role of these genes and their cumulative contribution to unexplained infertility is currently under investigation.

Additional risk factors of RPL include maternal thrombophilic conditions and a carrier status for pathogenic variants in the coagulation factor II gene (*F2*), coagulation factor V gene (*F5*), or the placental anticoagulant protein annexin A5 (*ANXA5*). Recurrent miscarriages in these women are commonly attributed to placental insufficiency.

9.9 Concluding Remarks and Future Perspectives

It is estimated that around 10% of women of reproductive age are affected by infertility. Etiology of female infertility is complex, with nearly 50% of patients having genetic defects. X chromosome aneuploidy and structural alterations, DNA copy number variations, monogenic (pathogenic sequence variants in a single gene), or polygenic (mutations among multiple genes) inheritance are the main categories among the genetic causes of infertility in females. Mammalian reproduction is regulated by many genes that orchestrate multiple processes on the intracellular and intercellular, tissue-specific, and organ-specific levels. A significant number of genes have been found to cause ovarian dysfunction and infertility in humans. With the implementation of exome, genome, and RNA sequencing in research and clinical practice, substantial progress has been made in the elucidation of genes causing ovarian dysgenesis, POI, alterations associated with early embryonic and fetal lethal phenotypes [27], parental abnormalities leading to implantation failures, and recurrent pregnancy losses. Increasingly, more women are seeking assisted reproductive technology (ART) as an option for infertility treatment. Identifying causes of infertility is important for the selection of an efficient treatment approach, management of general broader health implications associated with infertility, and understanding of the genetic determinants of this condition.

For many genes, autosomal recessive inheritance is associated with reproductive pathologies; however, the clinical consequences of heterozygous pathogenic variants in the same genes are poorly understood. A growing body of evidence suggests that a cumulative effect of less damaging heterozygous alterations among the genes involved in the same pathway may result in an outcome equivalent to a pathogenic homozygous change of a single gene, hence the term "polygenic disease."

Clinical laboratories have implemented next-generation sequencing (NGS) in a variety of medical fields as a tool for the diagnosis of heterogeneous genetic disorders. To date, NGS also can detect CNVs, structural chromosomal alterations, sex chromosome aneuploidies, and germline mosaicism. Despite several challenges associated with the implementation of genomic techniques in clinical practice and the interpretation of sequencing results, molecular diagnosis has become an integral part of ART and personalized reproductive precision medicine.

References

1. Yatsenko SA, Rajkovic A. Genetics of human female infertility. *Biol Reprod* 2019;101(3):549–66.
2. McGee EA, Hsueh AJ. Initial and cyclic recruitment of ovarian follicles. *Endocr Rev* 2000;21:200–14.
3. Pangas SA, Rajkovic A. Transcriptional regulation of early oogenesis: in search of masters. *Hum Reprod Update* 2006;12:65–76.
4. Choi Y, Rajkovic A. Characterization of NOBOX DNA binding specificity and its regulation of Gdf9 and Pou5f1 promoters. *J Biol Chem* 2006;281:35747–56.
5. Shin YH, Ren Y, Suzuki H, et al. Transcription factors SOHLH1 and SOHLH2 coordinate oocyte differentiation without affecting meiosis I. *J Clin Invest* 2017;127:2106–17.
6. Hanley NA, Ikeda Y, Luo X, Parker KL. Steroidogenic factor 1 (SF-1) is essential for ovarian development and function. *Mol Cell Endocrinol* 2000;163(1–2):27–32.
7. Eggers S, Ohnesorg T, Sinclair A. Genetic regulation of mammalian gonad development. *Nat Rev Endocrinol* 2014;11:673–83.
8. Weinberg-Shukron A, Rachmiel M, Renbaum P, et al. Essential role of BRCA2 in ovarian development and function. *N Engl J Med* 2018;379:1042–9.
9. Wood MA, Rajkovic A. Genomic markers of ovarian reserve. *Semin Reprod Med* 2013;31:399–415.
10. Rajkovic A, Pangas S. Ovary as a biomarker of health and longevity: insights from genetics. *Semin Reprod Med* 2017;35:231–40.
11. Richards JS, Ren YA, Candelaria N, Adams JE, Rajkovic A. Ovarian follicular theca cell recruitment, differentiation, and impact on fertility: 2017 update. *Endocr Rev* 2018;39:1–20.
12. Persani L, Rossetti R, Di Pasquale E, Cacciatore C, Fabre S. The fundamental role of bone morphogenetic protein 15 in ovarian function and its involvement in female fertility disorders. *Hum Reprod Update* 2014;20:869–83.
13. Zhao H, Qin Y, Kovanci E, et al. Analyses of GDF9 mutation in 100 Chinese women with premature ovarian failure. *Fertil Steril* 2007;88:1474–6.
14. Caburet S, Arboleda VA, Llano E, et al. Mutant cohesin in premature ovarian failure. *N Engl J Med* 2014;370:943–9.
15. MacLennan M, Crichton JH, Playfoot CJ, Adams IR. Oocyte development, meiosis and aneuploidy. *Semin Cell Dev Biol* 2015;45:68–76.
16. Wood-Trageser MA, Gurbuz F, Yatsenko SA, et al. MCM9 mutations are associated with ovarian failure, short stature, and chromosomal instability. *Am J Hum Genet* 2014;95:754–62.
17. AlAsiri S, Basit S, Wood-Trageser MA, et al. Exome sequencing reveals MCM8 mutation underlies ovarian failure and chromosomal instability. *J Clin Invest* 2015;125:258–62.
18. Desai S, Rajkovic A. Genetics of reproductive aging from gonadal dysgenesis through menopause. *Semin Reprod Med* 2017;35:147–59.
19. Yatsenko SA, Wood-Trageser M, Chu T, Jiang H, Rajkovic A. A high-resolution X chromosome copy number variation map in fertile females and women with primary ovarian insufficiency. *Genet Med* 2019;21(10): 2275–84.
20. Franco B, Ballabio A. X-inactivation and human disease: X-linked dominant male-lethal disorders. *Curr Opin Genet Dev* 2006;16:254–9.
21. Gupta SK. The human egg's zona pellucida. *Curr Top Dev Biol* 2018;130:379–411.
22. Feng R, Sang Q, Kuang Y, et al. Mutations in TUBB8 and human oocyte meiotic arrest. *N Engl J Med* 2016;374:223–32.
23. Shahine LK, Marshall L, Lamb JD, Hickok LR. Higher rates of aneuploidy in blastocysts and higher risk of no embryo transfer in recurrent pregnancy loss patients with diminished ovarian reserve undergoing in vitro fertilization. *Fertil Steril* 2016;106:1124–8.
24. Delhanty JD, SenGupta SB, Ghevaria H. How common is germinal mosaicism that leads to premeiotic aneuploidy in the female? *J Assist Reprod Genet* 2019;36(12):2403–18.
25. Dong Z, Yan J, Xu F, et al. Genome sequencing explores complexity of chromosomal abnormalities in recurrent miscarriage. *Am J Hum Genet* 2019;105(6):1102–11.
26. Dawes R, Lek M, Cooper ST. Gene discovery informatics toolkit defines candidate genes for unexplained infertility and prenatal or infantile mortality. *NPJ Genom Med* 2019;4:8.
27. Suo L, Zhou YX, Jia LL, et al. Transcriptome profiling of human oocytes experiencing recurrent total fertilization failure. *Sci Rep* 2018;8:17890.

Chapter 10

Preconception Genetics Analysis/Screening in IVF

Juan José Guillén and Rita Vassena

10.1 Introduction

The Human Genome Project officially began in the USA in October 1990 under the auspices of the Department of Energy and the National Institutes of Health (NIH) under the direction of Francis Collins. The objective was to build genetic and physical maps of the entire human genome, and at the same time to develop the technology needed to perform DNA sequencing on a large scale. Extensive international collaboration and advances in the field of genomics and bioinformatics enabled the first essentially complete version of the human genome (92.3% of the total) to be officially announced 13 years later, two years ahead of schedule, on April 14, 2003, with 99.9% reliability.

One of the consequences of the Human Genome Project was the development of technological advances that allowed for fast and comparatively inexpensive sequencing of genomic DNA. Ultimately these advances, refined and improved, ushered in today's genomic medicine, and are the base of current commercial sequencing services and products.

In reproductive medicine, genetic tests are used mainly with the goal of identifying the genetic causes of infertility (e.g. vas deferens agenesis related to pathogenic variants in the *CFTR* gene, early menopause in women with premutation in the *FMR1* gene; for a more comprehensive description of the genetic causes of infertility, see Chapters 7–9), assessing the risk of a healthy individual to pass on genetic diseases to their progeny (known as reproductive carrier screening), avoiding the transmission of genetic diseases to offspring by preimplantation genetic test of monogenic diseases (PGT-M), and detecting anomalies in the embryo by preimplantation genetic testing of aneuploidies (PGT-A; see Chapter 13 for a comprehensive description of PGT).

In this chapter we focus on carrier screening, a form of genetic testing that is used to determine a person's or a couple's risk of having a child with a recessive genetic disorder, when there is no a priori risk based on personal or family history.

10.2 Ethnicity-Based Carrier Screening or Population-Specific Screening

The concept of carrier screening is not new. In fact, for decades it has been a procedure offered to individuals and couples belonging to high-risk populations with an increased mutation frequency of certain autosomal recessive conditions because of their ethnicity, race, or geographic origin. Decades ago, when genetic studies were not yet available, carrier screening was carried out by means of hematologic parameters for hemoglobinopathies or by determining the levels of a certain enzyme in the blood in Tay–Sachs disease. When it became technologically possible, screening was carried out by analyzing the most frequent pathogenic variants, usually of a single gene such as that for cystic fibrosis.

More than half a century ago, Wilson and Jungner published a seminal work detailing 10 guiding principles when deciding whether to screen for a disease [1].

1. The condition sought should be an important health problem.
2. The natural history of the condition, including development from latent to declared disease, should be adequately understood.
3. There should be a recognizable latent or early symptomatic stage.
4. There should be a suitable test or examination.
5. The test should be acceptable to the population.
6. There should be an agreed policy on whom to treat as patients.
7. There should be an accepted treatment for patients with recognized disease.
8. Facilities for diagnosis and treatment should be available.

9. The cost of case-finding (including diagnosis and treatment of patients diagnosed) should be economically balanced in relation to possible expenditure on medical care as a whole.
10. Case-finding should be a continuing process and not a "once and for all" project.

These screening principles were set out as normative statements regarding what should be known in order to guide screening decisions. In the case of genetic disease, however, these principles should be complemented by the possibility of acting on the knowledge of parental carrier status, either by preimplantation testing or prenatal diagnosis with possibility of intervention. In all cases, people should be free to decide whether they wish to participate in the screening.

10.2.1 Population Screening for Hemoglobinopathies

Hemoglobinopathies are the most common monogenic diseases in the world and are prevalent in regions with a history of endemic malaria such as the Mediterranean, the Middle East, South-East Africa, and sub-Saharan Africa. Population screening for hemoglobinopathies has been carried out in the countries of the Mediterranean basin for more than 40 years, Sardinia and Cyprus being good examples.

Cyprus is an eastern Mediterranean island with a total area of 9251 km^2. Greek Cypriots represent about 98.8% of the inhabitants, the rest are Turkish Cypriots. An exceptionally high incidence of thalassemia on the island poses a major public health challenge. In the 1970s [2] the prevalence of thalassemia major was estimated to be around 1 in 1000, the carrier frequency of β-thalassemia was 1 : 4, and the annual birth rate of affected individuals was between 30 and 50 individuals. In 1978, Cyprus introduced a population screening for hemoglobinopathies based on blood test parameters, differences in mean corpuscular volume (MCV), and changes in hemoglobin electrophoresis indices, which combined allow for the identification of heterozygote carriers. In 1983, the Cyprus Church introduced a mandatory premarital certificate indicating that the couple seeking marriage had undergone screening for thalassemia and received appropriate counseling. As a result, in 1991, the annual birth rate of affected individuals had decreased to less than five affected children born annually on the island [3].

10.2.2 Program for the Prevention of Tay–Sachs Disease

Tay–Sachs disease (TSD) is a lethal neurodegenerative condition caused by mutations in the gene that encodes the enzyme hexosaminidase A (HEXA). In 1970, Michael M. Kaback from The Johns Hopkins Hospital (Baltimore) in collaboration with John O'Brien (University of California San Diego) and Ed Kolodny at the National Institutes of Health (NIH) showed that serum quantification of HEXA could identify healthy carriers for TSD, who almost always have lowers levels of HEXA than noncarriers. This made it possible to identify affected fetuses by enzymatic testing through amniocentesis, and to screen adults to identify carrier couples, who are at increased risk of having an affected child. Thus, the idea of community-based heterozygote screening was conceived.

After the establishment of standard carrier screening programs by Dr. Michael Kaback and others in 1971, the overall birth rate of children with TSD began to decline. However, this was not the case among the Orthodox Jewish community, due to a combination of very high carrier frequency in this community (1 : 30) and religious laws not allowing birth control or abortion. In 1983, Rabbi Josef Ekstein in collaboration with Dr. Robert Desnick, chairman of the Department of Human Genetics at the Mount Sinai School of Medicine (New York city), proposed to their community leaders that a screening program be established for TSD. They founded "Dor Yeshorim," the Committee for Prevention of Jewish Genetic Diseases. The implementation of this screening program among the Orthodox Jewish community had specific features: first, it would have a premarital approach, and was offered primarily to senior high school girls and boys in theological schools. Second, anonymity and confidentiality had to be preserved in order to prevent carriers or their families from being stigmatized or discriminated as undesirable marriage partners. To enforce this point, participants did not receive any results at the time of testing; rather, when they found a potential partner, the couple were informed if the marriage was feasible.

In 1970, before the program for the prevention of TSD was implemented, an estimated 50–60 infants with TSD were born each year in the USA and Canada. Of these, 40–45 infants were of Ashkenazi Jewish ancestry and 10–15 were of non-Jewish heritage. From 1983 to 1998 the average total number of new cases per year in the Jewish population was four, a 90% reduction in the incidence of TSD [4]. Moreover, the initiation in 1971 of a screening program to prevent TSD among Ashkenazi Jews in the USA led to the establishment in 1978 of a national carrier screening program in Israel under the support of the Health Department.

10.2.3 Consanguinity As Risk Factor for Recessive Conditions

Consanguinity and endogamy (marriage within a circumscribed community) both have well-known effects on the genetics of populations. Homozygosity for a pathogenic genetic variant can be diagnosed as early as three generations after its appearance in the first carrier and coinheritance of two or more rare recessive disorders is most frequent. For these reasons, consanguineous marriage has been identified as a risk factor for congenital malformations and genetic disorders. The rate of malformations is approximately two to three times higher among first cousins than in nonrelated couples from the same population [5]. In Bangladesh, the results from a cross-sectional study on carrier detection of β-thalassemia and hemoglobin E variants suggest that consanguinity contributes significantly to the increased thalassemia carrier frequency in young adults with consanguineous parents (23.5%) compared with nonrelated parents (11.4%) [6].

Israel, with 8 628 600 inhabitants in 2017, is characterized by a wide ethnic and religious diversity. This is a result of its history of migration waves and return since 1948; the population comprises approximately 75% Jew, 21% Arab, and 4% Bedouin. The Bedouins are a tribal nomadic population that live in the Negev region. Nowadays the Negev Bedouin communities remain very isolated, with a low socioeconomic and educational level, high birth rates, and a rate of consanguinity around 45% (second cousins or closer).

This is not an isolated occurrence; consanguineous marriage is overall common in the Arab world. The most frequent type of consanguineous marriage has been between first cousins and, among those, marriage between children of brothers (patrilineal parallel) [7]. Although first cousin marriages are in decline, they have mostly been replaced by marriage to more distant relatives, so endogamy remains high (>70%) [8]. Endogamy was also relatively common in Jewish communities.

10.2.4 The Israeli National Carrier Screening Program

An example of a comprehensive population screening program for reproductive purposes is the one in place in Israel since 2013, and includes screening for cystic fibrosis, fragile X syndrome, and spinal muscular atrophy free of charge for all citizens. Besides these common conditions, individuals are further screened according to their community of origin (Jews, Christian Arabs, Muslim Arabs, Negev Bedouins, and Druze) for all severe diseases with a carrier frequency of 1 : 60 or higher and/or disease frequency of 1 in 15 000 live births or higher according to the Israeli National Genetic Database (INGD) (http://INGD.huji.ac.il). This program is operated through sequential screening, where the woman in the couple undergoes testing first, and her partner is only tested if she carries a pathologic mutation. Couples that share a pathogenic variant, either the same one or different, in the same gene receive genetic counseling and reproductive options are discussed. The implementation of the Israeli National Population Program of genetic carrier screening has resulted in all but eradication of TSD, while thalassemia cases have been reduced by more than 80% [8].

Nowadays, national screening programs for high-risk populations to identify carriers of recessive genetic diseases have been adopted in countries such as the USA, the UK, Australia, and Cyprus [9] but none have been implemented as universally as in Israel.

10.2.5 Population Screening for Cystic Fibrosis

Cystic fibrosis (CF) is a disease with an autosomal recessive inheritance pattern; it is often severe and requires lifelong medical treatment. However, given its variable phenotype, a small percentage of individuals present moderate forms, compatible with a normal life. In 1989 Lap-Chee Tsui cloned the *CFTR* gene and a few mutations causative of this disorder. This, like other "rare diseases" do not affect many people, but globally they represent a not insignificant social and economic cost. The frequency of carriers in northern European countries and the USA is between 1 : 20 and 1 : 30, and therefore the risk of an unstudied couple having an affected child is estimated at 1 in 2300. Carriers are healthy individuals and approximately 90% of them do not have any affected family member.

In the early 1990s, the scientific community and health departments in some countries (USA, UK, Canada) began to consider whether universal population screening should be carried out for CF. The first pilot studies were carried out in the UK and the USA by general practitioners in primary healthcare who offered the test to any adult or pregnant woman. These initial studies concluded that CF met the necessary conditions for community carrier screening, and that it was feasible to inform the community of the test availability. On the other hand, several studies demonstrated that those who are found to be carriers do not appear to have undue anxiety, are not stigmatized, and are glad to have been tested. Considering the evidence, in 1997 the National Institutes of Health in the USA recommended that all couples planning a pregnancy be offered CF screening [10].

In summary, for more than 50 years the medical community has been working to inform prospective parents about their increased reproductive risk, with the final goal of preventing the birth of children with serious genetic diseases.

10.3 Expanded Carrier Screening

Thanks to the Human Genome Project (1990–2003) and its extraordinary advances in DNA sequencing, is it now possible to analyze hundreds of genetic diseases rapidly and cheaply from one single sample and in one sequencing run. Through sequencing panels, it is possible to interrogate thousands of genes for variants related to genetic diseases. This approach to carrier screening is called expanded carrier screening (ECS), which determines the carrier status of an individual for any of the diseases included in the panel. When performed in couples, ECS provides the opportunity to assess their reproductive risk (i.e. the risk of having a child affected by a severe genetic disorder). The ideal target for this test is healthy individuals who a priori do not have an increased genetic risk. In contrast to community screening programs focusing on one disease, the high number of genes analyzed in ECS makes it impractical and unnecessary to direct them to certain geographic areas or ethnicities. ECS should be offered to individuals prior to conceiving or in the early stages of pregnancy (recommended at 12 weeks' gestation or earlier); when offered preconceptionally, there are more reproductive options and less emotional distress in comparison to studies during pregnancy, when there are much fewer alternatives to avoid the birth of an affected child.

Several professional and scientific societies, such as the American College of Medical Genetics (ACMG) in 2013 and 2015 [11,12], the American College of Obstetricians and Gynecologists (ACOG) in 2015 and 2017 [12,13], the European Society of Human Genetics (ESHG) in 2016 [14], and the Spanish Fertility Society (SEF) in 2019 [15], among others, have outlined the pros and cons of ECS relative to more traditional approaches and have offered recommendations for its implementation.

10.3.1 ECS Is Complementary to Newborn Testing

Each test is designed to achieve a different goal. Neonatal testing programs are diagnostic studies whose objective is to detect, diagnose and, when appropriate, provide early treatment for individuals affected by genetic and/or endocrine diseases, which are rare but can be fatal or significantly compromise an individual's quality of life. This strategy makes it possible to establish treatment even before the symptoms appear, thus improving the prognosis or, for certain conditions, curing the affected individual. ECS screening studies, on the other hand, are designed to identify healthy carriers at an increased risk of transmitting a genetic disease to their offspring, and presenting carriers with actionable reproductive options to prevent the birth of sick individuals

To add to the complexity, neonatal testing programs are in place in most countries but there is wide heterogeneity in the diseases included. Congenital hypothyroidism, because of its high incidence, low cost of treatment, and good clinical results, is the only one that is usually included in neonatal testing programs worldwide [16]. Likewise, the composition of ECS panels can differ greatly depending on the company that commercializes them.

In any case, ECS and neonatal testing are complementary and not exclusive; in other words, performance of ECS does not preclude performance of neonatal testing.

10.3.2 What Should We Consider before Implementing an ECS Program?

As with any medical intervention or genetic test, pre- and post-test information is necessary. Counseling should at a minimum focus on the purpose of the test, its limitations, its risks, and the significance of the result, explained with language adapted to the circumstances of each patient. While it is not feasible, or even needed, to describe in detail all the diseases included in the panel, it is nonetheless mandatory that the information given includes the general nature of disease, and the consequences of the results that can be expected from the test. Further, users must understand that the results may have implications not only for them but also for their families and future children.

An expert, usually a genetic counselor, but, depending on the country and legal framework, also other trained healthcare workers, should explain both test and test results to the patients. To avoid a feeling of false security, it is very important to stress that a zero risk does not exist: even if the test result is negative, there will always be a residual and a reproductive risk. Residual risk is defined as the probability that an individual is a carrier even if the test is negative (i.e. even if the test has not detected any pathogenic variants). It is related to the limitations of the technique, the detection rate of the test for each disease, and the prevalence of the disease in the population to which the individual belongs (frequency of carriers). When both members of a couple are carriers of an autosomal recessive condition or a woman is a carrier of an X-linked condition, there is an increased risk of usually 1 : 4 of carrying an affected pregnancy. Couples or individuals with an increased reproductive risk have several options in order to either avoid or prepare for having a child affected depending on their choices, such as in vitro fertilization (IVF) with PGT-M, prenatal diagnosis or the use of donor gametes. For an exhaustive discussion of genetic counseling, see Chapter 11.

10.3.3 Is There Still a Role for Family History or Ethnicity-Based Screening?

Both family history and ethnicity-based screening were employed first when genetic testing for recessive and X-linked diseases was a long and costly proposition, in order to define a population with higher-than-average expectation for positive testing. However, the relevance of this approach can be discussed when testing becomes cheap and widely available.

The family history should include the ethnic background of family members as well as any known consanguinity. It will reveal whether an individual has an increased risk for a genetic disease with autosomal dominant or polygenic and multifactorial inheritance, or even for chromosomal alterations, but it will not provide much guidance on the existence of pathogenic variants related to autosomal recessive diseases. Carrier screening based on family history has some limitations; one of them is our inaccurate knowledge of the health conditions of generations past. Second, for a recessive disease to be expressed in an individual, it is necessary to have a pathogenic variant on each allele of the gene (two carrier parents). It may be that a relative several generations back was a carrier of the disease while successive generations would only present carriers. In fact, 80–90% of CF carriers do not report any known affected family member [17], while 95% of children with hearing loss detected by universal screening have both parents with normal hearing [18]. Finally, family history is often an unreliable source of information on genetic diseases, given the relatively recent development of genetic medicine and molecular diagnosis.

Carrier screening based on ethnicity also has some important limitations: self-reported ethnicity does not always correlate with genetic ancestry [19], society tends to be multiethnic with a mixed racial ancestry and an ever-increasing number of individuals do not know the exact origin of their family [20], and genetic conditions are not limited to ethnic groups [21].

Currently, ACOG [22] and ACMG [12] have recommended pan-ethnic screening for CF and spinal muscular atrophy. ACOG also recommend pan-ethnic preconception screening for hemoglobinopathies based on red blood cell indices. As discussed earlier, at present Israel is the only country in the world which provides free screening for CF, fragile X syndrome, and spinal muscular atrophy to the whole population.

There are arguments for and against the use of ECS for all couples. On the one hand, screening based on family history or ethnicity leaves the onus on patients and clinicians to recognize the individual's ethnicity, increases the chance of not detecting carriers, and favors ethnic stigmatization. On the other hand, the widespread use of ECS may lead to the stigmatization of couples not accepting ECS as careless and uninterested in the health of their offspring [23,24]. Further, the cost-effectiveness of population-wide ECS screening has not yet been established, and questions of access and equity still remain [25].

10.3.4 Should All Intended Parents Be Made Aware of ECS?

All humans are carriers of pathogenic variants in one or more genes associated with genetic diseases with autosomal recessive inheritance [26]. A variant is a stable change in the DNA sequence of an individual with respect to the sequence considered normal or reference. When the change is causative of an altered phenotype it is considered a pathogenic variant. Whole-genome sequencing of randomly selected individuals and their parents has demonstrated that every individual is born with 44–82 de novo single-nucleotide mutations and therefore, in a defined population, many of the newborns are carriers of new variants responsible for recessive diseases [8.]. Every year 8 million children are born with a serious defect of partial or total genetic origin [27]. EURORDIS (Rare Diseases Europe) estimates that 6–8% of the European population are affected by a rare disease (around 30 million citizens) and 80% of rare diseases have identified genetic origins (www.eurordis.org/content/what-rare-disease). Parents and families of children with health disabilities experience significant financial and psychological stress and often avoid having another child with the same condition [25]. It is estimated that about 2–3% of couples in the general population attempting pregnancy are at risk of having children with a serious genetic disease [28,29]. Apart from de novo mutations, which are not detectable in the parents as they appear for the first time in the fetus, this risk can be assessed for any individual or couple using ECS tests.

All these points would indicate that proposing reproductive carrier screening to couples intending to conceive is indeed a positive intervention that should be supported. However, there are still some issues to bear in mind. Firstly, there is no agreement among scientific societies: in 2011, the Human Genetics Commission (UK) [30] reported on prenatal genetic testing, where the following conclusion was reached: "Having considered the issues associated with preconception genetic testing, in our view, there are no specific ethical, legal or social principles that would make preconception genetic testing within the framework of a population screening programme unacceptable." In 2015, a joint statement of five American societies [12] stated: "This statement does not replace current screening guidelines, which are published by individual organizations to direct the practice of their constituents" and current screening guidelines only recommend pan-ethnic screening for spinal muscular atrophy and CF. The ESHG in 2016 stated "Governments and public health authorities should adopt an active role in discussing the responsible introduction of expanded carrier screening" [14]. ACOG in 2017 stated that ECS is an acceptable strategy for carrier screening [13]. Finally, the SEF in 2019 recommended that "patients should be informed of possible strategies to reduce the risk of having offspring affected by a serious recessive or X-linked genetic disease which if no preventive measures are taken, is estimated at about 0.5 %" [15]. Secondly, the cost of testing may hinder providers to adopt the new technologies and may promote economic disparities between users, with uneven access. Finally, a broad introduction of ECS would exacerbate the severe lack of clinical geneticists and genetic counselors worldwide, which will require a careful approach to the introduction of ECS, and urgent educational efforts to allow other health professionals to share part of the counseling burden.

10.3.5 How Can We Test Couples?

A couple can be screened by two main strategies: concurrent/simultaneous or sequential screening. In simultaneous screening both members are tested at the same time. In the sequential approach, the test is performed first in one of the spouses – screening the female first is usually recommended to allow screening for X-linked recessive disorders. If the woman is found to be a carrier for an autosomal recessive disorder, then the partner is also tested. This strategy may be cheaper, but has the disadvantage compared with concurrent screening that time is lost because study of the partner can only start when the results of the first individual are available. In order to make the process more comfortable for the couple, it is recommended that the samples are taken from both members of the couple at the same time, and that one sample will be processed while the other is stored for later analysis if necessary.

Normally, in the sequential approach, the partner should be screened using the same screening panel that was used to identify the variants in the person first tested, but some authors [31] have suggested that partners should rather be offered whole-gene sequencing to identify rare variants in the gene in which the first member screened was determined to be a carrier.

We think that for each positive test result, the residual reproductive risk needs to be evaluated and the appropriate decision can be reached in agreement with the couple.

On the other hand, as the purpose of reproductive carrier screening is to detect couples at risk for an affected child, individual knowledge about being a carrier is almost always irrelevant to the carrier themselves. However, this knowledge may still be important, as it allows relatives of the carrier to be informed about the existence of the variant in the family and to decide whether to be tested as well in what is known as cascade screening. In the future, when population screening programs will likely be more widespread, we can expect cascade screening to become less relevant.

10.3.6 Which Conditions Should Be Included in ECS Panels?

Usually, the conditions included in ECS panels are monogenic with an autosomal recessive or X-linked recessive pattern of inheritance and it is recommended that they have certain characteristics; the disease should present with a *well-defined phenotype* and should be *clinically severe*. Further, it is usually expected that the diseases are incurable and can only be treated symptomatically, and even then with significant loss in quality of life and/or life expectancy. Only variants for which the natural history of the disease is known should be included. How can we establish if an illness is severe? One way of defining a disease as severe is based on the prospective parents; in other words, according to ACOG, when most couples would be interested in prenatal diagnosis and potential pregnancy termination if the result is positive, then the disease can be considered severe [13]. A more objective categorization is based on the condition phenotype. A condition is usually considered severe if it is associated with intellectual disability, or if it requires lifelong medical or surgical treatment, or it diminishes the quality of life and manifests itself in the fetus, newborn, or early childhood [12], and if these diseases do not only have a serious impact on the individual but can also affect their relatives and offspring.

Further, a *highly accurate test* should be available, so that a positive result is predictive of disease and a negative result does not give a false sense of security. Therefore, a set of relatively well-characterized pathogenic variants should be tested for. The possibility to test for mutations causing only mild phenotypes should not be offered in the context of carrier screening for reproductive planning [32]. A path forward in the future would be to include, for each gene, all known pathogenic variants, as characterized via genomic analysis of patients affected by severe autosomal recessive diseases in the population. Nevertheless, results of whole-exome/genome sequencing show that some variants that were considered pathogenic, because detected in affected patients, have also been found frequently in homozygosity among unaffected individuals and are therefore probably benign [20].

The diseases included in the panel should be relatively *common*, but there is no agreed threshold of carrier frequency for condition inclusion. In 2010, the American College of Medical Genetics and Genomics defined the minimum carrier frequency as around 1 : 60 [33], whereas in 2015 a joint statement by experts defined a common condition as a carrier frequency of 1 : 50 or greater [22] but did not recommend this as a specific criterion for inclusion on an expanded carrier panel. In 2017 ACOG [13] and in 2019 the Association of Israeli Medical Geneticists [20] recommended including all severe diseases with a threshold at or above 1 : 100 in a pan-ethnic screening.

Another requirement that should be considered when evaluating inclusion of a disease in a panel is whether it can be *diagnosed prenatally*. Importantly, it would be useful to discuss openly with the intended target population in order to understand what they perceive to be an important disease to avoid [34].

10.3.7 Should There Be a Standard Recommended ECS Panel?

Currently, there is no consensus on how many of the more than 1000 known autosomal recessive conditions should be included in a panel, but several professional societies such as ACOG, ACMG, ESHG, and SEF have provided general guidelines about the requirements that the panels should meet. These requirements consider the clinical features of conditions (severe diseases), classification of variants (only pathogenic and probably pathogenic variants), the carrier frequency in the general population, penetrance, and value for reproductive decision-making or management.

Nowadays there are considerable differences in panel composition between laboratories. For instance, the Genetics Department of the University Medical

Centre Groningen (UMCG) in the Netherlands developed and validated a population-based ECS test for a limited set of 50 severe early-onset autosomal recessive conditions for which no curative treatment is available. From 2016, general practitioners in a pilot study offered this couple-based ECS to women and their partners from the general population at no financial cost [35].

In a study comparing 16 commercially available ECS tests, panel size ranged from 41 to 1792 conditions, with only three screened by all panels [36]. A recent study [37] found that screening solely for conditions with carrier frequencies of 1 : 100, which equated to variants in just 40 genes, would identify 76–97% of carrier couples. In several of their publications, Lazarin and colleagues argue that the 1 : 100 carrier rate threshold, although currently accepted, does in fact limit the detection of at-risk couples. They proposed the concept of a "panel carrier rate," defined as the proportion of patients who were carriers of at least one condition on the panel, because it describes how frequently partner testing is needed in a sequential-screening workflow, and thus reflects the logistical and economic costs incurred to reap the clinical benefit of identifying at-risk couples [38]. Their data show that for a panel including 18 conditions with a 1 : 100 cutoff, 61% of carriers and 84% of at-risk couples are detected, while with a 1 : 500 cutoff, including 91 conditions, 80.5% of carriers and 95% of at-risk couples are detected. Of course, they recognize that both the panel carrier rate and at-risk couple rate tend to become saturated if more rare diseases are included in ECS panels, and saturation is reached around the 1 : 1000 cutoff.

In the light of this consideration, it is evident that the model of inclusion "the more the better" is not the right one; nevertheless several testing companies tend to market to an uninformed medical establishment and patients alike, using the number of genes in their panel as proxy for quality. If we include hundreds of diseases with low frequency, we will of course increase the panel carrier rate (the number of individuals identified as carriers) but we will not improve the main goal significantly, because almost all carriers of extremely rare diseases will not have a matching carrier in their partner, and the residual risk of the affected child would have been therefore very, very low even without testing the partner; nevertheless, we will have increased the couple's anxiety, and the costs of the process, and the unnecessary use of scarce counseling resources.

For all these reasons, a better strategy for the future definition of panels should not focus on carrier frequency per se, nor attempt to include the highest number possible of diseases in order to increase the detection of at-risk individuals. Rather, it should focus on the false negatives resulting from the currently used panel, by analyzing the instances where children with a severe disease are born to parents who underwent ECS, and evaluate their inclusion in current panels. Nevertheless, the ESHG remarked in 2016 that "the effectiveness of carrier screening programs should be measured by assessing the extent to which it optimizes informed choice and reproductive decision making and not by demonstrating how much it reduces the birth prevalence of affected children" [39].

In summary, a standard ECS should benefit all individuals, regardless of them knowing their own ethnicity, should include severe diseases with a well-defined phenotype and high penetrance, and be able to identify high-risk couples in the tested population. Ultimately, the application of the knowledge of the molecular basis of disease, how many at-risk couples we are failing to detect, and patients' values and priorities are critical points to consider when designing an ECS program, rather than the number of diseases included in the panel.

10.3.8 Should We Use ECS on Gamete Donor Candidates?

The probability that an individual who wishes to be a gamete donor will be a carrier of a chromosomal abnormality or a pathogenic variant in a certain gene is the same as that of the rest of the general population. Nevertheless, most scientific societies agree that gamete donors should be screened for some genetic diseases with recessive inheritance that are common in the population to which they belong and should also have a chromosomal study performed before the donation; indeed, being a carrier for certain frequent diseases such as CF may in fact be a reason for exclusion from donation in certain countries. In Europe, the legislation is along these lines, and the European directive states that "The use of reproductive cells other than for partner donation must meet the following criteria: genetic screening for autosomal recessive genes known to be prevalent, according to international scientific evidence in the donor's ethnic background."

We have already argued that ethnicity reported by the tested individual should not be considered a useful

Table 10.1. Genetic tests recommended in different countries and by several recommending bodies for gamete donors

Country/ recommendation body	Test recommended for gamete donors
ASRM/SEF/UK/Italy/ France	Karyotype
ASRM/SEF/UK/Italy/ ACOG	Cystic fibrosis
ACMG/SEF/ACOG	Spinal muscular atrophy
ASRM/SEF	Fragile X syndrome
ACOG/SEF/UK	Hemoglobinopathies
Italy	Glucose-6-phosphate dehydrogenase deficiency
SEF	Nonsyndromic hearing loss

ASRM, American Society for Reproductive Medicine.

factor in deciding who should be studied or for which diseases they should be screened, and that the various scientific societies do not agree on which genetic studies should be carried out. There is no consensus among international bodies as to which genetic tests to carry out for donor screening (Table 10.1).

The request for genetic studies in gamete donors is determined by the impact of their use because of their wide distribution. The number of children that can be born from the same donor's gametes will often exceed the number that an individual could naturally have throughout his or her fertile life, especially sperm donors. Therefore, if the donor is a carrier of a pathogenic variant related to a recessive disease with a high frequency of carriers, the probability that it will manifest itself in his or her offspring will be higher. On the other hand, any case of an affected child, however isolated it may be, creates social alarm among not only the media but also the scientific community (e.g. "Sperm donor suffers years later from inherited disease" [40]. "Should gamete donors be screened for spinal muscular atrophy?" [41]. "Is sperm donor karyotype analysis necessary?" [42]). These situations, apart from possible lawsuits, confront the professionals and the couples involved with a series of logistical problems and ethical questions that are not always easy to resolve. Will it be possible to contact all the couples who have used the donor? When and how should we communicate this to the couples involved? When and how should the couples communicate it to their children? Should we carry out the relevant genetic study in the child or should we wait until the child reaches an age that allows him or her to make his or her own decision? Would this age correspond to the "legal age of majority"?

The way to minimize the risks does not involve further restricting the number of donations but rather legally establishing a maximum number of children per family and seeking continuous improvement in the study and selection processes of candidates for donation to make the process ever safer. This way, all gamete donors must undergo a genetic study that includes an assessment of their personal and family history, a karyotype, and analysis of at least pathogenic variants related to CF, spinal muscular atrophy and, in oocyte donors, molecular study of fragile X syndrome. Regarding the study of haemoglobinopathies, we consider it essential to perform a complete blood count (CBC) analysis, as almost all thalassemia carriers show microcytic hypochromic parameters with an apparently normal hemoglobin level. At this point it should be acknowledged that adding the analysis of pathogenic variants in the *HBB*, *HBA1* and *HBA2* genes will provide 100% sensitivity and 100% specificity in the diagnosis of thalassemia carriers, but the combination of CBC and hemoglobin electrophoresis tests showed 99.55 (95% CI 97.51–99.92) sensitivity and 99.82 (95% CI 99.47–99.94) specificity [6]. Some guidelines have recommended adding molecular analysis of *GJB2* [15] to detect carriers for autosomal recessive deafness 1A (DFNB1A). However, it seems that the best strategy is based on the performance of an ECS in all recipients whose gametes will be combined with the donors'. In this way, the recipient will be assigned a donor with whom they do not share pathogenic variants in the same gene.

Despite this effort, couples using ESC to assess their reproductive risk before becoming parents will have to understand the limitations of any study method used (e.g. germline mosaicism, late-onset genetic diseases, polygenic and multifactorial diseases) and of human reproduction itself (de novo mutations). In any case, the risk of having an affected child through one of these mechanisms is, at most, similar to that of natural reproduction due to the selection process that donors undergo.

10.3.9 Should Late-Onset Disorders Be Included in ECS Panels?

So far, all the published documents [12,31] agree that reproductive carrier screening is not aimed at predicting or planning for late-onset conditions, but that the goal is to enable couples to prevent the birth of affected children.

However, some authors disagree [43] and think that an ECS panel could be used not only for reproductive planning but also for early diagnosis and interventions, to screen presymptomatically for mutations in genes causing late-onset diseases such as the *BRCA1* and *BRCA2* genes that increase the risk for breast and ovarian cancers. These authors argue that this attitude "responds to what our patients have told us they want" and consider the result medically relevant for the patient and emphasize that in the pretest genetic counseling the patient has the option to decide what information they want or do not want to receive (e.g. accidental findings, variants of unknown significance, presymptomatic studies).

On the other hand, in the context of a gamete donors' program, clinical geneticists are beginning to ask that genetic studies of donor candidates consider the possibility of including diseases that are sometimes difficult to diagnose, such as neurofibromatosis or tuberous sclerosis, or of late onset, such as myotonic dystrophy type 1. The reason behind this request relates mainly to the relatively high number of couples that may use gametes from the same donor, especially sperm, and the proportional implication that an adverse result will have. Any implementation requires justification in terms of proportionality in which the advantages outweigh the disadvantages. In fact, discovering that one is going to develop a fatal neurodegenerative condition while undertaking gamete donation is much more than what the prospective donor had bargained for, and such discoveries will alter both the donor's life and the life of their family. While the expansion of screening tests could be beneficial, it also raises new ethical, psychological and societal issues that must be assessed by multidisciplinary committees of experts before a decision is taken.

10.3.10 If Over Time the Screening Panel Is Updated, Should We Test an Individual Again?

Carrier screening should not be repeated in subsequent pregnancies unless the family history has changed. It is currently not recommended to retest the patient. The reason for this recommendation is that the mutations responsible for most cases of a disease are usually known, and those that are added usually represent a minority of cases, which makes a new panel not cost-effective. It is also possible that in more recent panels there are no mutations or diseases present in older ones; this is because as the number of healthy individuals tested increases, some mutations that were previously considered pathologic because they were found in sick patients eventually turn out not to be related to the disease, and are simply non-pathologic variants of a gene.

10.3.11 All Carrier Screening Tests Use the Same Technology: Where Are We Going?

Although a technical discussion of the molecular biology behind ECS is beyond the scope of this chapter, we would like to point out a recent shift in ECS technology that may bring new possibilities and new challenges in the next few years. Currently, the vast majority of commercialized ECS panels use next generation sequencing (NGS) as the main technique for identifying genetic variants, while specific techniques are used to study certain genes or diseases such as spinal muscular atrophy, fragile X syndrome, and Friedreich's ataxia. There are two main NGS strategies used: one is targeted sequencing, where only certain portions of the genes known to harbor the mutations are tested. These portions are those where known pathogenic variants are found, or where the gene is known to accumulate mutations. The second strategy entails sequencing whole target genes, not just part of them. The rationale behind this strategy is that some new or unknown mutations may emerge anywhere in the protein-coding portion of the gene. Full gene sequencing is often done in genes with a known high number of mutations (e.g. *CFTR* gene).

While each strategy has pros and cons, both reach sufficient sensitivity and specificity in detecting pathogenic variants and are interchangeable for most screening instances. By its own nature, NGS (either targeted or whole gene) will detect any variation in the DNA and therefore the residual risks will be very reduced; on the other hand, it will identify variants for which there is incomplete or contradictory evidence about their pathogenicity, the so-called "variant of unknown significance" (VUS or VOUS), which have an uncertain effect on gene function and thus an uncertain relationship to clinical phenotype.

Currently, most scientific societies recommend that VOUS should not be reported [12].

A rapidly approaching technical development in the field of ECS is the adoption of exome sequencing. An exome is a collection of the exonic sequences of protein-coding genes in our genome, and while currently no full exome is used for ECS, several companies are starting to commercialize ECS panels based on partial exomes, sequencing several thousands of genes. It is important to understand that the panel reported remains relatively limited, and that the exome sequences are captured mostly as a cheaper alternative to targeted sequencing. However, exome sequencing raises some new issues and opportunities, such as the possibility of returning later to explore other sequenced genes at no extra cost, for instance to reassess newly identified genes/mutations or to evaluate other genes for different health reasons such as cancer, cardiovascular risk, or pharmacogenomics. However, the use of exome sequencing also raises issues such as the possibility of identifying serious medical problems, for instance breast and ovarian cancer predisposition and other late-onset diseases, which are definitely not in the scope of ECS.

10.4 Conclusion

In this chapter, we have reviewed the basic concepts related to ECS, its history, implementation, complexity, and opportunities. Since the 1970s, the medical community has demonstrated the usefulness of carrier screening as a tool to prevent the birth of children affected by certain prevalent and serious genetic diseases, such as the hemoglobinopathies, TSD, or CF (primary prevention). Technological developments in the field of medical genetics have made it possible to know the carrier status of an individual with respect to hundreds of serious genetic diseases with a recessive or sex-linked inheritance pattern (ECS). The comparison of results between two members of a couple contemplating a pregnancy allows us to evaluate the risk that the offspring will have of developing one or more of the diseases included in the panels.

Every human is a carrier of pathogenic variants in genes related to severe recessive diseases. Using family history or ethnicity as a criterion to define who should or should not undergo ECS leads to an obvious underestimation of risk. At present, the use of ECS as a population-based screening test has yet to be proven beneficial and cost-effective.

Our patients should be informed about the existence of carrier screening tests, their objectives, limitations, and the type of data they will provide. They must also know what options are available if their offspring are at risk of suffering from a recessive or X-linked genetic disease. The information facilitates reproductive decision-making.

ECS should be offered prior to conceiving, with mandatory pre- and post-test genetic counseling. It is important for couples to understand that a zero risk does not exist, and that there will always be a residual reproductive risk.

There is general agreement that a basic panel should include disorders with a well-defined phenotype, with a detrimental effect on quality of life, that cause cognitive or physical impairment, that require surgical or medical intervention, and that have an onset early in life and which should be able to be diagnosed prenatally to optimize reproductive outcomes. However, currently there is no consensus about a standard panel.

While there is no international consensus, it seems reasonable that gamete donors be assessed for their personal and family history, karyotype, evaluation of hemoglobinopathies, and analysis of pathogenic variants related to CF, spinal muscular atrophy and, in oocyte donors, fragile X syndrome carriership. A molecular study of *GJB2* should be assessed based on the prevalence of nonsyndromic sensorineural deafness in the population.

References

1. Wilson JMG, Jungner G. Principles and Practice of Screening for Disease. Geneva: World Health Organization, 1968.
2. Angastiniotis M, Kyriakidou S, Hadjiminas M. Comment on a endigué la thalassémie à Chypre. *Forum Mondial de la Santé* 1986; 7(3):308–15.
3. Angastiniotis M, Modell B, Englezos P, Boulyjenkov V. Prevention and control of haemoglobinopathies. *Bull World Health Organ* 1995;73(3):375–86.
4. Kaback MM. Screening and prevention in Tay–Sachs disease: origin, update, and impact. *Adv Genet* 2001;44:253–65.
5. Sagi-Dain L, Weissman I, Cohen-Kfir N, et al. Genetic counseling of high-risk isolated populations: a worldwide challenge. *Birth Defects Res* 2020;112(4):316–20.
6. Noor FA, Sultana N, Bhuyan GS, et al. Nationwide carrier detection

and molecular characterization of β-thalassemia and hemoglobin E variants in Bangladeshi population. *Orphanet J Rare Dis* 2020;15(1):15.

7. Zlotogora J, Shalev SA. The consequences of consanguinity on the rates of malformations and major medical conditions at birth and in early childhood in inbred populations. *Am J Med Genet A* 2010;152A:2023–8.
8. Zlotogora J, Patrinos GP. The Israeli National Genetic database: a 10-year experience. *Hum Genomics* 2017;11(1):5.
9. Delatycki MB, Alkuraya F, Archibald A, et al. International perspectives on the implementation of reproductive carrier screening. *Prenat Diagn* 2020;40(3):301–10.
10. National Institutes for Health. Genetic testing for cystic fibrosis. National Institutes of Health Consensus Development Conference Statement on genetic testing for cystic fibrosis. *Arch Intern Med* 1999;159:1529–39.
11. Grody WW, Thompson BH, Gregg AR, et al. ACMG position statement on prenatal/preconception expanded carrier screening. *Genet Med* 2013;15(6):482–3.
12. Edwards JG, Feldman G, Goldberg J, et al. Expanded carrier screening in reproductive medicine: points to consider. A joint statement of the American College of Medical Genetics and Genomics, American College of Obstetricians and Gynecologists, National Society of Genetic Counselors, Perinatal Quality Foundation, and Society for Maternal-Fetal Medicine. *Obstet Gynecol* 2015;125:653–62.
13. American College of Obstetricians and Gynecologists. Committee opinion no. 690: carrier screening in the age of genomic medicine. *Obstet Gynecol* 2017;129(3): e35–e40.
14. Henneman L, Borry P, Chokoshvili D, et al. Responsible implementation of expanded carrier screening. *Eur J Hum Genet* 2017;25(11):1291.
15. Castilla JA, Abellán-García F, Alamá P, et al. *Cribado* Genético *en* Donación *de* Gametos. Madrid: Grupo de trabajo de Donación de Gametos y Embriones de la SEF, 2019.
16. Therrell BL, Padilla CD, Loeber JG, et al. Current status of newborn screening worldwide: 2015. *Semin Perinatol* 2015;39:171–87.
17. Williamson R. Universal community carrier screening for cystic fibrosis? *Nat Genet* 1993;3(3):195–201.
18. Alford RL, Arnos KS, Fox M, et al. American College of Medical Genetics and Genomics guideline for the clinical evaluation and etiologic diagnosis of hearing loss. *Genet Med* 2014;16:347–55.
19. Shraga R, Yarnall S, Elango S, et al. Evaluating genetic ancestry and self-reported ethnicity in the context of carrier screening. *BMC Genet* 2017;18:99.
20. Zlotogora J. The Israeli national population program of genetic carrier screening for reproductive purposes. How should it be continued? *Isr J Health Policy Res* 2019;8(1):73.
21. Haque IS, Lazarin GA, Kang HP, et al. Modeled fetal risk of genetic diseases identified by expanded carrier screening. *JAMA* 2016;316(7):734–42.
22. Rink B, Romero S, Biggio JR Jr, Saller DN Jr, Giardine R. Committee opinion No. 691. Carrier screening for genetic conditions. *Obstet Gynecol* 2017;129(3): e41–e55.
23. Capalbo A, Chokoshvili D, Dugoff L, et al. Should the reproductive risk of a couple aiming to conceive be tested in the contemporary clinical context? *Fertil Steril* 2019;111(2):229–38.
24. Van der Hout S, Holtkamp KC, Henneman L, de Wert G, Dondorp WJ. Advantages of expanded universal carrier screening: what is at stake? *Eur J Hum Genet* 2016;25(1):17–21.
25. Rowe CA, Wright CF. Expanded universal carrier screening and its implementation within a publicly funded healthcare service. *J Community Genet* 2020;11(1):21–38.
26. Bell CJ, Dinwiddie DL, Miller NA, et al. Carrier testing for severe childhood recessive diseases by next-generation sequencing. *Sci Transl Med* 2011;3:65ra64.
27. Bonte P, Pennings G, Sterckx S. Is there a moral obligation to conceive children under the best possible conditions? A preliminary framework for identifying the preconception responsibilities of potential parents. *BMC Med Ethics* 2014;15:5.
28. Ropers HH. On the future of genetic risk assessment. *J Community Genet* 2012;3:229–36.
29. Plantinga M, Birnie E, Abbott KM, et al. Population-based preconception carrier screening: how potential users from the general population view a test for 50 serious diseases. *Eur J Hum Genet* 2016;24:1417–23.
30. Human Genetics Commission. *Increasing Options, Informing Choice: A Report on Preconception Genetic Testing and Screening.* London: HGC, 2011.
31. Gregg AR, Edwards JG. Prenatal genetic carrier screening in the genomic age. *Semin Perinatol* 2018;42:303–6.
32. Zlotogora J, Meiner V. Ashkenazi carrier screening for reproductive planning: is this what we planned for? *Genet Med* 2016;18(5):529.
33. Prior TW. Spinal muscular atrophy: newborn and carrier

screening. *Obstet Gynecol Clin North Am* 2010;37:23–36.

34. Archibald AD, Smith MJ, Burgess T, et al. Reproductive genetic carrier screening for cystic fibrosis, fragile X syndrome, and spinal muscular atrophy in Australia: outcomes of 12,000 tests. *Genet Med* 2018;20 (11):1485.
35. Schuurmans J, Birnie E, Ranchor AV, et al. GP-provided couple-based expanded preconception carrier screening in the Dutch general population: who accepts the test-offer and why? *Eur J Hum Genet* 2020;28: 182–92.
36. Chokoshvili D, Vears D, Borry P. Expanded carrier screening for monogenic disorders: where are we now? *Prenat Diagn* 2018;38:59–66.
37. Guo MH, Gregg AR. Estimating yields of prenatal carrier screening and implications for design of expanded carrier screening panels. *Genet Med* 2019;21(9):1940–7.
38. Ben-Shachar R, Svenson A, Goldberg JD, Muzzey D. A data-driven evaluation of the size and content of expanded carrier screening panels. *Genet Med* 2019;21(9):1931–9.
39. Henneman L, Borry P, Chokoshvili D, et al. Responsible implementation of expanded carrier screening. *Eur J Hum Genet* 2016;24(6): e1–e12.
40. Gebhardt DOE. Sperm donor suffers years later from inherited disease. *J Med Ethics* 2002;28 (4):213; discussion 214.
41. Tizzano EF, Cuscó I, Barceló MJ, Parra J, Baiget M. Should gamete donors be tested for spinal muscular atrophy? *Fertil Steril* 2002;77(2):409–11.
42. Weissenberg R, Litmanovitz T, Dekel M, et al. Is sperm donor karyotype analysis necessary? *Reprod Biomed Online* 2007;14 (6):724–6.
43. Baskovich B, Hiraki S, Upadhyay K, et al. Expanded genetic screening panel for the Ashkenazi Jewish population. *Genet Med* 2016;18(5):522–8.

Chapter 11

Genetic Counseling in Assisted Reproductive Treatment

Christine de Die-Smulders and Ron van Golde

11.1 Introduction

In the past decade the development of new technologies and innovations have revolutionized genetic testing. The ability to diagnose and prevent genetic disorders before an existing pregnancy and to detect embryonic and fetal genetic errors has dramatically increased, and has substantially changed the care for couples wanting a child.

In general, professionals and the public have become more aware of genetics and the possibilities of screening, action options, and treatments. Parents confronted with a high risk of genetic disease for their children have several new possibilities for overcoming the birth of affected offspring, such as invasive or noninvasive prenatal diagnosis (PND) or preimplantation genetic testing (PGT).

These advances are changing clinical practice and pose novel challenges for genetic and reproductive counseling [1]. The future is challenging both for the technical innovations and for the ethical compass while exploring new techniques. The aim of this chapter is to provide an overview of genetic counseling and how it is relevant to reproductive medicine practice and assisted reproductive treatment (ART).

11.2 General Principles of Genetic Counseling

Genetic counseling is traditionally defined as the process by which patients or relatives at risk of developing a potential hereditary disorder are advised on the consequences of the disorder, the probability of developing or transmitting it, and the ways in which the risk may be prevented, avoided, or attenuated. It is also defined as an educational process that seeks to assist affected and/or at-risk individuals to better understand the nature of a genetic disorder, its transmission, and the options open to parents in management and family planning.

History taking, application of tests, estimation and explanation of the risk for the patient or client and for offspring, and a discussion of the possibilities for preventing or treating the disease in the proband and offspring and of the consequences for family members are still standard elements in the process of genetic counseling.

The tremendous advances in molecular genetics have created great opportunities for diagnosing patients with rare diseases, who formerly went undiagnosed. Nowadays, a known or yet unknown clinical diagnosis is usually confirmed respectively elucidated by molecular testing. In general, these advances are perceived as very positive. However, the introduction of next generation sequencing (NGS) technology has also led to specific issues. The consequences of these technical improvements may be difficult to interpret and may pose professionals difficult dilemmas, such as the detection of unsolicited findings or of so-called variants of uncertain significance (VUS). It is almost unnecessary to mention that explaining the relevance of such unsolicited or indistinct findings to the patient poses a heavy burden on the treating physician and has deeply influenced the process of genetic counseling in recent years.

Usually genetic counseling is given by a genetic specialist (e.g. clinical geneticists or genetic counselors). Also, other (medical) specialists may be involved. In most clinics a multidisciplinary team is available for additional support. In particular, psychosocial support can be necessary, for example by helping clients to deal with their increased risk, supporting them when dilemmas arise in their wish for a child and their reproductive choices, and providing practical and emotional help with the genetic investigation of family members. The conversation with the genetic counseling specialist can be supported by the supply of specific online information or online decision aids [2].

Diverse factors, such as social and religious values, psychological and cultural circumstances, and prior experiences of the clients, have to be taken into account in every conversation. The application of genetic counseling should be specific for the context of the current culture of a society. In every setting, leading principles in counseling include respect for autonomy, privacy, and confidentiality.

Historically, nondirective counseling was mostly strived for in clinical genetics. In nondirective counseling, the counselor does not direct the decision-making process, but provides all the information needed for making a well-informed decision, with no pressure or guidance to choose one option over another. However, nondirective counseling is not always possible or desirable. An alternative framework is shared decision-making, defined as "an approach where clinicians and patients share the best available evidence when faced with the task of making decisions, and where patients are supported to consider options and to achieve informed preferences" [3]. Essential elements of shared decision-making include defining the decision to be made, conveying that the patient's opinion is important, presenting options and their pros and cons, discussing patient's values and preferences, supporting the patient in deliberation, making or explicitly deferring the decision, and arranging follow-up. Genetic counseling provides a natural context for shared decision-making because there are often multiple options for patients to choose. These decisions are complex with pros and cons, but with no clear choice that is clinically the best for all patients in all situations. Further, genetics-related decisions are frequently very much value-based and preference-driven. In the context of a pregnancy, such decisions translate into actions that matter most to future parents [3].

The success of treatment is measured by outcome. Despite advances, assessment of the effectiveness of genetic counseling is complicated by a lack of widespread agreement about the most appropriate outcome measures. Broadly speaking, genetic counseling can lead to increased knowledge, improvement of information recall, perceived personal control, positive health behaviors and improved accuracy in risk perception, as well as decreases in anxiety and worry and in decisional conflict. It improves psychological well-being, and is generally well regarded by patients [4].

The relatively new concepts of designing "healthcare pathways" and "patient journeys," which comprises active involvement of patients by listening to their experience and needs and their goal to improve care, may further add to desired improvements in genetic counseling in keeping with the spirit of the times [5].

11.3 Reproductive Options and Decision-Making

For couples of childbearing age who are confronted with an increased recurrence risk of a genetic disorder to their offspring, providing information on the reproductive options is essential. Fulfilling their wish for a child may feel pressing and urgent. The process of reproductive decision-making is generally perceived as difficult. It is therefore predictable that reproductive counseling sets special demands for doctor–patient communication.

Generally, the main options for consideration will include:

- Accepting the risk and taking the chance of a good ending with a pregnancy
- Refrain from (genetically related) children
- Adoption
- Gamete (sperm or egg) donation
- Prenatal diagnosis
- Preimplantation genetic testing.

For couples who prefer genetically related offspring, and want to prevent transmission of the familial genetic disorder, PND and PGT are the only available options.

11.3.1 Prenatal Diagnosis

PND includes the ultrasound anomaly scan as well as invasive testing (chorionic villus biopsy or amniocentesis) or noninvasive prenatal testing (NIPT) or noninvasive prenatal diagnosis (NIPD).

An ultrasound anomaly scan to detect fetal structural anomalies can be performed from 13 weeks on; the ideal time is between 18 and 20 weeks. Generally, it will not be possible to make a certain (etiologic) diagnosis using ultrasound. However, an abnormal ultrasound can be an indication for further genetic testing.

11.3.1.1 Invasive Prenatal Testing

Chorionic villus biopsy is possible from 11 to 14 weeks via the transcervical or transabdominal route. Amniocentesis is possible from 15 weeks on. Both are invasive procedures, probably carrying a small risk of miscarriage. Women pregnant after ART may feel that this miscarriage risk poses a serious obstruction to undergoing the prenatal test. If the couple faces a high risk of a serious genetic disorder in their child, and future parents are convinced that they want to prevent this, the decision to undergo the test may be relatively simple. However, if the risk for offspring is relatively low, for example a recurrence risk of a de novo mutation or a prenatal test meant to confirm the result of PGT, the risk of the PND procedure may be in the same range or even higher than the recurrence risk, complicating the decision to undergo the prenatal test. The familial molecular defect or structural chromosome anomaly is analyzed in the chorionic villi or amniotic cells. If the couple receives an adverse result, a decision has to be made whether to continue the pregnancy. It has to be emphasized that the aim of prenatal testing is to give the couple the opportunity to make their own informed reproductive decision, matching their personal and moral values. If they decide to continue the pregnancy, they can prepare for the birth of an affected child. If the couple chooses to terminate the pregnancy, a medical procedure burdening the woman is necessary. Although termination of pregnancy prevents a living affected child, the couple may go through a mourning process because of the loss of a highly desired son or daughter. This may be even more poignant for couples who have become pregnant by in vitro fertilization (IVF)/intracytoplasmic sperm injection (ICSI) after a period of infertility. These pregnancies may be perceived as more precious and the decision to terminate a highly desirable pregnancy will probably be more difficult if the couple knows that establishing a subsequent pregnancy may not be simple.

The variability of the phenotype or the late stage of onset of the condition may possibly add a complex dimension to the decision about the reproductive choice in the prenatal trajectory. Neurofibromatosis type 1 is one such condition where genotype confers no correlation with severity. A prenatal result will not confer information about the severity in the child. If a mild phenotype exists in the family or in the patient himself or herself, the couple may feel less able to terminate the pregnancy.

If the couple decides to continue the pregnancy and the particular disorder concerns a so-called late-onset disorder, future parents must realize that they deprive their child of an autonomous decision at adult age to opt for predictive DNA testing. The hope for a cure may be a prevalent issue as it concerns termination of pregnancy when a resultant child would not be affected until the fourth decade of life, such as in Huntington's disease.

11.3.1.2 Noninvasive Prenatal Testing

The identification of cell-free fetal DNA in the maternal blood paved the way for NIPT of the fetus. The test only requires a one-time blood collection, is risk-free, and not stressful for the pregnant women. From 9 weeks on the fetal sex and the rhesus D genotype can be established. Numerous mainly commercial tests are available for noninvasive testing of aneuploidy and a selected group of other chromosomal anomalies, such as microdeletions of unbalanced structural anomalies. Positive predictive value (PPV) varies depending on the test result. PPV is high for trisomy 21 and 18, but fairly low for several other trisomies such as trisomy 7 or 16. The latter are in most cases restricted to the placenta. Abnormal NIPT results must always be confirmed by an invasive test, before the meaning and the policy can definitely be determined [6].

There is an increasing need for universal approaches for NIPD for monogenic diseases. Recently a cost-effective, generic, cell-free fetal DNA haplotyping approach to screen the fetal genome for the presence of inherited monogenic disease was published [7]. However, this method has currently not yet been implemented on a larger scale.

11.3.2 Preimplantation Genetic Testing

Recently, the terms "preimplantation genetic diagnosis" (PGD) and "preimplantation genetic screening" have been replaced by the term "preimplantation genetic testing" (PGT) in order to ensure global consistency in terminology. In general, PGT comprises IVF followed by genetic testing of the embryos for the presence of a familial mutation and/or for a numerical or structural chromosomal abnormality.

PGT includes PGT for monogenic disorders (PGT-M), PGT for structural rearrangements (PGT-SR), and PGT for aneuploidy (PGT-A). PGD was introduced in the early 1990s and was developed as an alternative to prenatal testing. By performing a diagnosis before pregnancy a couple could avoid the difficult decision to terminate pregnancy. The indications at the start of PGD were mainly severe early childhood or lethal disorders. In recent years indications now include disorders with adult onset of the diseases and disorders with reduced penetrance, such as hereditary breast and ovarian cancer. Readers are referred to Chapter 13 for information on the technical application of PGT.

Couples with a high risk of affected offspring may consider PGT as the method of first choice. In most cases this choice is made to avoid termination of pregnancy or repeated loss of a fetus with an unbalanced karyotype due to a parental translocation or another structural rearrangement.

The decision whether to accept a couple for PGT is usually based on several criteria and taken by a multidisciplinary team of genetic and fertility specialists, possibly supplemented with other professionals. The general inclusion criterion for PGT-M and PGT-SR is the high risk of a serious disorder in the offspring. National legislation and local regulations may also apply. General principles for PGT are laid down in several ESHRE guidelines and task force reports [8].

A precondition for PGT is a confirmed clinical and molecular diagnosis in the proband and/or family members. Also, the couple must meet the conditions for an IVF treatment, with a reasonable chance of retrieving sufficient eggs and sperm. Safety issues regarding the IVF treatment itself and a future pregnancy and the well-being of the expected child will also be taken into account. Strong cooperation between the genetic department and the fertility clinic to make balanced choices is warranted in order to guarantee optimal selection and guidance of couples.

General issues to be discussed with the couple include their motivation for PGT, their chance of an ongoing pregnancy after PGT taking into account maternal age, and possible risks during pregnancy especially when the genetic condition influences maternal health. Furthermore, timelines, costs and their alternatives to healthy offspring should be discussed. The practical and psychological impact needs to be considered, especially in couples already responsible for the care of an affected child. The counseling is preferably aimed at enabling the couple to make an informed reproductive choice.

In cases where the genetic condition influences the general health of one of the parents, the impact on the future well-being of the child should be evaluated. In cases where there are serious concerns about the safety and well-being of the child, a multidisciplinary team should discuss if treatment is going to be offered.

If the couple makes a definite decision to embark on PGT, extensive discussion of treatment-related issues will follow. All information should be given by qualified persons in understandable language tailored to the educational level of the couple. Information leaflets and specific web-based information can support the informational process. The patients will sign a written informed consent to confirm that they agree to the procedure.

11.3.2.1 Specific Information Related to the Genetic Analysis

The principle of the test should be explained and that, depending on the indication, blood or DNA samples and genetic reports from the couple and relevant family members are required. The expected time frame for the laboratory work-up is discussed. Depending on the indication and mode of inheritance, the number and type of transferable embryos will differ. The couple should be aware of the possibility that embryos may show an inconclusive result and of having no embryos for transfer if they are all genetically unsuitable. The possibility of a misdiagnosis and the possible explanations and level of error rates need to be explained as well as the recommendation for prenatal diagnosis for confirmation of the PGT result. The local policy regarding the follow-up of pregnancies and children born from PGT should be mentioned.

11.3.2.2 Specific Information Related to the Mode of Transmission

Autosomal Dominant Disorders

Autosomal dominant disorders are among the most frequent indications for PGT [9]. Assuming that one of the future parents is affected, the risk for offspring will be 50%. Reduced penetrance and a variable expression are characteristic for many autosomal dominant disorders. The variable complaints and uncertainty about the severity in their future child are difficult to deal with for most couples. On the one hand, the fear for severely affected offspring and the fact that they will burden the child with the same problem as they face may be the reason to ask for PGT. On the other hand, PGT and selection against affected embryos may feel a denial of their own life. It is not uncommon that, if PGT does not result in a pregnancy, the couple accepts the risk and attempts natural conception.

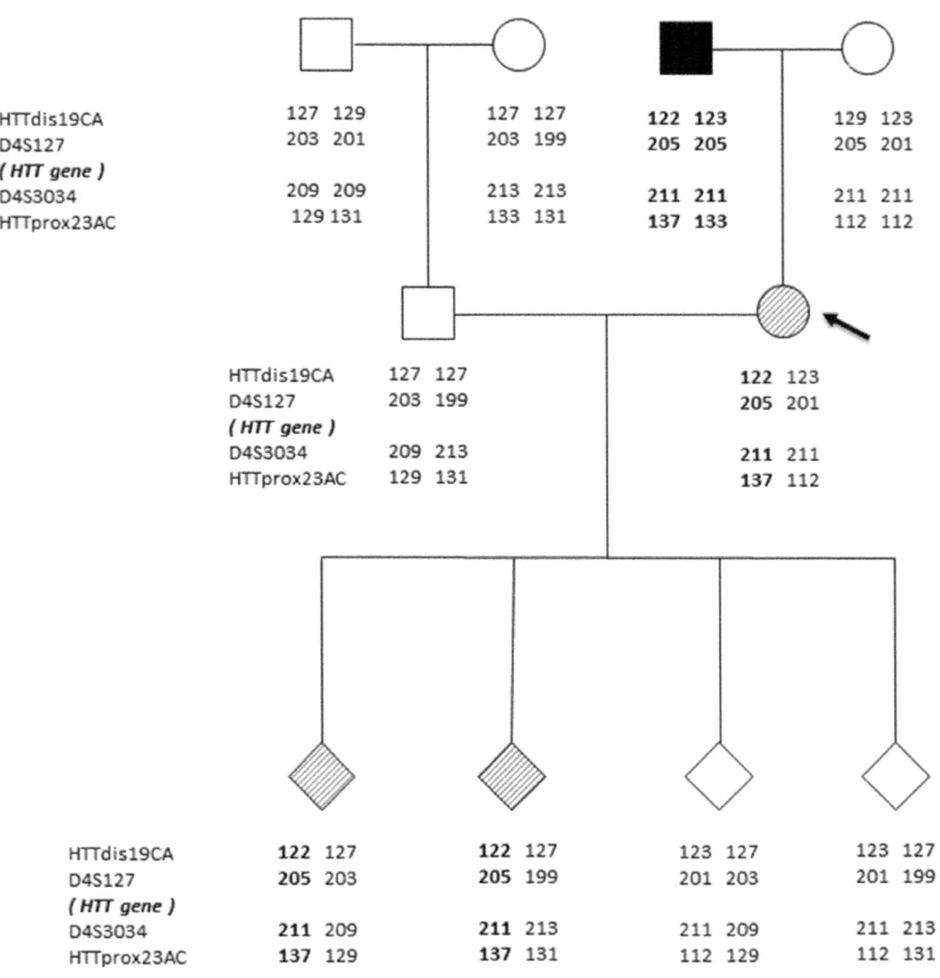

Figure 11.1 Pretest work-up for exclusion testing in Huntington's disease. The numbers indicate the alleles for the polymorphic markers around the *HTT* gene and are listed in the pedigree. The at-risk haplotypes of the affected father are in bold. The at-risk person does not want to know her carrier status. The left chromosome of the grandfather is transmitted to the at-risk patient as an example, but no information is given on the genotype of the *HTT* gene. The four possible haplotypes in the embryo are depicted in the lower row. Embryos with any of the two grandpaternal haplotypes are at 50% risk for Huntington's disease, like the mother, and will be discarded. Only embryos with the grandmaternal haplotype will be transferred.

For dominant disorders caused by repeat expansions, such as Huntington's disease and myotonic dystrophy, the couple should be informed that the length of the repeat will also be tested in the healthy partner in the pretest work-up. Although the a priori risk of a repeat expansion in the pathogenic range is low, this risk is certainly not zero.

An alternative approach for individuals at risk for an autosomal dominant disorder, but who do not want to know their carrier status, is so-called exclusion testing [10]. Classically, exclusion testing was applied in Huntington's disease, but nowadays it can also be an option in other severe late-onset neurodegenerative disorders, such as hereditary forms of early dementia. Exclusion testing is based on identifying the grandparental origin of the two alleles and only embryos with the allele from the nonaffected grandparent are considered transferable (Figure 11.1). The reason to apply exclusion testing is that the person at risk fears the burden of knowing his or her carrier status, while he or she wants to prevent the disease in offspring. Disadvantages are that statistically half of the treatments are in fact useless as the person at risk is not affected, but this is unknown to the patients and the team. Another drawback is that DNA of the affected parent of the person at risk should be available. Finally, an argument against this practice is that in one case out of two, the at-risk parent will develop Huntington's disease. However, this is still more true with PGT for Huntingtons disease carriers, where the risk of developing symptoms is almost 100%. Nevertheless, experience has taught that exclusion testing certainly meets a need.

Autosomal Recessive Disorders

In autosomal recessive disorders homozygous or compound heterozygote embryos for the mutation (s) involved are not transferable. As the carriers have a normal phenotype, there is no medical need to select against carriers. Also, the risk for a carrier having affected offspring is generally low, given the relatively low population frequency of most autosomal recessive disorders. However, couples may ask for preferential transfer of noncarriers. Whether this so-called "ranking" can be honored will depend on the specific situation of the couple and on local regulations and rules. If the couple completely rejects the transfer of carrier embryos, 75% of the embryos will not be transferable and the chance of pregnancy will decline notably.

With the recent introduction of whole-exome sequencing in clinical genetic practice, many previously undiagnosed cases can now be elucidated and novel diagnoses made. Among them are a couple of rare autosomal recessive disorders. Consanguineous couples are at particular risk for offspring with such rare autosomal recessive disorders. Until recently these couples were usually identified as carriers after the birth of an affected child, and could then opt for PGT or prenatal testing. However, a recently introduced whole-exome sequencing (WES)-based preconception carrier test (PCT) provides the opportunity for genome-wide carrier screening for recessive mutations and thus of identifying the future parents as carriers before conception, so before the birth of affected children (see Chapter 10 on PCT). Subsequently, PGT can be requested. A practical point to consider is that in such situations a reference (i.e. an affected child) for the development of a haplotyping-based PGT test is lacking, which may complicate the pretest work-up. Another problem complicating the pretest work-up PGT for consanguineous couples, particularly highly consanguineous couples, is the absence of informative short tandem repeat (STR) markers of haplotypes, inherent to the fact that the couple shares common ancestors. In most cases the unaffected alleles will be different, and may be used for discriminating the various haplotypes.

The PCT test can also be offered to known (consanguineous) carriers of rare autosomal recessive diseases to test them for carriership of additional autosomal recessive mutations. If a second autosomal recessive disorder is identified, it can be established if PGT for the two indications is technically possible and a desirable approach for both the couple and their caregivers.

X-Linked Disorders

The first applications of PGT, at the beginning of the 1990s, concerned the exclusion of an X-linked disease through polymerase chain reaction (PCR) or later fluorescent in situ hybridization (FISH)-based sexing of embryos [11]. Female carriers of an X-linked disorder could prevent the birth of an affected boy by choosing for a daughter as all male embryos, affected or unaffected, were discarded. Nowadays, specific mutation detection is the standard, and male embryos with the mutation will not be transferred. The transfer policy of female embryos with the mutation (i.e. female carriers) has to be discussed. In the decision to transfer a female carrier embryo, two factors may play a role: the expected health of the female carrier and her reproductive risk.

Most X-linked disorders are denominated as X-linked recessive. Examples include Duchenne muscular dystrophy and hemophilia A/B. Most female carriers are asymptomatic. However, some may have complaints related to the mutation, such as dilated cardiomyopathy or mild muscle weakness in carriers of Duchenne or Becker muscular dystrophy, or bleeding tendency in hemophilia carriers. These complaints are usually much milder than in affected males, but these aspects could be important for carrier women undergoing IVF and during their pregnancy.

In some other X-linked disorders, such as Alport syndrome or adrenoleukodystrophy, complaints in female carriers are more frequent and more serious. Kidney insufficiency in female carriers of Alport syndrome is not rare. This is the reason why Alport syndrome may be designated an X-linked dominant disorder rather than X-linked recessive.

The relatively frequent fragile X mental retardation syndrome is also X-linked dominant. About half of the females with a full mutation are severely affected equal to males with the mutation.

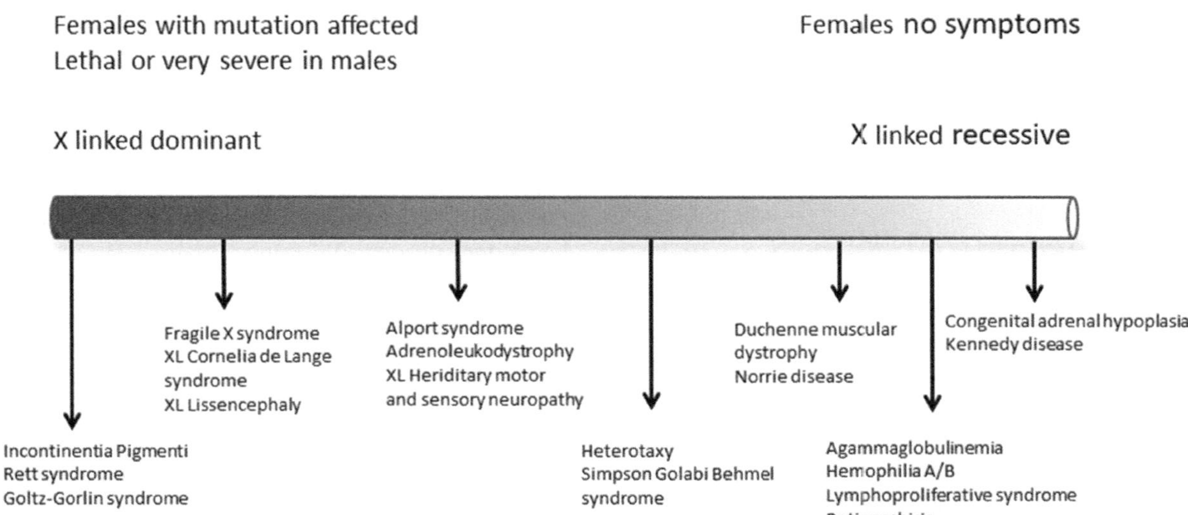

Figure 11.2 The health of female carriers in X-linked disorders: a sliding scale from dominant to recessive. Most severe phenotypes are on the left, milder phenotypes on the right. Examples of X-linked disorders are under the bar. The position on the bar is not exact but approximate.

In fact, making a distinction between an X-linked recessive and dominant condition is difficult in practice and it is probably better to start from the concept of a sliding scale (Figure 11.2) for counseling purposes.

What is wise regarding transfer policy in PGT for X-linked disorders strongly depends on the indication. Counseling should be adapted accordingly and the personal experience of the female carrier and her family members may also be taken into account. Of course, transfer policy should also be in accordance with national and local regulations.

For X-linked recessive diseases or situations where specific mutation detection is not possible, the pros and cons of embryo sexing should be discussed, since all male embryos, affected or unaffected, will be discarded and carrier females cannot be distinguished from unaffected female embryos. Of course, this strategy implies that girls born after PGT with sexing face a 50% risk of being a carrier and the same reproductive risk as their carrier mothers; this also applies to carrier girls born after mutation-focused PGT.

A special application of PGT is the male patient suffering from an X-linked disease who prefers male progeny only and asks for sex selection to avoid the birth of a daughter, who would thus be an obligate carrier. His daughters will usually be healthy (or mildly affected) carriers. In this situation the procedure would thus not be done to prevent severe disease in the offspring, and PGT for such "transgenerational risk" is still a subject of discussion in many countries.

Mitochondrial Disorders

Mitochondrial diseases are, in about 15% of cases, caused by exclusively maternally inherited mitochondrial DNA (mtDNA) mutations. The phenotype is variable but may be severe, with involvement of many organs, an unpredictable course, and lack of effective treatment. Recurrence risk for offspring of carrier females may be high, and therefore couples wish to prevent transmission to their offspring. Prenatal diagnosis is problematic for several reasons, and concerns the often poor correlation between mutation percentages and disease severity and the uncertainties about the representativeness of a fetal sample. PGT may circumvent these problems by transferring an embryo below the threshold of clinical expression [12].

MELAS (mitochondrial encephalopathy, lactic acidosis and stroke-like episodes) syndrome is probably the most frequently observed mitochondrial disorder in PGT practice. Females with the m.3243A>G mutation may be asymptomatic or may show a variety of signs and symptoms, varying from deafness only to life-threatening cerebral and other complications, which may in itself be a contraindication for IVF/PGT treatment. Risk of recurrence for the child is higher for the future mother with a high mutation load. The threshold for transfer after PGT has to be set before the PGT treatment. For the m.3243A>G mutation a threshold of 15% has been suggested, but will be different for other mitochondrial diseases. It has been shown that PGT provided women with MELAS syndrome the opportunity to conceive healthy offspring [12]. See also Chapter 12.

Structural Chromosomal Abnormalities

It is estimated that 1 in 625 individuals carries a balanced chromosomal rearrangement, mostly a reciprocal or Robertsonian translocation. The risk for offspring varies from very low to high (incidentally as high as 50%) and depends on the segments involved and the size of these segments. Carriers are at risk for repeated miscarriage, male and female infertility, or unbalanced offspring. PGT can be offered to avoid (another) miscarriage or pregnancy termination. Different technologies can be applied to diagnose chromosomal imbalance in the embryos. Initially FISH was the method of choice, but has now been mostly replaced by more comprehensive genome-wide methods.

Collated data from the ESHRE PGD consortium [9] show that reciprocal translocations are the most frequent indication in this group, followed by Robertsonian translocations in males and lastly in females. It has to be realized that a high number of embryos will not be suitable for transfer, in particular in the reciprocal translocation group. The relatively low clinical pregnancy rate reflects the low proportion of chromosomally normal embryos available for transfer.

For carriers of balanced complex chromosomal rearrangements involving at least three chromosome breakpoints on two or more chromosomes, the probability of conceiving a normal balanced pregnancy is even lower and PGT may be the only option enabling them to have genetically related offspring, although the chance of pregnancy may also not be high. For structural chromosomal rearrangements, it is important to discuss that the applied technology may not allow discrimination between normal and balanced results (see Chapters 3 and 7).

As for other PGT indications prenatal testing is recommended in pregnancy, in particular when one of the parents carries a translocation with small segments involved and a high a priori risk of an ongoing pregnancy of an unbalanced conceptus. These small unbalanced segments may also be more easily missed in PGT.

If the child is healthy, it is recommended to inform and test the child born after PGT for carriership of the balanced rearrangement at a later age.

11.3.2.3 PGT with HLA Typing

PGT for HLA typing aims to establish a pregnancy that is HLA-compatible with a sibling who requires hematopoietic stem-cell transplantation. HLA typing alone can be applied when the affected child has a nongenetic disorder such as leukemia. Most cases concern genetic disorders in the affected sibling, such as one of the hemoglobinopathies; β-thalassemia is the most frequent indication [9]. The first cases of HLA typing, reported in 2001, were controversial, but it is now a well-established method. If a couple applies for PGT with HLA typing, the chance of finding a transferable embryo should be discussed. This chance will of course be lower than with PGT for the genetic disorder only, as just one of four embryos will have the same HLA type as the affected sibling. This results, for example, in a probability of 18.8% (3 of 16) of identifying a transferable embryo when PGT is concurrently administered for an autosomal recessive or X-linked recessive disorder.

Some parents will ask for ranking, with transfer of nonaffected HLA-suitable embryos first, followed by transfer of non-HLA-matched embryos. The policy regarding the transfer of unaffected non-HLA-matched embryos should be agreed upon, according to local and national regulations, before the start of treatment.

In a recent multicenter cohort study [13], the clinical utility of PGT-HLA was evaluated. Pregnancy rate per embryo transfer was 41.3% whereas pregnancy rate per started cycle was only 23.3%. The study also underlines the need for close collaboration between specialists of the genetic, IVF, and PGT departments and the transplant units involved in the procedure. The couple should be aware that the procedure is highly complex and time-consuming. The couple should also be counseled on issues related to hematopoietic stem cell transplantation, such as potential stem cell source, timing, expected success rate, and other possible treatment options for their affected child.

11.4 Assisted Reproductive Techniques

The growth in our knowledge of the genetic origin of disease, accelerated by revolutionary advances in genomic techniques, and the continually improving technology of ART has transformed medical options. Since the introduction of IVF in 1978, many patients can now be treated effectively for their infertility. In particular, since the introduction of ICSI more couples have been able to achieve a pregnancy even in the case of severely impaired semen quality. Of course, patients using assisted reproductive methods such as IVF and ICSI should have a clear indication like tubal blockage or severe male factor subfertility [14]. In this section we describe how these improvements have opened up opportunities in patient care and how this knowledge can be implemented.

Before starting a fertility treatment, it is important to collect information on possible genetic risks in the couple and their desired offspring. In the distant past, infertile couples were in fact not at risk of transmitting a genetic disease, because they did not reproduce. With the introduction of ART couples could have children and could subsequently pass a possible genetic disease to their offspring. Especially in men with severely impaired semen quality there was a new paradox of transmitting the genetic disease or even transmitting their infertility. Therefore, during history taking one should focus on abnormalities related to a genetic condition such as congenital anomalies, mental retardation in the couple themselves or their families, fertility problems (especially premature ovarian insufficiency and male subfertility), and recurrent miscarriage in the couple or their family members.

Furthermore, the presence of malignancies occurring at a young age, thrombosis and cardiovascular disease are as important for the identification of genetic factors as for risks related to the IVF treatment itself. For instance, in women with an increased risk of thrombosis, ovarian hyperstimulation syndrome causes a hyperestrogenic state that increases the risk of a thromboembolic event. Patients with an increased (genetic) risk of thromboembolic processes should be counseled by a specialist in order to formulate a plan to prevent such an event.

Women with an increased risk of complications should be counseled on preventive measures and, if possible, treatment during pregnancy, delivery and after childbirth. These preventive measures are even more important if the treated patient already has symptoms due to her genetic disease. Women with an autosomal dominant or X-linked disorder may undergo ART to perform genetic selection for PGT. For each patient it should be clear before starting ART how the treatment and pregnancy can influence the symptoms of the genetic disease and vice versa. This also accounts for patients undergoing oocyte donation treatment. In particular, patients with Turner syndrome are at increased risk of an aortic dissection and rupture with a risk of dying during pregnancy.

In general, the risks of ART due to ovarian stimulation or oocyte retrieval are limited. Ovarian hyperstimulation increases circulating estrogen levels and could thereby induce or accelerate estrogen-sensitive tumors of endometrium or breast. The hyperestrogenic state can also affect coagulation as stated above.

During ovum pick-up, patients should have adequate anesthesia. The risks of ovum pick-up and the use of certain medications for pain relief should be evaluated, especially for particular genetic conditions such as myotonic dystrophy type 1, mitochondrial diseases, and cardiogenetic diseases.

11.5 The Subfertile Couple

Patients facing a fertility problem and therefore probably requiring fertility treatment could also be at increased risk of a genetic trait. Firstly, a genetic origin of their fertility problem could be present and the genetic disease may potentially also bear an increased risk for their offspring. In addition, a genetic condition in the family or physical examination of the partners may lead to the diagnosis of a genetic disorder in the couple. Subsequently, patients and their prospective children can be at increased risk for several problems complicating IVF treatment and pregnancy. These include:

- Genetic abnormalities leading to decreased ovarian reserve. This can, for example, be the case in women carrying a fragile X premutation or having (mosaic) Turner syndrome.
- Genetic causes leading to male infertility such as cystic fibrosis, leading to an obstructive azoospermia due to congenital (bilateral) absence of the vas deferens.
- A chromosome rearrangement leading to impaired spermatogenesis, or deletions on the Y chromosome leading to azoospermia.
- Chromosome abnormalities such as reciprocal or Robertsonian translocations leading to infertility or recurrent miscarriage.
- An increased risk of a genetic disorder in the offspring.
- Diverse gene panels for female and male infertility, if proven to be of clinical impact. When

a genetic abnormality is found, patients should be counseled on the genetic disorder and its consequences. If there is a risk for the patient and possible family members facing premature ovarian insufficiency, they can be counseled on timely fertility preservation.

11.5.1 Male Subfertility Genetics and ART

Male subfertility is mostly caused by a diminished number of spermatozoa and a decrease in motility and is often accompanied by a higher number of morphologically abnormal spermatozoa. The possibility of treating severe male subfertility by ICSI has revolutionized ART. However, the introduction of ICSI has raised concerns about the health of babies born after such treatment, with theoretical risks of transcending natural selection and making infertility paradoxically a hereditary disease, accompanied by possible other mental or physical disorders. Discussion about the safety of ICSI for offspring is still ongoing.

It is known that in male subfertility the incidence of genetic abnormalities increases when more semen parameters are impaired. The number of chromosomal abnormalities in severe oligoasthenozoospermia is 10–15% and is even higher in azoospermia [15]. More details are described in Chapter 8.

If a chromosomal abnormality is detected as the cause for the male subfertility, PGT can be offered in addition to ICSI. The goal of PGT is to shorten time to pregnancy and prevent a miscarriage or a newborn with an unbalanced chromosomal abnormality [16].

Another genetic cause of male subfertility is a deletion in the AZF region of the Y chromosome. This is important to recognize, as (1) men can be informed about the origin of their infertility, (2) using sperm from a man with an *AZFc* deletion will probably lead to male subfertility in sons born after ICSI, and (3) spermatogenesis is absent in males with an *AZFa* or *AZFb* deletion. So in these men it will not be possible to collect sperm from the testicles using testicular sperm extraction.

In addition, sperm parameters can be impaired in men with a monogenic disorder such as myotonic dystrophy [17].

11.5.2 Female Subfertility Genetics and ART

About 10% of women of reproductive age are unable to conceive or carry a pregnancy to term. Female factors alone account for at least 35% of all infertility cases and comprise a wide range of causes affecting ovarian development, maturation of oocytes, and fertilization competence, as well as the potential of a fertilized egg for preimplantation development, implantation, and fetal growth. Genetic abnormalities leading to infertility in females mainly comprise abnormalities of the X chromosome, and rare DNA sequence variations in the genes that control numerous biological processes implicated in oogenesis, maintenance of ovarian reserve, hormonal signaling, and anatomic and functional development of the female reproductive organs. Despite the high number of genes implicated in reproductive physiology, known from the study of animal models, only a subset of these genes is associated with human infertility. Most genetic causes, such as a fragile X (pre)mutation or galactosemia, will lead to premature ovarian insufficiency, and these women can choose oocyte donation. In case patients decide to request female relatives as their oocyte donor, one should be aware of the same genetic risk in these women. If there is adequate ovarian reserve, the genetic origin of female infertility and the possible increased risks for their offspring could, if desired, be prevented using PGT. Chapter 9 describes the genetics of female subfertility in more detail.

11.6 The Subfertile Couple with a Genetic Disease, ART, and Pregnancy

Before starting IVF treatment with or without PGT the couple should be advised on their chances of pregnancy, treatment-related risks, and the possible risks during pregnancy (Table 11.1). We describe here the most common diseases encountered in IVF centers. In general, counseling of the couple with an increased genetic risk of an affected offspring starts with a detailed discussion of the pros and cons of adding PGT to a regular IVF treatment. The couple should be informed that by adding PGT the chance of pregnancy is generally lower than in IVF without selection of embryos. The couple may also opt for IVF followed by prenatal testing and termination of pregnancy to prevent the birth of an affected child. Others will express that their wish to have offspring is even more important than to have genetically healthy offspring and thus accept the genetic risk for their future child. Furthermore, policy should be defined regarding preventive measures or treatments during

Table 11.1. Genetic diseases, ART and pregnancy

Genetic disease	ART	Pregnancy
Oncogenetic diseases • Hereditary breast–ovarian cancer syndromes (HBOC) • Familial adenomatous polyposis (FAP) • Hereditary nonpolyposis colorectal cancer (HNPCC) • Neurofibromatosis type 1 (NF1) **Neuromuscular diseases** • Myotonic dystrophy type 1 (MD1) • Duchenne/Becker muscular dystrophy **Hematologic diseases** • Hemophilia A/B • Thalassemias **Renal disorders** • Polycystic kidney disease • Mitochondrial diseases **Connective tissue disorders** • Marfan syndrome • Ehlers–Danlos syndrome vascular type	• Ovarian reserve especially after gonadotoxic treatment • Hyperstimulation and possible risks of hyperestrogenic state • Ovum pick-up possibilities after surgery • Neurofibroma in uterus or vascular complications • Semen analysis in men with MD1 • Anesthetics in women undergoing ovum pick-up • Cardiomyopathy in female Duchenne carriers • Risk of bleeding during ovum pick-up for carrier woman • Renal and hepatic cysts, influenced by hyperestrogenic state	• Risks of recurrence of malignancy • Risk of malabsorption in case of bowel surgery • Risk of obstetric and peripartum complications due to medication or history of surgery • Increased risk of obstetric complications, e.g. growth retardation, premature birth, and preeclampsia • Pulmonary function in patients with myotonic dystrophy • Increased risk of postpartum hemorrhagic bleeding • Cardiomyopathy or rhythm disturbances in pregnancy or during delivery • Peripartum management for preventing hemorrhage • Risks of anemia and iron overload after treatment • Impaired renal function • Hepatic cysts • Diabetes, nephropathy influencing pregnancy outcome, epilepsy • Aortic aneurysm, spontaneous rupture of the uterus or vessel malformation

IVF or pregnancy. In particular, this applies for women affected with a (genetic) disorder. If the woman is affected with a (genetic) disease, the discussion must consider the options should the disease worsen due to the IVF treatment or during pregnancy.

11.6.1 Oncogenetic Diseases

11.6.1.1 Hereditary Breast and Ovarian Cancer

Hereditary breast and ovarian cancer (HBOC) is a cancer predisposition syndrome mostly caused by mutation in the *BRCA1* or *BRCA2* gene. Women with a mutation in one of these genes face elevated risks of breast, fallopian tube, and ovarian cancer. Since the *BRCA1* and *BRCA2* genes have autosomal dominant inheritance, both male and female mutation carriers have a 50% risk of transmitting the mutation to their offspring.

The reproductive decision-making process for couples deciding on PGT for a *BRCA1/BRCA2* mutation may be even more complex than for other genetic disorders. Female mutation carriers may have a history of breast cancer and experience fear of treatment-related decreased ovarian reserve or recurrence of their malignancy. Women with a *BRCA1/BRCA2* mutation who develop breast cancer and (still) have the wish to become pregnant can consider fertility preservation to prevent infertility using emergency IVF or cryopreservation of ovarian tissue. For some couples the use of PGT for future children is an extra argument for fertility preservation [18]. Another argument for fertility preservation for female *BRCA1/BRCA2* carriers could be the limited time available to have children due to their increased risk of breast or ovarian cancer and to necessary preventive measures like ovariectomy before finalizing their child wish.

PGT seems to be a suitable reproductive option for both asymptomatic male and female *BRCA1/BRCA2* mutation carriers and female breast cancer survivors, yielding good pregnancy rates.

11.6.1.2 Neurofibromatosis Type 1

Neurofibromatosis type 1 (NF1) is an autosomal dominant disease characterized by café-au-lait spots of the skin and neurofibromas, which can occur in different sites on the body thereby causing diverse problems. Severity greatly differs from one patient to another. Neurofibromas in the uterus may impair fertility, but probably only when they influence the uterine cavity. During IVF treatment there are no additional risks, but during pregnancy there may be an increase in the number and size of neurofibromas, with possible spinal expansion. PGT can be offered to patients with neurofibromatosis [9].

11.6.2 Neuromuscular Diseases

Neuromuscular diseases are a major indication for performing PGT. Especially for women with a genetic disease, this may impact the IVF treatment necessary for PGT and may also subsequently influence the pregnancy. This can also be the case in asymptomatic carriers or persons with minor symptoms.

11.6.2.1 Myotonic Dystrophy Type 1

Myotonic dystrophy type 1 is one of the most common hereditary adult muscular diseases with autosomal dominant inheritance. In addition to muscle involvement (myotonia and muscle weakness), many other organs may be affected, such as the heart, digestive tract, and central nervous system. It can take various forms: a mild type with onset from 50 years of age with cataract and mild muscle weakness to the most severe congenital type with onset before birth, breathing and swallowing problems, muscle weakness, intellectual disability, and often death at young age. Men with myotonic dystrophy type 1 can suffer from subfertility and sperm parameters can be so low that testicular biopsy to retrieve sperm is necessary.

Before PGT treatment, a plan should be formulated by a multidisciplinary team and it is important to discuss the risks with the couple:

- Increased risk of maternal complications with IVF/ICSI treatment, especially around ovum pick-up.
- Possible risk of maternal complications during pregnancy.
- IVF/ICSI/PGT can only proceed if there are no contraindications; therefore all women with myotonic dystrophy are referred to gynecology and obstetrics for preconception screening.
- If necessary, offer psychosocial/medical social counseling.

11.6.2.2 Duchenne and Becker Muscular Dystrophy

Duchenne muscular dystrophy (DMD) and Becker muscular dystrophy (BMD) are X-linked recessive disorders that both lead to progressive muscle disease starting proximally, calf hypertrophy, and dilated cardiomyopathy; BMD is the milder form.

Symptoms in women with a mutation in the *DMD* gene (carriers) are mostly mild. It is estimated that 22% of carriers, with an age of onset averaging 34 years, have exercise-related muscle pain and cramps, mild muscle weakness, and heart complaints such as dilated cardiomyopathy and ECG abnormalities.

11.6.3 Hematologic Diseases

11.6.3.1 Hemophilia A and B

Hemophilia A and B are caused by a mutation in the factor VIII (*F8*) or factor IX (*F9*) gene. In men with a mutation this will lead to an increased bleeding tendency. Carrier women may have a risk of reduced factor VIII or IX resulting in a bleeding tendency, but which is less severe than in men with a mutation.

In carrier women who are planning to become pregnant a hematologist should evaluate the patient and advise her on the measures taken during pregnancy and delivery. If the couple is planning to use IVF with PGT they should also be advised on preventive measures during treatment and ovum pick-up. If the factor VIII/IX level is seriously diminished, advice is given about coagulation correction. Hemophilia A/B carriers should have regular follow-up by a gynecologist during pregnancy in combination with the hematologist and should deliver in a hospital familiar with the disorder, preferably under the supervision of a multidisciplinary team with expertise on bleeding disorders.

11.6.3.2 Thalassemias

Thalassemias are genetic disorders comprising two main types, α- and β-thalassemia. These can result in fetal death or severe or mild anemia, caused by increased degradation of erythrocytes with abnormal globin. Patients can further suffer from an enlarged spleen and icterus. Blood transfusions, with an additional risk of iron overload or bone marrow transplantation, are necessary in many cases. Thalassemias are common in people of Mediterranean, Middle Eastern, South Asian, and African descent. Female patients with symptoms of thalassemia should be counseled before conception on possible risks during the IVF treatment and subsequent pregnancy.

11.6.4 Renal Disorders

11.6.4.1 Polycystic Kidney Disease

The most common hereditary kidney disease is the autosomal dominant form of polycystic kidney disease (ADPKD). It is a late-onset disorder characterized by progressive development of bilateral kidney cysts during the lifetime of the patient, eventually resulting in end-stage renal disease. Patients may also display polycystic liver disease and arterial hypertension, with an increased risk of intracranial aneurysms. Most women with ADPKD have no specific fertility problems, whereas male patients can have impaired sperm parameters due to cyst formation in the reproductive tract and/or structural abnormalities of the spermatozoa. A recent study has shown that male infertility accompanying ADPKD does not substantially affect the clinical outcome of IVF/PGT cycles. It was also concluded that affected men who suffer from infertility should seek treatment in time to improve their chances of conceiving a child [19].

Female ADPKD patients can be at increased risk for severe polycystic liver disease, arterial hypertension, and increased risk of intracranial aneurysms, so ovarian hyperstimulation syndrome and multiple pregnancy should be avoided. Furthermore, in patients with polycystic liver disease, cysts could grow during ovarian hyperstimulation due to increased levels of estrogens. During pregnancy the patient's blood pressure and kidney function should be closely monitored.

11.7 Future Perspectives

The most important drawback of ART, with or without PGT, is the burden of treatment and the limited chance of an ongoing pregnancy. Regarding the last point, better selection of embryos and enhancing the chances of a successful implantation are challenges for the near future. The upcoming noninvasive techniques for embryo testing are promising and can be applied on a larger scale, if proven effective.

The role of genetic treatment of embryos with genetic disorders, for example by means of CRISPR/Cas9 techniques, thereby increasing the number of transferable embryos, is awaited (discussed in Chapter 15).

Should PGT also be applied to other patient groups in the near future? Nowadays PGT is available for couples with an already existing increased genetic risk. A rapidly upcoming group comprises couples without a family history of genetic disease who are identified by PCT to be both carriers of an autosomal recessive disease; this mainly applies to consanguineous couples. PCT combined with PGT prevents the birth of a first affected child for these couples.

However, de novo mutations, which are the cause of the majority of the cases of severe mental retardation, will still not be prevented. Diagnosing de novo mutation in an embryo will require complex techniques and of course a thorough ethical evaluation. A step further is the screening of embryos for polygenic traits such as intelligence, which was recently described as being of limited utility [20].

A wish for the future is that burdensome IVF therapies may not be required. Selection of sperm before fertilization could be an option. Stem cell technology could replace IVF treatments if oocyte and/or sperm could be derived from stem cells.

References

1. Vermeesch JR, Voet T, Devriendt K. Prenatal and pre-implantation genetic diagnosis. *Nat Rev Genet* 2016;17:643–56.
2. Reumkens K, Tummers MHE, Gietel-Habets JJG, et al. Online decision support for persons having a genetic predisposition to cancer and their partners during reproductive decision-making. *J Genet Couns* 2019;28:533–42.
3. Birch PH, Adam S, Coe RR, et al. Assessing shared decision-making clinical behaviors among genetic counsellors. *J Genet Couns* 2019;28(1):40–9.
4. Resta RG. What have we been trying to do and have we been any good at it? A history of measuring the success of genetic counseling. *Eur J Med Genet* 2019;62:300–7.
5. Matt Bolz-Johnson M, Meek J, Nicoline Hoogerbrugge N. Patient journeys: improving care

by patient involvement. *Eur J Hum Genet* 2020;28:141–3.

6. van der Meij KRM, Sistermans EA, Macville MVE, et al. TRIDENT-2: national implementation of genome-wide non-invasive prenatal testing as a first-tier screening test in the Netherlands. *Am J Hum Genet* 2019;105:1091–101.
7. Che H, Villela D, Dimitriadou E, et al. Noninvasive prenatal diagnosis by genome-wide haplotyping of cell-free plasma DNA. *Genet Med* 2020;22: 962–73.
8. Harton G, Braude P, Lashwood A, et al. ESHRE PGD consortium best practice guidelines for organization of a PGD centre for PGD/preimplantation genetic screening. *Hum Reprod* 2011;26:14–24. Erratum in *Hum Reprod* 2012;27:2569.
9. Coonen E, van Montfoort A, Carvalho F, et al. ESHRE PGT Consortium data collection XVI–XVIII: cycles from 2013 to 2015. *Hum Reprod Open* 2020;2020(4): hoaa043. https://doi.org/10.1093/hropen/hoaa04
10. de Die-Smulders CE, de Wert GM, Liebaers I, et al. Reproductive options for prospective parents in families with Huntington's disease: clinical, psychological and ethical reflections. *Hum Reprod Update* 2013;19:304–15.
11. Handyside AH, Pattinson JK, Penketh RJ, et al. Biopsy of human preimplantation embryos and sexing by DNA amplification. *Lancet* 1989;1(8634):347–9.
12. Sallevelt SC, Dreesen JC, Drüsedau M, et al. Preimplantation genetic diagnosis in mitochondrial DNA disorders: challenge and success. *J Med Genet* 2013;50:125–32.
13. Kakourou G, Kahraman S, Ekmekci GC, et al. The clinical utility of PGD with HLA matching: a collaborative multi-centre ESHRE study. *Hum Reprod* 2018;33:520–30.
14. Evers JLH. Female subfertility. *Lancet* 2002;366:151–9.
15. Oud MS, Volozonoka L, Smits RM, et al. A systematic review and standardized clinical validity assessment of male infertility genes. *Hum Reprod* 2019;34:932–41.
16. De Krom G, Arens Y, Coonen E, et al. Recurrent miscarriage in translocation carriers: no differences in clinical characteristics between couples who accept and couples who decline PGD. *Hum Reprod* 2015;30:484–9.
17. Hortas ML, Castilla JA, Gil MT, et al. Decreased sperm function of patients with myotonic muscular dystrophy. *Hum Reprod* 2000;15:445–8.
18. ter Welle ter-Butalid E, Vriens I, Derhaag J, et al. Counseling young women with early breast cancer on fertility preservation. *J Assist Reprod Genet* 2019;36: 2593–604.
19. Berckmoes V, Verdyck P, De Becker P, et al. Factors influencing the clinical outcome of preimplantation genetic testing for polycystic kidney disease. *Hum Reprod* 2019;34:949–58.
20. Karavanu E, Zuk O, Zeevi D, et al. Selecting human embryos for polygenic traits has limited utility. *Cell* 2019;179:1–12.

Chapter 12

Mitochondrial Genetics in Reproductive Medicine

Claudia Spits and Filippo Zambelli

12.1 Mitochondria and Their Genome

Mitochondria are typically described as the powerhouse of the cell, because they are the cytoplasmic organelles responsible for the production of ATP through oxidative phosphorylation. Over the years, it has become clear that their function within the cell is more complex as they are also involved in numerous other processes, including lipid and carbohydrate metabolism, heme biosynthesis, apoptosis, and calcium homeostasis [1]. Human cells contain multiple mitochondria, with the exception of red blood cells that have none. The numbers, mass, morphology, and distribution vary greatly across different cell types, generally depending on the energy demands of the tissues. For instance, sperm contain 20–75 mitochondria in their midpiece, while hepatocytes and muscle cells contain thousands.

Mitochondria have a double membrane that forms two compartments, an inner matrix and an intermembrane space. The inner membrane has low ion permeability and is folded into numerous cristae that protrude into the matrix. The protein complexes needed to carry out oxidative phosphorylation are located on this inner mitochondrial membrane and are composed of five subunits (I–V) and two electron carriers (coenzyme Q and cytochrome C) (Figure 12.1).

Mitochondria likely originated by endosymbiosis of a prokaryotic entity, and still have their own small circular multicopy genome that resides in the inner matrix. Again, the numbers of mitochondrial DNA molecules vary greatly across cell types. Human cells contain between 500 and 10 000 mitochondrial DNA molecules, with the exception of mature oocytes, which generally have between 100 000 and 200 000 mitochondrial DNA copies. The mitochondrial DNA is composed of a heavy strand and a light strand and is organized in structures named nucleoids. In the nucleoids, single copies or small clusters of mitochondrial DNA molecules organize with regulatory proteins and is where both replication and transcription occur. Conversely, nucleoids are not bound by histones and do not possess the extensive machinery for DNA repair present in the cell nucleus. This, in combination with its close exposure to the byproducts of energy production (i.e. reactive oxygen species, molecules with high reactivity to DNA, lipids and proteins), makes it particularly susceptible to spontaneous mutagenesis, resulting in a mutation rate that is at least 5–10 times higher than in the nuclear DNA.

The human mitochondrial DNA comprises 16 569 base pairs and is characterized by the absence of introns, with some regions coding for parts of two different proteins. In this small genome, there are only a few noncoding regions, mainly located in the so-called "D-loop" or "hypervariable" region. This sequence contains the origin of replication of the heavy strand and is the target of the replication and transcription factors. The rest of the molecule encodes for 13 proteins of the oxidative phosphorylation complexes I, III, IV and V, two ribosomal RNAs, and 22 tRNAs exclusively active within the mitochondria [2,3] (Figure 12.2a). The remaining approximately 1500 genes required for the oxidative phosphorylation complexes and for mitochondrial homeostasis are encoded by the nuclear genome, making mitochondria fully dependent on the nucleus for their survival and activity. In fact, during evolution, the majority of the mitochondrial genes migrated from the mitochondrial to the nuclear genome. This is thought to be the consequence of the higher mutation rate of the mitochondrial DNA, which drove the transfer of genes to the more stable nuclear genome [4]. This dependency also means that the correct regulation of oxidative phosphorylation is governed by mitonuclear interactions.

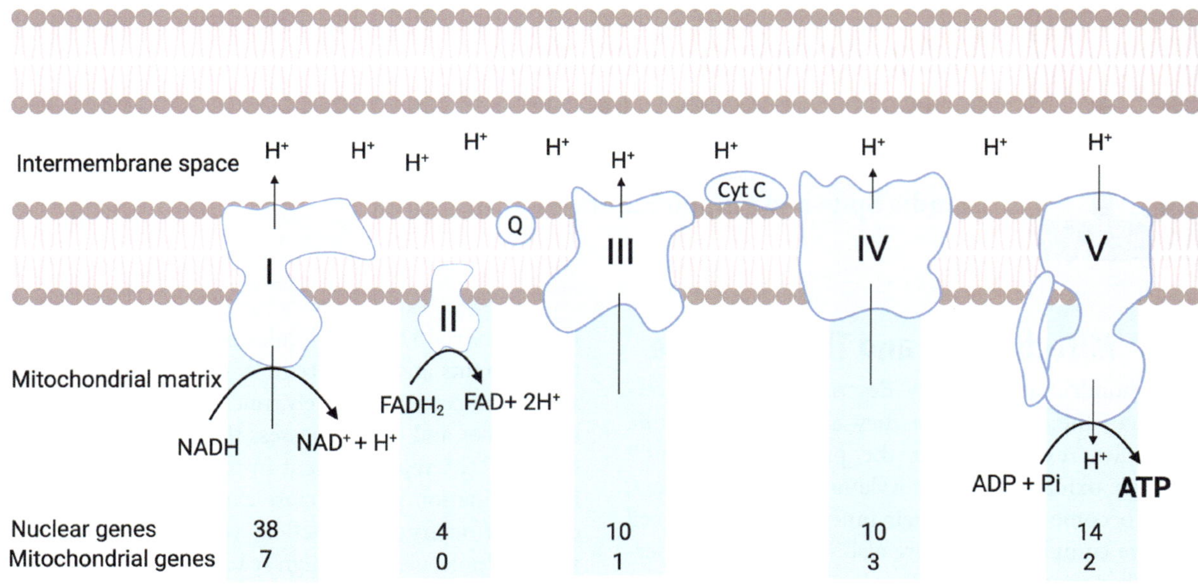

Figure 12.1 Schematic overview of the electron transport chain. Complexes I–IV generate a proton gradient in the intermembrane space that is used by complex V to convert ADP into ATP. Each complex is composed of multiple proteins. The number of genes encoded by the nuclear and mitochondrial genome for each complex is shown in the green boxes.

The shape, size and, number of mitochondria in a cell are regulated by cycles of fission and fusion. During these processes, the organelles can exchange their membrane components, as well as soluble parts such as their DNA. During fission, a single mitochondrion will divide in two by cleavage of both membranes, while these are joined during fusion. While the remodeling of mitochondrial structure and the control of their numbers is not strictly connected to the replication of mitochondrial DNA, both respond to physiologic and environmental cues, and serve to adjust to the demands on mitochondrial function during the different stages of development and in the homeostatic and correct function of different tissues under different physiologic conditions [5]. While mitochondrial DNA replication does not coincide with the cell cycle and occurs independently of nuclear DNA replication, many of the factors required for replication are encoded by the nuclear genome, giving the nucleus a key role in the control of mitochondrial copy number. The best-characterized components of the replication machinery include the DNA polymerase γ (*POLG*), which also carries out repair activities; the mitochondrial single-stranded DNA-binding protein (mtSSBP), and Twinkle, the mitochondrial helicase, which work together to achieve helix destabilization during replication; and *TFAM*, an essential component of nucleoid structure that is necessary for the initiation of transcription and replication; all of them are nuclearly encoded [6]. The dynamic nature of the mitochondrial network, the link between mitochondrial DNA copy number and mitochondrial activity, and the relative ease of quantifying the mitochondrial DNA copy number in contrast to counting the mitochondria per cell explains why many researchers refer to mitochondrial DNA copy number in their studies rather than number of mitochondria per cell.

A key aspect of mitochondrial genetics is the multicopy nature of this genome, and the fact that each cell contains numerous mitochondria. This results in two different situations, depending on whether a variant is present in all the mitochondrial DNA copies or in only a fraction. This is termed "homoplasmy" and "heteroplasmy," respectively (Figure 12.2b).

All mitochondrial genomes carry homoplasmic variants that are stably transmitted through the generations, and which have been classically used to study human genealogy and evolution [7]. Homoplasmic variants are the basis of mitochondrial haplogroups, in which a group of specific variants cosegregate. For instance, haplogroup H is defined by the presence of the variants m.2706G>A and m.7028T>C. Since mitochondrial genome variance

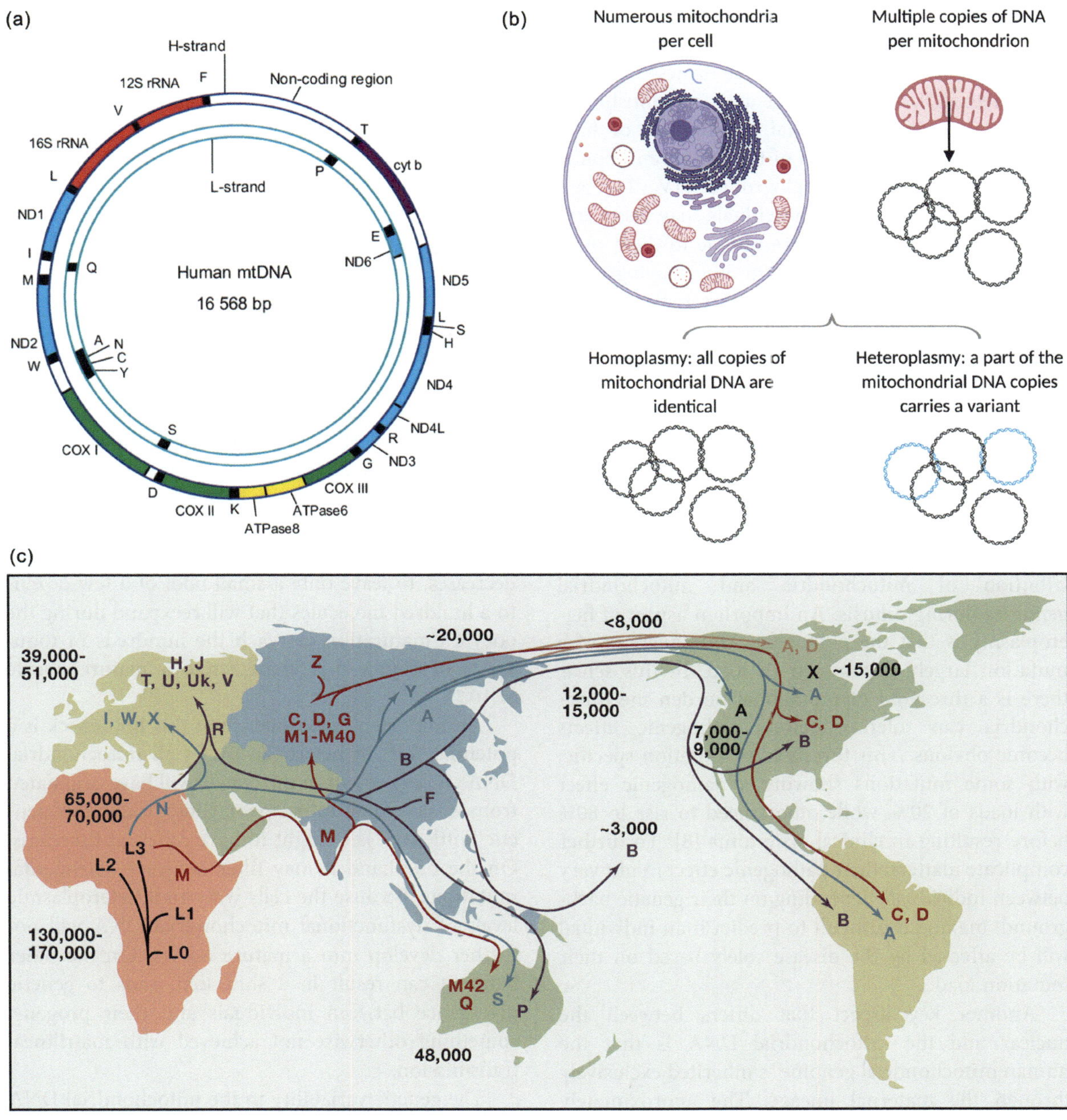

Figure 12.2 (a) The human mitochondrial DNA gene map. The human mitochondrial DNA is a double-stranded circular molecule of 16.6 kb. Single letters indicate the positions of the corresponding tRNA genes. ND, NADH dehydrogenase genes; cyt b, cytochrome b gene; COX, cytochrome c oxidase genes; A6/8, ATP synthase genes 6 and 8; 12S/16S, ribosomal RNA genes. (b) Homoplasmy versus heteroplasmy. Cells contain numerous mitochondria, and each mitochondrion carries multiple copies of its own DNA. The term "homoplasmy" applies to copies of this genome that are identical (i.e. carry the same variants). "Heteroplasmy" refers to the situation in which only a fraction of the DNA molecules carry a variant. Its frequency is termed "heteroplasmic load." (c) Human migration and regional distribution of mitochondrial DNA haplogroups. All African mitochondrial DNAs cluster under macrohaplogroup L and are assumed to have a common origin about 130 000–170 000 years ago. The M and N lineages emerged from the sub-Saharan African lineage L3 in northeastern Africa, and successfully left Africa, giving rise to macrohaplogroups M and N. N haplogroups radiated into European and Asian populations, while M haplogroups were confined to Asia. Haplogroups A, C, and D became enriched in northeastern Siberia and crossed the Bering Land Bridge 20 000 years ago with the founders of the Native American population. Additional Eurasian migrations brought to the Americas haplogroups B and X. Finally, haplogroup B colonized the Pacific Islands. On the map, the letters represent the haplogroups and the numbers indicate years before present. Sources: (a) adapted with permission from Spelbrink [30]; (c) reproduced with permission from Wallace [7].

has evolved as a result of the sequential addition of mutations, the by now over 5500 haplogroups follow a phylogenetic tree (the tree can be browsed at www.phylotree.org/). This phylogenetic relationship is also reflected in the regional distribution of haplogroups. For instance, 90% of the European population belongs to macrohaplogroups HV, JT, and U (Figure 12.2c). Finally, individuals may also carry homoplasmic variants that are unique to them or to their family and do not appear in other haplogroups, as well as variants that are characteristic of haplogroups other than their own.

With regard to heteroplasmic variants, their relative proportion in the cell/tissue/individual is termed "heteroplasmic load." The load of an inherited heteroplasmic variant may differ significantly across different tissues within one individual and between members of the same family. This diversity is caused by different mechanisms, including somatic and germline bottlenecks and by random drift in the distribution of mitochondria and mitochondrial genomes during mitosis. An important aspect of heteroplasmy is that the pathogenic consequences of a mutation largely depend on its load. In this sense, there is a threshold of mutational burden the mitochondria can tolerate before pathogenic effects become obvious. This threshold is mutation specific, with some mutations showing a pathogenic effect with loads of 20%, while others need to rise to 80% before resulting in clinical symptoms [8]. To further complicate matters, their pathogenic effect might vary between individuals depending on their genetic background, making it difficult to predict if an individual will be affected by the disease solely based on their mutation load.

Another key aspect that differs between the nuclear and the mitochondrial DNA is that the human mitochondrial genome is inherited exclusively through the maternal lineage. The approximately 100 000–200 000 copies of mitochondrial DNA contained in the oocyte will represent the source of all mitochondrial DNA of the future individual. In contrast, the few paternal mitochondria and their DNA are targeted for active degradation. The dogma of the exclusive maternal inheritance has only recently seriously been challenged. In 2018, the work of Luo and colleagues showed that, in some rare instances, some members of one family showed evidence of carrying the paternal mitochondrial genome in a heteroplasmic state, while their siblings showed the standard maternal inheritance pattern [9]. The work of Luo and colleagues is the first to extensively document this phenomenon, and although it opens new interesting research questions, their findings remain to be replicated.

12.2 Mitochondria in Gametes and Development

12.2.1 Mitochondrial DNA Transmission and Germline and Somatic Bottlenecks

During primordial germ cell specification, the mitochondrial genome undergoes what is termed a "genetic bottleneck" (Figure 12.3a). This bottleneck starts with a passive reduction of the number of mitochondrial DNA copies due to a halt in mitochondrial DNA replication. During the subsequent cell divisions, the mitochondrial DNA copy number significantly decreases, to leave only a small pool of a few dozens to a hundred molecules that will reexpand during the oocyte's maturation to reach the hundreds of thousands of copies that characterize a mature oocyte [4,10,11].

An important consequence of this bottleneck is a potential shift in heteroplasmy, as all mitochondrial DNA copies present in the oocyte will have originated from a very small initial pool. Biologically, this genetic bottleneck is thought to have different functions. On the one hand it may filter out very detrimental mutations, because the cells with high heteroplasmic levels of dysfunctional mitochondrial DNA will not further develop into a mature oocyte. On the other hand, it can result in a shift that leads to genetic divergence between individuals and their progeny, something otherwise not achieved with matrilineal transmission.

The genetic variability in the mitochondrial DNA composition of primordial germ cells is reflected in mature oocytes, which are known to carry very variable loads of mutations and harbor different variants within the same cohort. This renders the prediction of the mitochondrial genotype of an individual oocyte almost impossible, and creates its own set of problems when trying to prevent the transmission of mitochondrial diseases. An additional problem is caused by the tendency of some pathogenic mutations to very strongly skew in heteroplasmic load while progressing through development: it has been observed in both

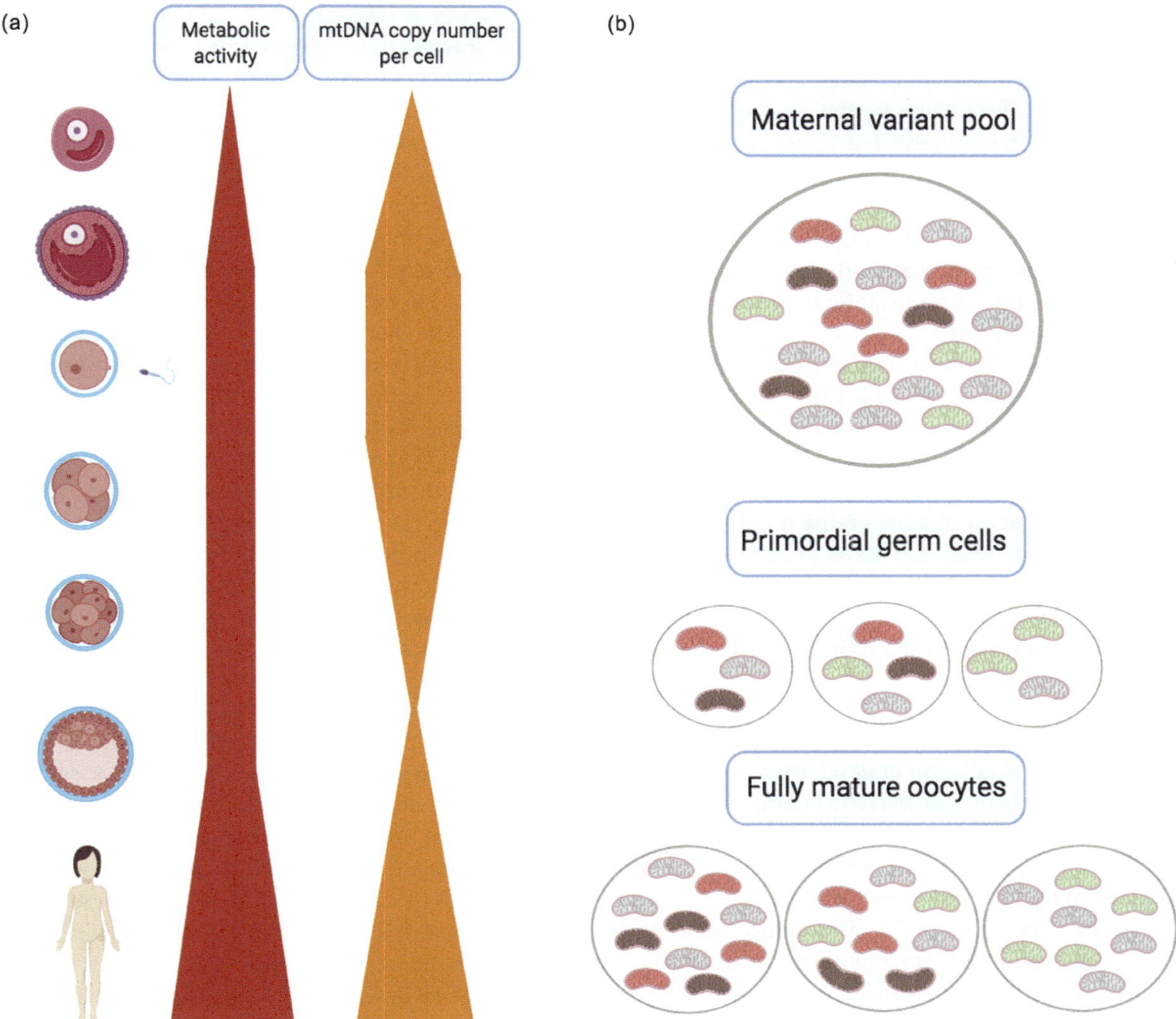

Figure 12.3 Mitochondrial activity and mitochondrial DNA segregation during oocyte maturation and early embryo development. (a) Schematic representation of the variations in metabolic activity and mitochondrial DNA copy number per cell from primordial follicles to blastocyst and adulthood. (b) Germline bottleneck: mitochondrial replication arrest during primordial germ cell (PGC) specification will cause a differential segregation of the variants present in the maternal pool (represented as differentially colored mitochondria), inducing considerable diversity in the oocytes from the same cohort, and significantly complicating the reproductive options for preventing disease transmission.

animal models and humans that specific mutations appear at frequencies that are close to 100% or 0% in gametes, while via random drift they should present with a more normalized distribution. Examples of these are mutations causing Leber hereditary optic neuropathy (m.11778G>A) and Leigh syndrome (m.8993T>G).

The replication of mitochondrial DNA is active throughout oocyte maturation, and it stops at full maturity. During the initial cleavage divisions of the embryo no replication occurs, and the number of copies of mitochondrial DNA per individual blastomere will decrease. Only at the blastocyst stage will replication resume, initially in the trophectoderm and later in the inner cell mass. This probably reflects the energy demands of the different lineages, as the trophectoderm requires more energy earlier on to guarantee rapid proliferation and successful implantation and invasion.

Once cells start dividing, we can observe a differential segregation of mitochondrial genomes in the daughter cells whether or not there is active replication of the mitochondrial DNA. This is what causes the mosaic nature of the mitochondrial genome in adult cells and tissues, and it represents a second type of bottleneck, the so-called "somatic bottleneck." The somatic bottleneck is still poorly understood, as illustrated by findings in mouse models, where some mutations skew their segregation based on factors independent from the dependency on oxidative phosphorylation or the levels of replication, and accumulate recurrently in specific tissues by unknown mechanisms.

In the case of mitochondrial disease, the somatic bottleneck can determine which tissue will be affected by mitochondrial dysfunction, and greatly complicates the diagnostic procedures when dealing with a patient suspected of mitochondrial disease. The preferential segregation of a high burden of the mutation in a tissue that does not have extensive needs for oxidative phosphorylation might attenuate the severity of the disease, or make the individual healthy but able to transmit the disease to its children.

12.2.2 Mitochondria in Oocytes

The metabolic regulation of developing oocytes in the human is still being uncovered (Figure 12.3a), as for years the only available data were from animal models such as mice, pig, and cow. The intrinsic characteristics of primordial germ cells in humans, such as their very early specification in development and their scarce number, has severely limited their study. Work on animal models has shown that mitochondria in primordial germ cells and early oocytes (before the antral phase) are generally inactive, and present an immature morphology, with a globular shape and few cristae with a nonpolarized membrane. When final maturation begins, mitochondria are located in the perinuclear area of the mouse oocyte, and when the germinal vesicle undergoes breakdown they migrate toward the outer area of the ooplasm to finally occupy the subcortical area in the mature metaphase II oocyte. The mitochondria residing in the inner area are not polarized and are therefore considered inactive, while the subcortical ones have a polarized membrane. Overall, the membrane potential of the oocyte will not significantly increase during the final meiotic maturation phases, although in bovine oocytes an increase in ATP content has been observed during in vitro maturation. Also in the bovine model, a better cumulus–oocyte complex morphology is associated with a higher ATP content and a specific mitochondrial distribution. Work in the human suggests a similar pattern of polarized mitochondria in mature oocytes, but further research is needed in earlier developmental stages to fully understand mitochondrial regulation during human oocyte maturation.

The different cellular functions taking place in, or being controlled by, mitochondria are increasingly being linked to the acquisition of developmental competence [12]. For instance, a lower mitochondrial potential has been observed in oocytes with failed fertilization, and women with reduced ovarian function, whether because of premature ovarian insufficiency or advanced maternal age, present oocytes with a decreased metabolic signature and an altered morphology. Despite the suggestive evidence, it is not clear if the mitochondrial dysfunction is directly causing the lower developmental competence or if it is just part of the oocyte's overall reduced fitness. More recently, it has been shown in cattle and mice that supplementation of the culture media used for in vitro maturation with NAD^+ metabolic precursors, capable of replenishing intracellular NADPH availability, ameliorates the quality of the oocytes, suggesting a direct effect of mitochondrial activity on the establishment of developmental competence.

Regarding the mitochondrial genome, work on different vertebrates has shown that a sufficient number of mitochondrial DNA molecules must be present to allow fertilization and support embryo development. Research on human oocytes has found a smaller number of mitochondrial genomes in oocytes that fail to fertilize and in degenerated oocytes when compared with normally fertilized ones. In mice oocytes, it has been calculated that 50 000 copies of the mitochondrial DNA allow proper embryo development, while a much lower number are needed to ensure fertilization. In analogous fashion to what was observed for the membrane potential, patients with premature ovarian insufficiency have a reduced number of mitochondrial DNA copies in their oocytes, with the same reduction observed in women of advanced maternal age. Finally, patients with mitochondrial DNA depletion caused by mutations in the mitochondrial DNA polymerase *POLG* show a higher incidence of primary ovarian insufficiency. This suggests not only an association but also a functional link between mitochondria and ovarian physiology.

12.2.3 Mitochondria in Sperm

Spermatogenesis is an energy-demanding process that proceeds continuously in postpubertal men, generating a high number of spermatozoa on a daily basis. The mitochondrial contribution of sperm to embryos is not fully understood, and it is generally accepted that the mitochondria from the paternal side are degraded after fertilization and have little or no contribution to embryonic development [13,14].

However, mitochondria and oxidative phosphorylation are active in mature sperm, and participate in the process of spermatogenesis . Mitochondrial metabolism is associated with correct maturation of spermatozoa precursors, as mitochondria in developing spermatozoa have different shapes, are characterized by specific protein isoforms at different maturation stages, and their activity is modified throughout spermatogenesis. In spermatogonia, mitochondria have the lowest activity. During spermatogenesis, the mitochondria will condense and progressively increase their activity concomitant with sperm maturation.

While mitochondria are active in mature spermatozoa, their contribution to energy production in sperm appears to be rather limited and still needs to be conclusively proven. Flagellar movement, for example, was thought to be the most obvious function of mature sperm requiring the energy provided by mitochondria. However, this function probably relies on glycolysis rather than mitochondrial metabolism for its ATP needs. While inhibition of oxidative phosphorylation does not influence sperm motility, the inhibition of glycolysis, even when oxidative phosphorylation is actively increased, causes a significant decrease in motility. This suggests that the amount of ATP required for the spermatozoa's movement is higher than that provided by the small number of mitochondria in a mature sperm [14]. It is estimated that only 20–75 mitochondria are stored in the midpiece of an individual sperm cell. They are organized in tubular structures and protected by the mitochondrial capsule. The existence of this specific structure that guarantees mitochondrial survival and activity suggests an important role in correct sperm function, albeit probably not related to ATP production but rather to signaling. Another aspect that has been investigated is the expression of mitochondrial proteins. A reduced expression of these proteins has been correlated with bad semen quality and compromised embryo development. Semen of asthenozoospermic patients has also been found to have low levels of mitochondrial proteins involved in energy production and reactive oxygen species scavenging, along with downregulation of several transcripts, some of them being specific to sperm function (e.g. *ANXA2* and *BRD2*) and others being related to mitochondrial metabolism (e.g. *ND2* and *ND3*). Taken together, while these results are suggestive of a role for mitochondrial DNA, and for mitochondrial function in general, in correct spermatogenesis and sperm function, no concrete factors have been identified and validated that could have diagnostic value.

12.2.4 Mitochondria in Embryos

The role of mitochondria during fertilization and early embryo development is not yet fully understood, but research from the past decade suggests that correct embryo development requires finely regulated mitochondrial activity [15]. In the case of mitochondrial DNA, its copy number has been found to play an important role in both fertilization and early embryo development of several vertebrate models, and humans seem to follow the same trend (Figure 12.3b). When the oocyte completes its maturation, it will have numerous mitochondria with a large number of copies of mitochondrial DNA, but it will not show a high metabolic activity, and most of the mitochondria will not have a high membrane potential and present with a globular shape and inactive appearance.

Fertilization does not trigger mitochondrial activation, as no differences in activity are seen between mature and fertilized oocytes. Their low oxidative metabolism is probably a reflection of the anoxic environment that preimplantation embryos encounter during development, where they mainly rely on glycolysis to produce ATP. Initial embryo cleavages will not change the level of oxidative phosphorylation, which will remain low as reflected also by unaltered oxygen consumption, and replication will still be halted. This causes a reduction in the number of mitochondrial DNA molecules per cell, as they simply become segregated in the dividing blastomeres while not increasing in number. When reaching peri-implantation stages, synchronic activation of mitochondrial metabolism and replication will occur in trophectoderm cells, and mitochondria will mature and increase their number, size, and activity. Conversely, no mitochondrial activation is seen in the inner cell mass. These differences are likely due to the differences in energetic needs between trophectoderm and inner cell mass, with the first having to proceed through implantation and invasion, and the latter still having to start the differentiation steps leading to gastrulation. Later, when the inner cell mass begins gastrulation, the mitochondria will be activated, both in terms of activity and of replication. At this point in development all the cells switch their glycolytic metabolism to oxidative metabolism, and this will be maintained through development and adulthood.

12.2.4.1 Mitochondrial DNA as a Marker of Embryo Quality

In Embryos

Recently, the absolute quantity of mitochondrial DNA present in embryonic cells has been suggested to be predictive of the embryo's quality and implantation capacity, and to its potential to result in a live birth [16].

The background for this idea is the "quiet embryo hypothesis," which is based on two aspects of mitochondrial biology during early development: (1) the fact that mitochondrial DNA replication is halted during early embryo development, and (2) that oxidative phosphorylation does not have major increases during the cleavage stage and only increases in trophectodermal cells before implantation. This has led to the hypothesis that the developmental capacity of the embryo is associated with its oxidative phosphorylation activity, with higher metabolic levels resulting in a lower developmental and implantation potential.

The first step to test this has been the identification of the correct sample to analyze. Polar bodies are not reliable in representing mitochondrial DNA copy number of the oocyte, in mice or in humans, and single nucleotide variants identified by sequencing of the polar bodies do not match the ones found in the oocyte. In the same fashion, analysis of free mitochondrial DNA molecules in the spent culture media of oocytes and embryos yields a poor correspondence and suffers from contamination from the cumulus cells. Therefore, most of the methods developed for mitochondrial DNA quantification and its association with in vitro fertilization (IVF) outcomes rely on the analysis of blastomeres from eight-cell embryos and trophectoderm cells.

The first reports of this type of testing were in line with the core hypothesis, and by using both blastomere and trophectoderm biopsy it was found that embryos that were not able to implant had relatively high mitochondrial DNA copy numbers. These promising findings attracted a wide number of independent research groups that tried to replicate these results in order to develop a minimally invasive tool to improve embryo selection. However, the results of these replicative studies have been conflicting. Different groups did not find the same association, with mitochondrial DNA levels not associating with any of the measured IVF outcomes. Some studies associated lower levels of mitochondrial DNA with a higher quality of embryos, rather than a higher implantation potential, while others failed to detect this link. These differences might be due to different technical and biological factors, making the standardization of this approach challenging and suggesting that the practical predictive value of mitochondrial DNA copy number for the improvement of IVF outcomes is limited [16].

In Cumulus Cells

The cells forming the follicle, namely granulosa and cumulus cells, have been extensively investigated in the hope of finding a noninvasive marker able to predict the developmental potential of the enclosed oocytes. Despite the fact that these cells are in direct contact with the oocyte and provide a significant amount of proteins and molecules responsible for correct maturation during the latest steps of oogenesis, no reliable markers for embryo/oocyte quality have yet been identified in such cells. During the last steps of follicular development, cumulus cells (and in particular theca cells) provide many different molecules to the oocyte, including ATP. For this reason, analyzing the metabolic profiles of cumulus cells appears an attractive approach to infer the metabolic state of the oocyte and the developmental potential of the resulting embryo in a noninvasive way. In the case of mitochondrial DNA, it has been observed that its decreased abundance in cumulus cells correlates with a diminished ovarian reserve, together with downregulation of key mitochondrial nuclear-encoded genes such as *PPARGC-1A*. In terms of embryo quality and implantation potential, a higher mitochondrial DNA copy number has also been associated with more favorable IVF outcomes [17]. Surprisingly, in the same studies, there is no association with the developmental capacity of the oocytes or their capacity to be correctly fertilized. However, the preliminary prospective studies published so far have not established the potential of this analysis in improving the selection of competent embryos, and this will have to be examined further in larger cohorts.

12.3 Inherited Mitochondrial DNA Disease

Mitochondrial diseases have an estimated prevalence of 5–15 cases per 100 000 individuals. They are characterized by a wide range of clinical manifestations, typically affecting organs or systems with high dependency on aerobic metabolism, such as the nervous, endocrine, muscular, gastrointestinal, and cardiovascular systems [3]. This heterogeneous group of diseases can be caused by different types of mutations, such as single nucleotide variants, and rearrangements such as insertions and deletions of varying size. The type of mutation and its localization will determine the disease, but symptoms and disease progression can greatly vary across different individuals. Mitochondrial disease can be caused by mutations in both nuclear and mitochondrial DNA, as well as by depletion of mitochondrial DNA itself (i.e. reduction of its copy number). Mutations in the nuclear DNA can have autosomal recessive, dominant, and X-linked inheritance patterns. While mutations in mitochondrial DNA will mostly affect the electron transport chain and therefore impair oxidative phosphorylation, nuclear mutations can act at the level of mitochondrial DNA replication, protein translation, and mitochondrial homeostasis.

Categorization of mitochondrial diseases is based on a combination of clinical features and the causal mutations, both of which are often very diverse within one syndrome. Also, there is a weak correlation between genotype and phenotype, with patients with the same mutation showing different clinical manifestations and disease progression, further complicating the correct diagnosis of mitochondrial disease. This variable clinical manifestation appears to be only partially related to the levels of heteroplasmy, and likely other yet to be identified factors play an important role, including the nuclear genetic background.

Leigh syndrome (OMIM #256000) is a childhood-onset disorder characterized by global developmental delay and/or regression, hypotonia, ataxia, dystonia, and basal ganglia and brainstem lesions. Causal mutations have been identified in more than 75 genes, both in nuclear and mitochondrial DNA. Other clinical manifestations of Leigh syndrome form a clinical continuum with those of NARP (neurogenic muscle weakness, ataxia, and retinitis pigmentosa, OMIM #551500). NARP can have a childhood or adult onset, and is characterized by pigmentary retinopathy, sensorimotor neuropathy, and ataxia. It is caused by mutations in mitochondrial DNA, of which the most common is the m.8993T>G mutation. Congenital lactic acidosis is another syndrome that can be caused by mutations in both nuclear and mitochondrial DNA. It is a childhood-onset, often fatal disease, with progressive neuromuscular weakness and accumulation of lactate in the blood (acidosis), urine and/or cerebrospinal fluid.

Pearson syndrome (OMIM #557000) is a childhood sideroblastic anemia associated with exocrine and/or endocrine pancreatic dysfunction, pancytopenia, and renal tubulopathy. It is caused by single, large-scale mitochondrial DNA deletions or rearrangements. Single large deletions also cause Kearns–Sayre syndrome (OMIM #530000), which is of adult onset. It is characterized by pigmentary retinopathy, progressive external ophthalmoplegia, cerebellar ataxia, cardiac conduction abnormalities, myopathy, diabetes mellitus, deafness, bulbar weakness, and dementia.

Leber hereditary optic neuropathy (OMIM #535000), MELAS (mitochondrial myopathy, encephalopathy, lactic acidosis, and stroke-like episodes, OMIM #540000), and MERRF (myoclonic epilepsy with ragged red fibers, OMIM #545000) are adult-onset syndromes exclusively caused by point mutations in the mitochondrial genome. Leber hereditary optic neuropathy is characterized by subacute bilateral visual loss during midlife. MELAS is characterized by myopathy, encephalopathy, lactic acidosis, and stroke-like episodes and MERRF is associated with progressive myoclonic epilepsy, ataxia, and weakness.

A large number of nuclear DNA mutations have been associated with childhood- and adult-onset syndromes. *POLG*-related mutations cause four childhood-onset syndromes, dominated by neurologic and muscular symptoms. Alpers–Huttenlocher syndrome (OMIM #203700) is characterized by intractable epilepsy, psychomotor retardation, and liver disease, and is often lethal before the age of 3 years. Patients with childhood myocerebrohepatopathy spectrum (OMIM #613662) suffer from myopathy, developmental delay or deterioration of intellectual function, and liver disease. Ataxia neuropathy spectrum (OMIM #607459) is characterized by sensory ataxic neuropathy, dysarthria, and ophthalmoparesis. Finally, patients with myoclonic epilepsy myopathy sensory ataxia (OMIM #607459) have similar symptoms on the ataxia neuropathy spectrum, with a higher frequency of migraine headaches and seizures.

Sengers syndrome (OMIM #212350) is a childhood-onset disorder caused by mutations in the *AGK* gene. The patients have congenital cataracts, hypertrophic cardiomyopathy that may be the cause of early death, skeletal myopathy, exercise intolerance, and lactic acidosis. MEGDEL syndrome (OMIM #614739) is caused by mutations in the *SERAC1* gene. It is characterized by childhood onset of delayed psychomotor development or psychomotor regression, sensorineural deafness, spasticity, dystonia, hepatopathy, and lactic acidosis.

Chronic progressive external ophthalmoplegia (CPEO, OMIM #157640) is an adult-onset disorder that can be caused by a large spectrum of mutations in nuclear genes, such as *POLG1*, *POLG2*, and *C10orf2* (encoding Twinkle). The clinical manifestations vary depending on the mutation, but include myopathy, resulting in muscle weakness, progressive external ophthalmoplegia and hearing loss. Mutations in some of the same genes that cause CPEO (specifically *TYMP*, *RRM2B*, and *POLG*) result in mitochondrial neurogastrointestinal encephalopathy (MNGIE, OMIM #603041) syndrome. The symptoms are similar, but also include gastrointestinal dysmotility.

12.4 Preventing the Transmission of Mitochondrial Disease

The absence of a cure for mitochondrial diseases has made the prevention of disease transmission currently the only option for patients wanting to avoid having affected children. However, the wide spectrum of mutations causing mitochondrial disorders and the nature of mitochondrial DNA itself can have a significant impact on the available treatments and their success rates (Figure 12.4) [18,19].

Mitochondrial disorders caused by mutations in the nuclear genome are inherited in an X-linked, autosomal dominant or autosomal recessive fashion, and their transmission can be prevented in the same manner as other monogenic disorders (i.e. by prenatal diagnosis or preimplantation genetic testing). Prenatal diagnosis requires the difficult decision of pregnancy termination if the fetus is affected, while preimplantation genetic testing is based on the diagnosis of embryos in vitro and the selective transfer of unaffected embryos. This ensures that if the treatment results in a pregnancy, it will be of an unaffected child (see Chapter 13).

More challenging is when the mutation is located in the mitochondrial genome, and this for multiple reasons. First, different types of mutations have different rates of recurrence. For example, single large

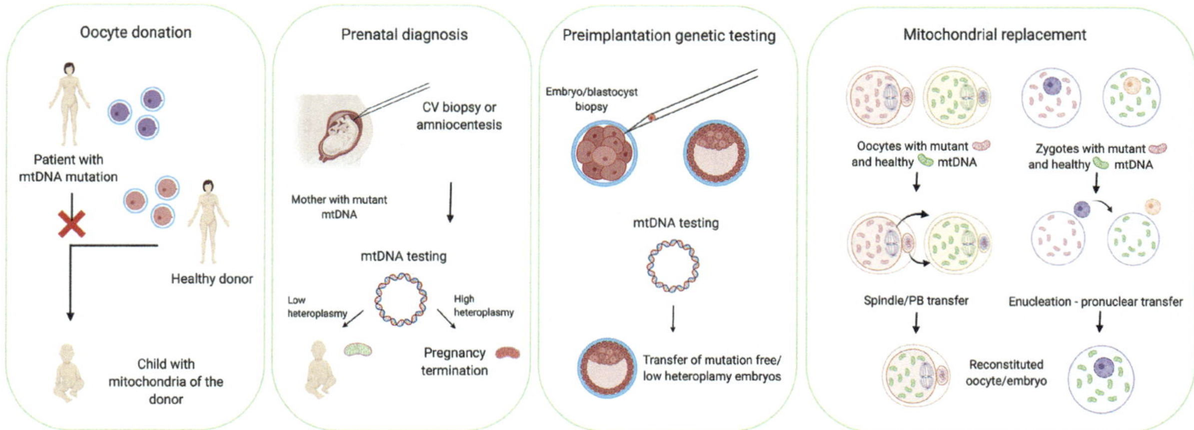

Figure 12.4 Reproductive options for preventing the transmission of mitochondrial diseases. Different types of mutations and their heteroplasmic or homoplasmic condition call for different reproductive options. Patients with homoplasmic mutations, or who do not produce oocytes with sufficiently low heteroplasmic loads, can resort only to oocyte donation and mitochondrial replacement, while patients carrying heteroplasmic mutations can undergo prenatal diagnosis and preimplantation genetic testing. CV, chorionic villi; mtDNA, mitochondrial DNA.

deletions have a recurrence rate of around 1 in 25. When the deletions in the mitochondrial DNA are multiple, they can be caused by a mutation in the nuclear gene for the mitochondrial polymerase *POLG*, which can be transmitted in an autosomal dominant or recessive fashion. Subsequent to the hurdles in correct counseling for the risk of recurrence, there are also technical difficulties associated with the reliable diagnosis of heteroplasmic mutations. The material on which to carry out the diagnosis has to be carefully chosen to ensure that it is representative of the mutational burden of the fetus. This is not a trivial issue given the significant variation in terms of mutation load that is found across tissues within one individual. Finally, the diagnosis is not based on absence or presence of an affected allele, but on the heteroplasmic load of a mutation. Given that the pathogenic threshold varies across different mutations, it is necessary to establish a safe heteroplasmic load for each mutation. Furthermore, due to the somatic bottleneck, the heteroplasmic load identified in an embryo may change during fetal development. This results in prenatal or preimplantation diagnosis that is not able to fully exclude the disease in the offspring, but to very significantly reduce its risk.

Currently, women carrying a mutation in their mitochondrial DNA can opt for oocyte donation, prenatal testing, or preimplantation genetic testing. A recent development that is still in an experimental stage is mitochondrial replacement therapy, and in the future mitochondrial DNA editing may also become a possibility (Figure 12.4).

12.4.1 Oocyte Donation

The first and most straightforward strategy to prevent the transmission of mitochondrially encoded mutations is to use donated oocytes in an assisted reproduction treatment. With this method, the transmission of any mitochondrial DNA mutation will be prevented, and the high success rates of assisted reproduction with donated oocytes makes it an effective solution for women of all reproductive ages, independent of their ovarian reserve. The main drawback of such a technique is the lack of genetic link between the prospective mother and the child. The approach also requires medically assisted reproduction, which is not only an invasive medical procedure but is also costly in countries where it is not covered by public healthcare. In addition, oocyte donation is not allowed in all countries, and can add to the financial burden of IVF as well.

12.4.2 Prenatal Testing

Prenatal testing is performed on a biopsy of chorionic villi obtained between weeks 11 and 14 of gestation, or on a sample of amniotic fluid collected between weeks 15 and 20. While prenatal testing for nuclear DNA mutations can be carried out as for any monogenic disorder, the diagnosis of heteroplasmic mitochondrial DNA mutations is more challenging. This is mainly because the procedure relies on the analysis of a sample that might not represent the mutation loads in the different tissues of the developing fetus. In addition, the association between mutation load and diseased phenotype is not always certain, especially for mid-range frequencies (around 50%), complicating the counseling process. Despite the mosaic nature of the placenta in terms of the mitochondrial DNA, data on pathogenic mutations such as the ones causing NARP/Leigh syndrome show good correspondence between the extraembryonic and fetal tissues, although with some exceptions. Overall, few data are available on the outcomes of prenatal testing for heteroplasmic mutations, and in combination with the technical challenges and the difficult counseling of cases with mid-range heteroplasmic loads, it is still an approach with limited applicability.

12.4.3 Preimplantation Genetic Testing

During preimplantation genetic testing, oocytes are fertilized in vitro, biopsied at the cleavage or blastocyst stage, and only embryos free of the disease under consideration are transferred to the uterus (see Chapter 13). In the case of mitochondrial DNA mutations, the diagnosis is not only based on the absence or presence of the mutation, but also on being able to accurately establish the mutation load in the embryo. In this sense, preimplantation genetic testing for heteroplasmic mitochondrial DNA mutations has similar technical hurdles to prenatal testing.

A first issue is the question of whether all cells of the embryo are equivalent in terms of mutation load, and whether it is best to biopsy at the cleavage or at the blastocyst stage. As stated before, the sample we will use to make the diagnosis must be informative of the whole embryo, and therefore we must know how cell divisions will change the mitochondrial DNA presence in different cells of the same embryo. Different groups have shown relatively homogeneous heteroplasmic loads across the blastomeres of cleavage-stage embryos. Some exceptions were found, typically patient or mutation specific, and generally due to one of the blastomeres of the embryo being an outlier as compared with the rest of the cells of the same embryo [8,20].

In blastocysts, the correspondence between the mutation load in trophectoderm biopsies and inner cell mass still needs to be proven. The trophectoderm undergoes a higher number of cell divisions, and the analyzed biopsy is just a small part of the total cell pool, and therefore it is potentially subject to more variation. At present, the biopsy of one blastomere (or even two in case of novel mutations) is the most used approach. However, only a handful of groups worldwide have been reporting healthy live births after preimplantation testing for mitochondrial disease, and no guidelines or recommendations have yet been developed.

The second point to address is the method used for the quantification of the heteroplasmic load. Several methods have been employed, such as semi-quantitative fluorescence polymerase chain reaction (PCR) with restriction enzyme digestion or allele refractory mutation system (ARMS)-based quantitative PCR, but no consensus has been reached yet on the optimal strategy for heteroplasmy quantification.

The last problem concerns the threshold for calling an embryo transferrable. As previously explained, the clinical manifestation of a mitochondrial mutation is very variable, and is not strictly associated with a high mutation load. In addition, every mutation appears to have a different threshold for inducing a phenotype, and the genetic background of each individual can change the appearance and severity of the pathology. Nevertheless, despite the very complex evaluation of the pathogenicity of mitochondrial DNA mutations, it has been seen that the majority of patients (>95%) that had a mutation load of 18% or below in their muscle tissue did not develop any pathologic symptoms, irrespective of the characteristics of the mutation itself. Based on this, and assuming that the mutation load detected in the blastomere is drifting significantly in the different tissues of the fetus, it has been suggested that 18% can be considered as the upper threshold for considering an embryo suitable for transfer [8].

Despite the critical points listed above, preimplantation genetic testing for mitochondrial disease has been performed by several different groups, and these have reported live births after preimplantation genetic testing for mutations causing MELAS, Leigh syndrome, NARP, and Leber hereditary optic neuropathy. The newborns tested disease-free and displayed a mutation load in line with that of the biopsied embryo, which was reassuring regarding the reliability of the technique. However, attention must still be given, and children extensively tested to achieve a full validation of the method. One case has been reported where a child displaying symptoms of mitochondrial disease was born after preimplantation genetic testing on trophectoderm biopsy despite the relatively low amount of mutated mitochondrial DNA in the transferred embryo. The embryo was diagnosed as carrying a load of 12% of the m.3243A>G mutation, but testing of the child at 6 weeks and 18 months of age showed that it carried heteroplasmic loads of 47% and 46% in blood, and 52% and 42% in urine, respectively [21].

12.4.4 Mitochondrial Replacement Therapy

In the case of homoplasmic mutations or for women who do not produce oocytes with low heteroplasmic loads, techniques relying on the selection of mutation-free embryos are not useful. An alternative approach that has been recently developed and which is still at an experimental stage is mitochondrial replacement therapy [22]. Mitochondrial replacement therapy is based on nuclear transfer, where the nuclear genome is transferred from the oocyte of the patient to an enucleated donor oocyte that carries a normal mitochondrial genome. This results in an oocyte with the genetic content of the prospective mother (and of the father in the case of pronuclear transfer) and the healthy mitochondria from the oocyte donor; children born after such a technique have been named "three-parent babies." Several strategies can be used to perform the transfer of the nuclear DNA, such as transfer of the metaphase spindle, transfer of the pronuclei of a fertilized zygote, or transfer of the second polar body extruded after fertilization. All these procedures, with varying efficiency, have proven to be effective in the generation of competent oocytes carrying a minimal carryover of mutant mitochondrial DNA, first in animal models and later in humans. Despite its obvious theoretical advantages, this technique has many bottlenecks in terms of efficiency and its safety and is still being debated. First of all, the interactions between mitochondrial and nuclear DNA (mitonuclear interactions) are disrupted when performing nuclear transfer. Reconstituted oocytes will have a different match between the mitochondrial and nuclear genomes, and in animal models this has been shown to have a deep impact on their fitness, including regulation of their metabolism and longevity [23,24].

Despite the fact that mitonuclear interactions must de facto be tolerant to different nuclear counterparts (due to the paternal contribution in sexual reproduction, which is normally an "unknown" nuclear DNA to the maternal mitochondrial DNA), there is increasing evidence that human populations maintain similar allelic combinations of mitochondrial DNA and nuclear DNA variants during evolution, and that there may be mitonuclear linkage disequilibrium throughout the whole genome [25]. Although the association of specific mitochondrial and nuclear variants is not very strong, the consequences of modifying these interactions remain poorly understood. To prevent detrimental interactions, the matching of the nuclear content of the patient and the mitochondrial background of the oocyte donor will minimize the genetic distance between the different genomes, and avoid the occurrence of undesired problems. Since a match between the different nuclear genomes seems unlikely, matching of the haplogroup between donor and recipient appears the best option.

Another problem is the potential carryover of mutated mitochondrial genome when the nuclear content is transferred. Most of the techniques mentioned above have a very low carryover of mitochondrial DNA (2–5%), but it has been observed that a small pool of mutated mitochondrial DNA can quickly take over during development and become predominant. Different groups have observed that when mouse embryonic stem cell lines are derived from an embryo generated by mitochondrial replacement therapy and left in culture, in a small but relevant proportion of them the mutant mitochondrial DNA progressively increases its load and becomes the most abundant mitochondrial genome after only a few passages. The reason for this is not yet known, although some groups have speculated that some specific mitochondrial genomes might contain variants that confer a replicative advantage, but it raises some concerns about the possible reversion of the mutated mitochondrial DNA in the developing fetus.

The issues surrounding this technique have instigated most European countries and the USA to impose a moratorium on the use of this technique while safety concerns are addressed. At the same time, the first "three-parent babies" have already been born in different countries such as Mexico, where the regulations did not specifically forbid this kind of experiment. The UK was the first country to allow the option of mitochondrial replacement therapy at the national level, although very strict regulations are in place regarding which cases qualify to receive the treatment.

12.4.5 Editing of the Mitochondrial DNA

The advent of novel gene-editing techniques such as CRISPR/Cas and its more recent variations has promising translational potential in the clinic, as the editing of embryos, oocytes, and spermatozoa to prevent the transmission of hereditary diseases has become a concrete possibility [26]. Nevertheless, despite the variable success in the editing of the nuclear genome, dealing with the mitochondrial DNA is a different challenge. The main problem is that it is located inside the mitochondrial double membrane. Therefore, the editing machinery must contain a specific mitochondrial targeting sequence, which implies additional modifications that are difficult to achieve in a consistent way. Additionally, the number of copies of the mitochondrial DNA within a cell requires a very high editing efficiency, since the majority of the molecules of the cell must be edited to achieve the desired switch in phenotype. Despite the difficulties, several techniques have been optimized for the genetic editing of the mitochondrial genome. Among the ones reported to date, transcription activator-like effector nucleases (TALENs) and zinc finger nucleases (ZFNs) have shown the greatest efficiency in editing of the mitochondrial DNA, in both somatic and germline cells, where a significant reduction in the presence of pathogenic variants specifically targeted by the nucleases has been achieved. Recent work has demonstrated the first successful in vivo treatment for the reduction of the heteroplasmic levels of a disease-causing mutation in mice hearts, paving the way to preclinical testing of TALEN-based therapies. A potential problem when aiming to edit the mitochondrial DNA of oocytes and embryos is the very low rate of mitochondrial DNA replication. If the editing strategy is based on the removal of the mutant mitochondrial DNA, lack of replication will result in an inability to compensate for the removal of the mutant molecules by replicating the wild-type genomes. This, in turn, would result in mitochondrial DNA depletion, which would be potentially highly detrimental for embryonic development.

This problem might have been recently solved by the development of a more refined editing method allowing actual editing of the mitochondrial DNA sequence, rather than its elimination. The new system includes a TALEN protein bound to a mitochondrial localization tag and a cytosine base editor derived from a bacterial toxin, and called DddA-derived cytosine base editor (DdCBE). This editing method has been shown to efficiently edit human cell lines and to reduce their heteroplasmy level without reducing the copy number [27]. However, the current potential of the technique is limited to the editing of C bases preceded by a T, and relies on mitochondrial DNA replication, so its efficiency in gametes and embryos must be proven.

Despite these shortcomings, the promising results reported so far indicate that the editing of mitochondrial genes encoded by nuclear DNA will be successfully achieved by CRISPR/Cas, while TALEN- and ZFN-based methods may be eventually successful in the editing of mitochondrial DNA.

12.5 Mitochondrial DNA Variation and Fertility

Because of the important role of mitochondria in the development and function of the reproductive system, the potential link between infertility and mitochondrial dysfunction has been the subject of much speculation. Conversely, studies directly investigating the link between human infertility and mutations in nuclear-encoded mitochondrial components or in the mitochondrial genome itself are rare [28,29].

Work on animal models has provided interesting insights on a potential link between genetic-based mitochondrial dysfunction and infertility. For instance, in the mouse, disruption of the nuclear-encoded adenine nucleotide transferase *Ant4*, which is selectively expressed in testicular germ cells and the developing ovary, results in complete loss of male fertility, with the germ cells arresting during meiosis. *Immp2l* is also a nuclear-encoded gene that encodes for a part of the mitochondrial inner membrane peptidase complex. Among other symptoms, *Immp2l* mutations affect reproductive function in males and females differently. The males have erectile dysfunction and a significant reduction in germ cells, while females have defects in folliculogenesis and ovulation.

Another example are mice with a large mitochondrial DNA deletion present at high heteroplasmic levels, but insufficient to result in an overt phenotype of multisystem mitochondrial disease. These animals have abnormal sperm morphology, number, and motility, caused by the arrest of spermatocytes at the meiotic stage and subsequent clearance by apoptosis.

Finally, a broadly used animal model in aging research is a mouse strain with a mutation that ablates the proofreading function of the mitochondrial DNA polymerase subunit. These mice show a premature aging phenotype, with high levels of somatic mutations in their mitochondrial DNA. They also present reduced fertility in both sexes. Females become infertile by 20 weeks of age and male show severe oligozoospermia by 40 weeks.

In the human, inherited mitochondrial disorders can present with fertility-related symptoms, the most frequent being hypogonadism. It is worth noting, though, that this is not the case for all patients nor all disorders. For instance, patients with Kearns–Sayre syndrome caused by large-scale deletions of mitochondrial DNA will frequently also present hypogonadism and irregular menses. Patients with MELAS commonly have endocrine dysfunction, including diabetes and growth hormone deficiency, and a number of studies report low levels of gonadotropins and estradiol, and hypogonadism due to defects of the hypothalamic–pituitary axis. Another example is the association between a variant in the overlapping section of *MT-ATP6* and *MT-ATP8* and a phenotype that includes hypergonadotropic hypogonadism, ataxia, peripheral neuropathy, and diabetes mellitus.

With regard to nuclear-encoded genes, *POLG* mutations have been associated with premature menopause, hypergonadotropic hypogonadism, testicular atrophy, and ovarian dysgenesis, although not all *POLG* syndromes show these symptoms. Other examples are mutations in *RRM2B*, *C10orf2*, and *TYMP*, which cause a severe multisystemic phenotype in some of the patients, including hypogonadism.

The findings in animal models and the association between some cases of inherited mitochondrial disease and hypogonadism make compelling arguments for studying the role of nonsyndromic variants in human infertility. Few studies have investigated this potential link, and most of the results remain inconclusive.

In mice, mitochondrial haplogroups and their matching to the nuclear genome have been proven to have a very significant impact on the general health status of the animal and to its rate of aging [23]. In the human, a growing number of studies show an association of haplogroups with health parameters in general or to predisposition to specific disorders (e.g. association with longevity, cancer predisposition or diabetes [7]). Conversely, in the field of reproductive medicine, the current evidence for a link between haplogroups and reproductive function is very limited. Haplogroups have been suggested to play a role in ovarian reserve, fertilization failure, and sperm quality. For instance, in a study on Caucasian women, haplogroup JT had a lower prevalence in women with diminished ovarian reserve, suggesting a possible protective role for this haplogroup. Haplogroup H has been reported to be more frequent in men with normal sperm motility, while haplogroup T would be more common in men with asthenozoospermia. Another study focusing exclusively on haplogroup U found that there are significant differences in sperm parameters according to sub-haplogroups. For instance, haplogroup U4 had the lowest sperm counts, highest incidence of abnormal sperm cells, and lowest motility while individuals with U5 haplogroup showed the best motility grades of the group. Nevertheless, these associations have been the subject of debate, as other studies have failed to identify a similar link.

Large meta-analyses of genome-wide association studies suggest that there may be an association between nonsyndromic *POLG* variants and both the age of menopause and early menopause. Conversely, other studies that have studied common *POLG* variants in patients with premature ovarian insufficiency have not identified a link between the two. Along similar lines, several studies testing an association of a polymorphic CAG repeat in the *POLG* locus with male infertility have found contradictory results. Also, a small number of studies have reported links between mitochondrial DNA variants and premature ovarian insufficiency, sperm quality and fertilization failure, but their results are sometimes contradictory and remain to be replicated in larger populations.

With regard to mitochondrial DNA copy number, a link has been suggested between mitochondrial DNA depletion in peripheral blood and premature ovarian insufficiency and unfavorable IVF outcomes. Not only the causes of the depletion remain to be elucidated, but other studies have failed to observe a significant correlation between mitochondrial DNA copy number in blood and gamete quality, making these observations of limited clinical value. In men, higher mitochondrial DNA copy number appears to be associated with poor-quality sperm. Interestingly, studies in mouse models support these findings, but the mechanisms behind this link still remain poorly understood. Similarly, mitochondrial DNA deletions have been linked to reduced sperm motility and quality, although it is unclear if the deletions are the cause, or secondary to mitochondrial dysfunction of other origin.

12.6 Concluding Remarks

Mitochondria are key to cells in all stages of development, and genetic variation in the nuclear or mitochondrial DNA can have subtle or drastic impacts on their functionality. Benign variation will modulate aspects of human health such as athletic performance, while pathologic variants can result in severe genetic disease. There is much suggestive evidence of a role for mitochondrial DNA variation in modulating reproductive health at multiple levels. Conversely, there are still few consolidated data linking specific mitochondrial DNA variation to specific phenotypes in early human development, gametogenesis, or infertility, and the conflicting results highlight the need for larger and better controlled studies to provide definitive evidence and to support their potential diagnostic value.

With regard to the prevention of transmission of mutations encoded in the mitochondrial genome, prenatal testing is not broadly used and for preimplantation genetic testing, only a limited number of centers have been successful in the development and clinical application of this challenging diagnosis. Mitochondrial replacement therapy has been developed for those patients for whom preimplantation genetic testing is not an option, although its application is still very experimental and controversial. Alternatively, researchers are working on genome-editing tools that can effectively and safely target the mitochondrial genome, with promising results, although they are still a long way from safe clinical application.

References

1. Spinelli JB, Haigis MC. The multifaceted contributions of mitochondria to cellular metabolism. *Nat Cell Biol* 2018;20:745–54.
2. Chen XJ, Butow RA. The organization and inheritance of the mitochondrial genome. *Nat Rev Genet* 2005;6:815–25.
3. Gorman GS, Chinnery PF, DiMauro S, et al. Mitochondrial diseases. *Nat Rev Dis Primers* 2016;2:16080. DOI 10.1038/nrdp.2016.80. PMID: 27775730
4. Otten ABC, Smeets HJM. Evolutionary defined role of the mitochondrial DNA in fertility, disease and ageing. *Hum Reprod Update* 2015;21:671–89.
5. Fenton AR, Jongens TA, Holzbaur ELF. Mitochondrial dynamics: shaping and remodeling an organelle network. *Curr Opin Cell Biol* 2021;68:28–36.
6. Holt IJ, Reyes A. Human mitochondrial DNA replication. *Cold Spring Harb Perspect Biol* 2012;4(12):a012971.
7. Wallace DC. Mitochondrial DNA variation in human radiation and disease. *Cell* 2015;163:33–8.
8. Hellebrekers DMEI, Wolfe R, Hendrickx ATM, et al. PGD and heteroplasmic mitochondrial DNA point mutations: a systematic review estimating the chance of healthy offspring. *Hum Reprod Update* 2012;18:341–9.
9. Luo S, Valencia CA, Zhang J, et al. Biparental inheritance of mitochondrial DNA in humans. *Proc Natl Acad Sci USA* 2018;115 (51):13039–44.
10. St John J. The control of mtDNA replication during differentiation and development. *Biochim Biophys Acta* 2014;1840:1345–54.
11. Van den Ameele J, Li AYZ, Ma H, Chinnery PF. Mitochondrial heteroplasmy beyond the oocyte bottleneck. *Semin Cell Dev Biol* 2020;97:156–66.
12. Gu L, Liu H, Gu X, et al. Metabolic control of oocyte development: linking maternal nutrition and reproductive outcomes. *Cell Mol Life Sci* 2015;72:251–71.
13. Ramalho-Santos J, Varum S, Amaral S, et al. Mitochondrial functionality in reproduction: from gonads and gametes to embryos and embryonic stem cells. *Hum Reprod Update* 2009;15:553–72.
14. Amaral A, Lourenço B, Marques M, Ramalho-Santos J. Mitochondria functionality and sperm quality. *Reproduction* 2013;146:163–74.
15. Dumollard R, Duchen MR, Carroll J. The role of mitochondrial function in the oocyte and embryo. *Curr Top Dev Biol* 2007;77:21–49.
16. Cecchino GN, Garcia-Velasco JA. Mitochondrial DNA copy number as a predictor of embryo viability. *Fertil Steril* 2019;111:205–11.
17. Taugourdeau A, Desquiret-Dumas V, Hamel JF, et al. The mitochondrial DNA content of cumulus cells may help predict embryo implantation. *J Assist Reprod Genet* 2019;36:223–8.
18. Smeets HJM, Sallevelt SCEH, Dreesen JCFM, de Die-Smulders CEM, de Coo IFM. Preventing the transmission of mitochondrial DNA disorders using prenatal or preimplantation genetic diagnosis. *Ann NY Acad Sci* 2015;1350:29–36.
19. Burgstaller JP, Johnston IG, Poulton J. Mitochondrial DNA disease and developmental implications for reproductive strategies. *Mol Hum Reprod* 2015;21:11–22.
20. Sallevelt SCEH, Dreesen JCFM, Drüsedau M, et al. Preimplantation genetic diagnosis in mitochondrial DNA disorders: challenge and success. *J Med Genet* 2013;50:125–32.
21. Mitalipov S, Amato P, Parry S, Falk MJ. Limitations of preimplantation genetic diagnosis for mitochondrial DNA diseases. *Cell Rep* 2014;7:935–7.
22. Craven L, Alston CL, Taylor RW, Turnbull DM. Recent advances in mitochondrial disease. *Annu Rev Genomics Hum Genet* 2017;18 :257–75.
23. Latorre-Pellicer A, Moreno-Loshuertos R, Lechuga-Vieco AV, et al. Mitochondrial and nuclear DNA matching shapes metabolism and healthy ageing. *Nature* 2016;535:561–5.
24. Reinhardt K, Dowling DK, Morrow EH. Mitochondrial replacement, evolution, and the clinic. *Science* 2013;341:1345–6.
25. Sloan DB, Fields PD, Havird JC. Mitonuclear linkage disequilibrium in human populations. *Proc R Soc B Biol Sci* 2015;282:20151704.
26. Gammage PA, Moraes CT, Minczuk M. Mitochondrial genome engineering: the revolution may not be CRISPR-Ized. *Trends Genet* 2018;34:101–10.
27. Mok BY, de Moraes MH, Zeng J, et al. A bacterial cytidine deaminase toxin enables CRISPR-free mitochondrial base editing. *Nature* 2020;583:631–7.
28. Demain LAM, Conway GS, Newman WG. Genetics of mitochondrial dysfunction and infertility. *Clin Genet* 2017;91:199–207.
29. Luo SM, Schatten H, Sun QY. Sperm mitochondria in reproduction: good or bad and where do they go? *J Genet Genomics* 2013;40:549–56.
30. Spelbrink JN. Functional organization of mammalian mitochondrial DNA in nucleoids: history , recent developments , and future challenges. *IUBMB Life* 2010;62:19–32.

Chapter

13

Preimplantation Genetic Testing

Jan Traeger-Synodinos and Carmen Rubio

13.1 Introduction

Preimplantation genetic testing (PGT), until recently known as preimplantation genetic diagnosis (PGD), is an early form of prenatal testing for couples at high risk of transmitting a genetic condition to their offspring, either for monogenic disorders (PGT-M) or chromosomal structural rearrangements (PGT-SR). The goal is to test for the specific genetic status in cells biopsied from oocytes/zygotes or embryos obtained in vitro through assisted reproductive technology (ART) and, following analysis, to transfer to the uterus only those embryos identified as genetically suitable relative to the condition under consideration. The selective transfer of unaffected embryos to the uterus for implantation means that PGT minimizes the need to consider the termination of affected pregnancies. This advantage of PGT means that it has become a widely acceptable alternative to conventional prenatal diagnosis. Of course, PGT is not applied without some ethical concerns. In patients with high risk for transmitting specific genetic conditions to their offspring, including serious single gene disorders and chromosomal rearrangements, there is usually little ethical debate. However, the increased availability and emergence of new uses, such as PGT for autosomal dominant late-onset disorders, cancer predisposition syndromes, PGT for histocompatibility (HLA) typing, and mitochondrial DNA (mtDNA) mutations, are associated with greater ethical controversy and important ethical questions (see Chapter 15).

The feasibility of PGT was facilitated by developments in reproductive medicine, genetics, and biotechnology methods. Edwards and Gardner successfully performed the first known biopsy on rabbit embryos in 1968. In humans, the first clinical application of PGT was reported in 1990 and described the exclusion of an X-linked disease through polymerase chain reaction (PCR)-based sexing of embryos [2]. In the mid-1990s the use of in situ hybridization with chromosome-specific probes was adapted to be used on single interphase nuclei [3]. This method, termed fluorescence in situ hybridization (FISH), was initially used for gender determination. With the development of multicolor FISH, the simultaneous detection of several chromosomes in a single cell became possible. It was the latter technique that was introduced in the setting of ART to select for chromosomally normal embryos in an effort to increase the rates of implantation and successful pregnancy. This procedure has been designated as preimplantation genetic screening (PGS), or more recently preimplantation genetic testing for aneuploidy or PGT-A and is considered "low-risk" preimplantation genetic testing (see Chapter 4).

A first step in both clinical PGT-M and PGT-SR is extensive counseling and genetic work-up of the couple and often their first-degree relatives, assuming that the couple agree to this and that the relatives are available. In addition, it must be ensured that there is a robust genetic test available for testing the cell samples biopsied from each embryo. Particularly for PGT applied to preclude rare (private) genetic conditions (monogenic or chromosomal) the tests themselves may need to be "personalized," although this has become less pertinent with the recent availability of generic universal methods. Furthermore, clinical PGT requires close collaboration between experts in ART and genetics and involves many stages, including evaluation and counseling relative to the genetic and reproductive status of the couple, all stages of ART, zygote/embryo biopsy, the genetic testing and, if implantation occurs after embryo transfer, follow-up of pregnancy and baby (or babies) delivered. The stages of embryo biopsy and genetic analysis must be performed with the highest precision, and both aspects have been continuously improved over the years (Figure 13.1).

The most challenging step in PGT remains the stage of genetic analysis. Since the first report of PGT-M by Handyside et al. in 1990 [2], two major

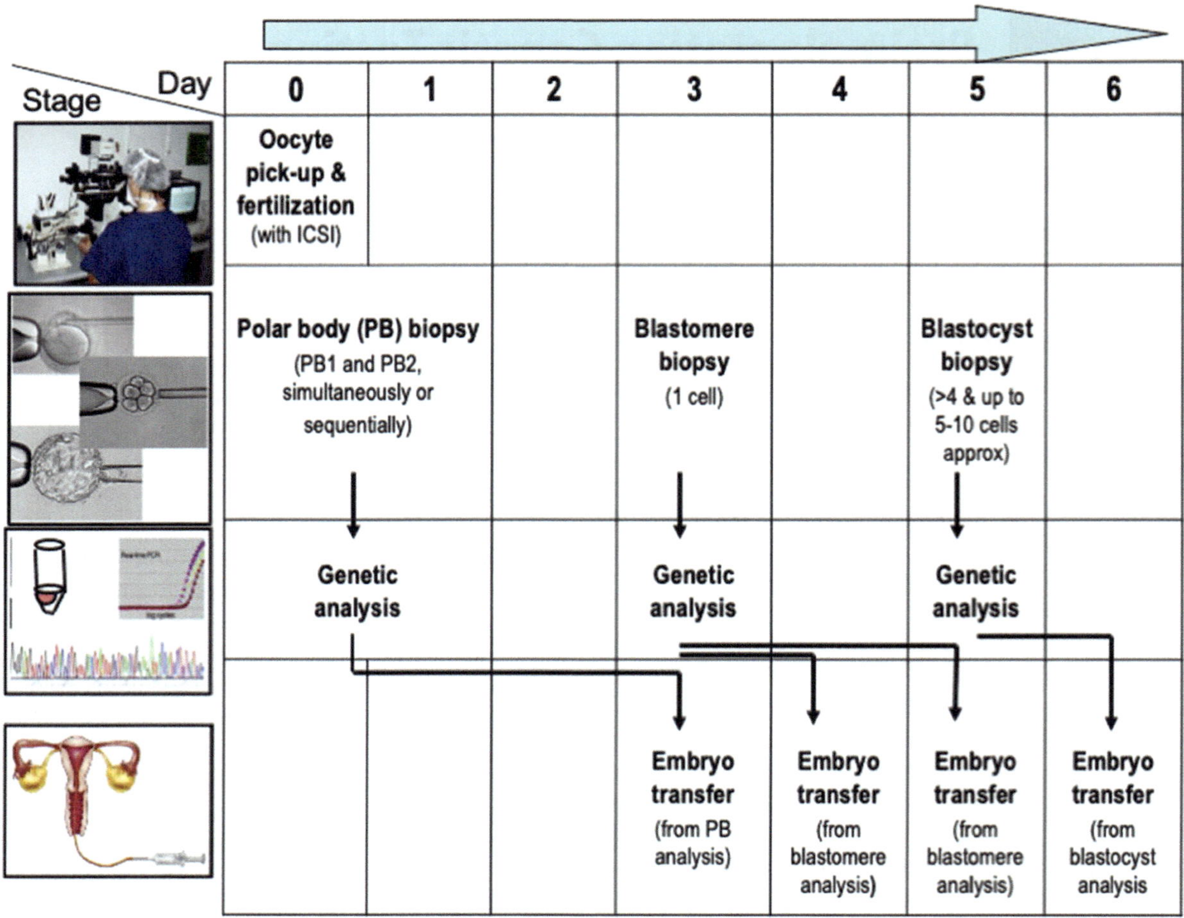

Figure 13.1 Representation of the procedure of preimplantation genetic diagnosis (PGT) for fresh cycles, including the option of the three developmental stages at which cells that are suitable for PGT analysis can be biopsied, and the timing of genetic analysis and embryo transfer. However, oocyte and embryo cryopreservation have become routinely incorporated into assisted reproductive procedures, circumventing the need for such precise timing of stages. PB, polar body; ICSI, intracytoplasmic sperm injection.

diagnostic methods have been internationally applied in centers performing in vitro fertilization (IVF) and PGT. They include PCR-based methods, mainly used for diagnosis in monogenic diseases, while FISH-based techniques have been widely used to analyze chromosomes for patients carrying chromosome abnormalities or for embryo sexing for patients carrying X-linked diseases. Around the year 2000 a newer molecular cytogenetic technique called comparative genome hybridization (CGH) was introduced, facilitating simultaneous evaluation of the copy number of all the chromosomes in a cell. Methods based on analysis of single nucleotide variants (traditionally known as single nucleotide polymorphisms or SNPs) using "SNP arrays" and, more recently, massive parallel sequencing or MPS (also referred to as next-generation sequencing, NGS) have been developed for PGT applications, potentially supporting the simultaneous analysis of monogenic and chromosome abnormalities in a single test, although there are some limitations which remain to be resolved.

This chapter presents an overview that highlights many of the technical and practical issues associated with performing PGT as part of a clinical service to support couples who are at risk for transmitting a genetic condition to have healthy offspring.

13.2 ART and Sources of Genetic Material for PGT

Assisted reproductive technology is an intrinsic part of PGT. With the exception of the procedure for oocyte/zygote or embryo biopsy, ART within PGT is essentially the same as for infertility treatment, irrespective of whether the couple is fertile or not. Controlled ovarian hyperstimulation supports the production of multiple mature oocytes to create sufficient zygotes/embryos and increases the chance of identifying embryos with the desired genetic status. For PGT methods involving a DNA amplification step, all cumulus cells should be removed from the aspirated oocytes. Furthermore, fertilization should be through intracytoplasmic sperm injection (ICSI) rather than IVF to preclude contamination by residual sperm associated with the latter.

A prerequisite for the genetic analysis in PGT is the acquisition of genetic material for analysis through a biopsy step. Needless to say, this step should be performed by a highly skilled embryologist to ensure successful biopsy while maintaining embryo viability.

There are essentially three developmental stages at which cells suitable for PGT analysis can be biopsied. These include polar bodies (PBs) from the oocyte/zygote stage, blastomeres from cleavage-stage embryos, or trophectoderm cells from blastocysts (Figure 13.1). Both cleavage-stage biopsy and PB biopsy have been in use for over 20 years, whereas blastocyst biopsy has been more recently introduced. The first step of any biopsy procedure is to breach the zona pellucida, which surrounds the oocyte or embryo until the expanded blastocyst stage, by mechanical means, chemical means (with acid Tyrode's solution), or subsequently with laser. Each zona breaching method and biopsy stage has relative advantages and disadvantages, and selection is made based on the indication for which PGT is to be performed, as well as the number and quality of zygotes/embyros that are available. For zona breaching acid Tyrode's solution was initially most used, but since 2003 laser breaching has become the preferred method, as it reduces manipulation time of embryos and improves the efficiency. The majority of PGT cycles performed to date have used cleavage-stage biopsy [4,5], although currently blastocyst biopsy is mostly performed.

The recent observations of DNA in the blastocoel fluid (BF) of blastocysts has led to some interest in the potential of developing methods to support noninvasive PGT approaches. However, these studies have highlighted several technical limitations related to the low quantity and overall quality of the DNA in the BF, as well as whether the DNA represents the genetic status of the embryo from which it has been isolated, or just a subset of cells. More research is required to evaluate whether BF DNA constitutes a valid source for PGT. Another recent approach is the analysis of the embryonic cell-free DNA secreted into the culture media, without any previous manipulation of the embryo or the blastocyst [6].

13.2.1 Polar Body Biopsy

Polar bodies are produced in the first and second meiotic division as oocytes complete maturation upon fertilization. In PGT the genotype status of the oocyte should be based on analysis of both the first and second PBs to preclude misdiagnosis, which may arise from recombination or allele dropout (ADO) (monogenic diseases), or by nondisjunction or meiotic errors (chromosomal analysis).

Timing of PB biopsy is critical to support optimal biopsy success and diagnostic result. Biopsy of both PBs may be done sequentially or simultaneously. Sequential biopsy disadvantageously requires two manipulations, whereas with simultaneous biopsy the first PB may degenerate, precluding completion of an accurate diagnosis, and furthermore there may be difficulties distinguishing each PB. The optimal time for biopsy of the first PB is considered to be 4–12 hours after ICSI and 8–16 hours for the second PB [4].

For PB biopsy, zona breaching is performed using either mechanical or laser methods, since acid Tyrode's solution may adversely affect subsequent oocyte development. Although laser biopsy offers the advantage of speed, recent evidence indicates that it may have a negative effect on subsequent embryo quality. Relative to legal, ethical, and safety issues, PB biopsy is advantageous since manipulations involve oocytes rather than embryos; moreover, the genetic material removed is not destined to become part of the developing embryo. In fact, PB biopsy is the only option in countries where legislation prevents embryo biopsy. In addition, PB biopsy allows more time to complete a genetic diagnosis. However, PB analysis is only useful for excluding mutations or aneuploidies of maternal origin, and the genetic contribution from the father remains unknown. Another practical disadvantage of PB biopsy for PGT is that diagnosis of both PBs doubles the number of samples for analysis. Furthermore, since some fertilized oocytes fail to develop, analysis of all PBs may be considered a waste of time and resources, unless the biopsy samples are frozen/preserved until embryo development can be assessed, removing the time advantage of PB biopsy as compared to cleavage-stage embryo biopsy.

13.2.2 Cleavage-Stage Embryo Biopsy

Biopsy of blastomeres from cleavage-stage embryos is performed on day 3 about 66–72 hours following ICSI, when the early embryo has around 6–10 cells. On the third day post insemination the cells are still totipotent and the cells of the embryos are usually not yet adhering to one another (compacting). In the event that compaction has started, brief exposure of the embryos to Ca^{2+}/Mg^{2+}-free media will reduce adherence between cells, facilitating the removal of a blastomere. Mechanical, chemical, and laser are all suitable for zona breaching when performing cleavage-stage biopsy [4].

Preimplantation genetic diagnosis may be based on analysis of either one or two biopsied blastomeres from a single embryo. In earlier years, PGT analysis of two cells presented a means to potentially improve accuracy of diagnostic outcome. However, with remarkable improvements in single-cell genetic testing technologies, this is no longer an issue. Furthermore, results of recent studies confirm that removal of two cells at the cleavage stage is more detrimental to pregnancy outcome than removal of only a single cell.

The analysis of blastomeres is suitable for all PGT indications. The major disadvantage of cleavage-stage biopsy is the limited amount of material present in a single blastomere for analysis. In addition, high rates of mosaicism are observed in embryos at this early stage of development, although for monogenic PGT, errors attributed to false-positive or false-negative results may be minimized by applying appropriate diagnostic strategies and interpretation of results.

13.2.3 Blastocyst-Stage Biopsy

The blastocyst develops about 5–6 days post insemination and contains approximately 100 cells that can be characterized into two cell lineages, the outer trophectoderm (TE) and the inner cell mass (ICM). Zona breaching may be performed by mechanical means or laser and it is usually done in the TE region furthest away (opposite) from the ICM.

With blastocyst biopsy several cells may be removed for analysis, a potential advantage compared to blastomere biopsy for subsequent diagnosis. However, many embryos fail to reach this stage of development. The fact that TE cells do not contribute to the embryo but eventually form the placenta and other extraembryonic tissue (comparable to an early chorionic villus sampling) partly reduces ethical considerations.

The first clinical PGT cycles based on blastocyst biopsy were reported in 2005 with high survival and pregnancy rates [7]. Blastocyst biopsy is becoming more widely used, supported by improvements in culture medium, although with current protocols only about 40–50% of preimplantation embryos develop to this stage in vitro, which limits the application of this biopsy method for PGT. In addition, the time available to complete diagnosis based on blastocyst biopsy is very limited if fresh embryo transfer is to be performed by day 6. The option of cryopreservation may address this issue, allowing transfer of embryos in a subsequent cycle.

13.2.4 Embryo Cryopreservation in PGT

Embryo cryopreservation has become a routine step in many assisted reproductive procedures. There are several situations in PGT when it is considered appropriate to freeze embryos. Besides cases when ovarian hyperstimulation syndrome has occurred, cryopreservation may provide more time to complete a diagnosis. In addition, there are often "unaffected" supernumerary embryos in PGT cycles. Preimplantation embryos usually undergo biopsy prior to cryopreservation and thus do not have an intact zona pellucida, although current cryopreservation protocols do not generally differ between biopsied versus intact embryos [4].

Reports of cycle outcomes for cryopreserved PGT embryos are not entirely conclusive [4]. Generally intact embryos survive better, and vitrification generally shows higher survival rates compared with slow freezing. The approach that is currently most widely practised is to perform embryo biopsy on day 5 or 6 blastocysts, followed by vitrification of all blastocysts; genetic testing is performed following vitrification, facilitating transfer of genetically suitable embryos in a subsequently planned cycle.

13.3 PGT for Single-Gene Disorders (PGT-M)

13.3.1 Indications for Monogenic PGT

To date PGT-M has been reported for hundreds of different, mainly severe, monogenic conditions, with autosomal recessive, autosomal dominant, or X-linked transmission [5]. Not surprisingly, the majority of PGT cycles have been applied for the common autosomal recessive disorders, including cystic fibrosis and the β-hemoglobinopathies (β-thalassemia major and sickle cell syndromes), followed by spinal muscular atrophy. For autosomal dominant disorders, the largest numbers of cycles have been reported for myotonic dystrophy type I and Huntington's disease (HD). For the X-linked disorders, fragile X syndrome is the most common indication, followed by Duchenne muscular dystrophy and hemophilia A and B. Preimplantation genetic testing for most other single-gene disorders is less frequently requested and performed.

The fact that PGT precludes the need to consider pregnancy termination means that its application has also been adopted for many conditions that are not traditionally considered for conventional prenatal diagnosis. Besides late-onset disorders (e.g. HD) or high penetrance cancer predisposition such as familial adenomatous polyposis I, the use of PGT has also been extended to cancer predisposition syndromes with lower penetrance (e.g. breast cancer caused by *BRCA1* mutations), and even conditions that are not life-threatening such as nonsyndromic deafness. Furthermore, the analysis of embryos for nonmedical reasons is used for the selection of a histocompatible sibling to facilitate a bone marrow transplant in a sick child. This can be done for HLA matching alone (e.g. to treat a child affected with leukemia) or in combination with PGT-M to exclude the familial monogenic disease (e.g. to treat a child with β-thalassemia major). For both applications, there was initially much ethical debate, but certainly HLA combined with PGT-M is now widely accepted, as evidenced by the steadily increasing number of requests and cycles each year for HLA-PGT [5]. Finally in the context of mitochondrial disorders, a non-Mendelian indication, PGT potentially offers a superior alternative to conventional prenatal diagnosis, apparently substantially reducing the risk of transmitting the mitochondrial disorder, although only a few cases have been reported to date. However, recent progress in nuclear transfer techniques potentially offers an alternative to PGT for preventing mitochondrial disorders, although further evaluation relative to safety and clinical efficacy is indicated [8].

13.3.2 Diagnostic Strategies for PGT-M

Preimplantation genetic diagnosis is feasible for any single-gene disorder, provided either the disease-causing mutation and/or the disease-associated phase within a family is confirmed. This information is mandatory to support the selection and application of an appropriate protocol with a high diagnostic value. The molecular genetics associated with a disorder largely determines the strategy and method(s) selected for performing the PGT-M (see below). Factors that should be considered include the mode of inheritance within the family and the nature of the mutations involved; most single-gene disorders are caused by variations of single or up to a few nucleotides and larger gene rearrangements (usually deletions) are rare. In the early years the strategies for PGT-M tended to focus on the detection (or exclusion) of the disease-causing mutation(s) alone, or for X-linked disorders, the detection (or not) of Y-specific amplification products. With the evolution of molecular technologies, most notably improved *Taq* polymerases, instrumentation, and accumulated experience, along with the observed risks of failure to amplify targeted alleles (i.e. ADO), current recommendations include the analysis of multiple targets across the genomic region associated with the disease in the format of linkage analysis of polymorphic markers. In this way, linkage analysis, with or without concurrent analysis for the disease-causing mutation (s), supports confirmation of the diagnosis through results from more than a single genomic region. The main limitation of this approach versus direct genotyping alone is the need to perform a family analysis.

Polymorphic markers include SNPs or microsatellite repeats, otherwise known as short tandem repeats (STRs). Until the introduction of newer technologies such as SNP arrays and NGS, both of which support parallel analysis of many SNPs simultaneously, STRs were preferably used in PGT-M protocols. SNPs usually only have two alternative alleles, reducing the likelihood of finding informativity in order to distinguish the high- versus low-risk allele in a family; in a typical couple, about 20–40% of all SNPs will be informative. In contrast, each STR usually has several potential allelic variants with a higher chance of finding informative STRs in a couple requesting PGT-M, and thus STRs are considered more useful for PGT-M applications. Additionally, STR analysis is easier than SNP analysis when using

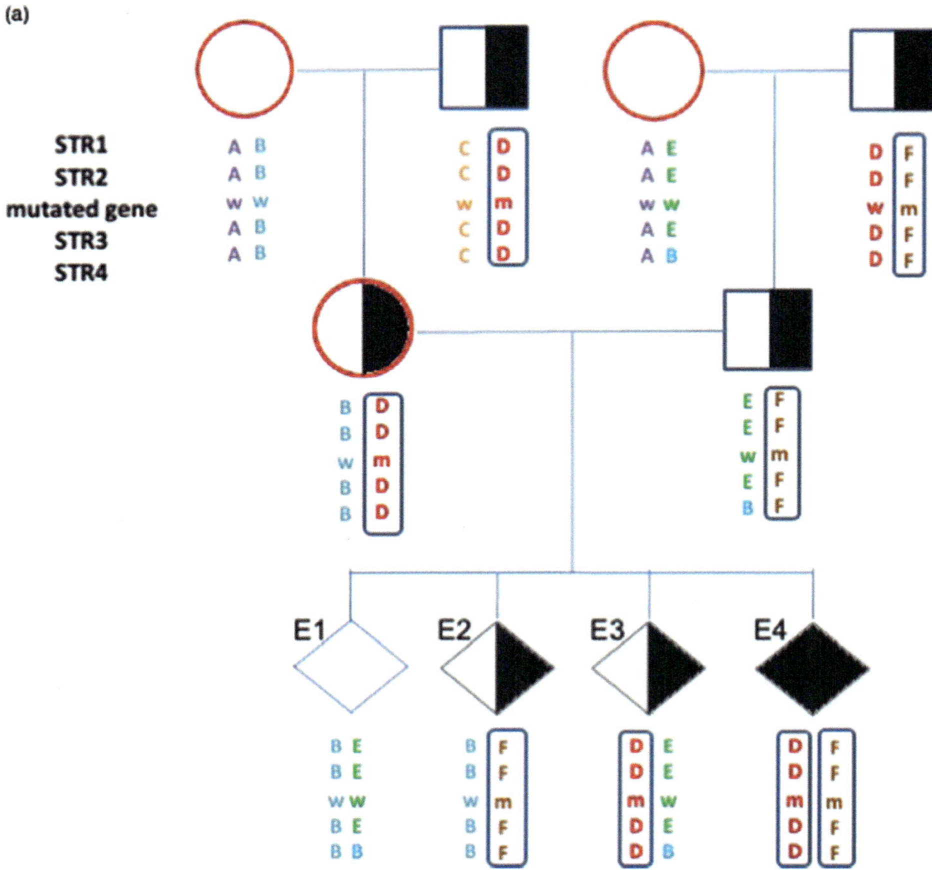

Figure 13.2 (a) The phase of informative short tandem repeats (STRs) syntenic to a disease locus in an imaginary family, illustrating the haplotypes on each allele. (b) The results of sizing of the informative parental STRs following polymerase chain amplification of STR markers and analysis on automatic sequencer.

direct PCR-based methods, through simply sizing the previously amplified fluorescently labeled repeats on an automated genome analyzer (sequencer) (Figure 13.2).

13.3.3 Principles of Amplification-Based PGT-M and Protocol Work-Up

The first step in setting up an amplification-based PGT-M protocol for any disorder involves locating the gene and/or disease-associated mutations in silico through any of the many databases for human genetic information available in the public domain (NCBI, Ensembl, UCSC, Blast). These databases will also support the identification of closely located STRs (ideally within 1 Mb distant from the disease-associated gene) and also facilitate the design of primer sets for direct mutation detection (if appropriate) and the amplification of the STRs. If possible, informative STRs should be selected (ideally at least two to four) to flank the disease variant(s) at a maximum distance of about 1 Mb to preclude misdiagnosis in the event of recombination.

During the work-up of a couple, informative STRs linked to the disease-causing gene are identified by testing appropriate family members with known genetic status, such that during the PGT-M application embryo diagnosis can be based on the detection of high-risk and low-risk alleles at each linked STR (see Figure 13.2). The PGT-M can be done either in the form of a multiplex PCR (see Section 13.3.4) or as multiple independent PCR assays if preceded by a whole-genome amplification (WGA) step. For PGT-M to preclude transmission of dominant disorders, where one of the partners is the only family member affected and the de novo germline mutation is known,

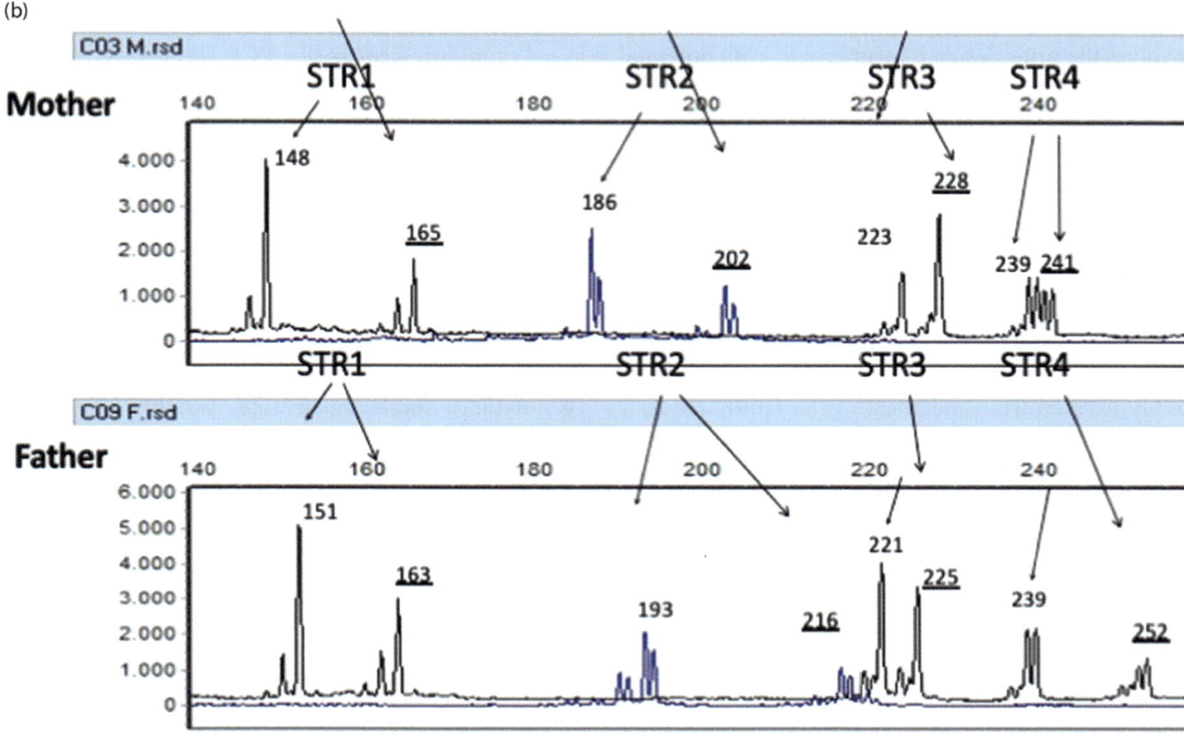

Figure 13.2 *(cont.)*

the PGT-M protocol can be based on an assay for the mutation alone. However, to prevent misdiagnosis caused by ADO, it is preferable to support the mutation diagnosis with linked markers. For de novo mutations that have arisen in the male partner, individual sperm can be tested to establish the phase of high-risk and low-risk STRs; if it has arisen in the female partner, then first and second PBs can be tested. This can be done as part of the first PGT-M cycle, with PB biopsy at day 0 and embryo biopsy at day 3 or 5. However, this approach is not ideal due to the negative consequences of multiple biopsy on embryo survival.

The huge spectrum of disease indications, along with family-specific differences, even for the same disease indication, means that there are very few completely generic PGT-M protocols, with the possible exception of those involving a prior step of WGA coupled with the analysis of large panels of STR markers (see Section 13.3.5 and Chapter 2).

Preimplantation genetic testing for single-gene disorders requires an initial amplification step, essential with current technologies to facilitate the analysis of the tiny amount of DNA sample, often no more than that in a single cell. Until recently PCR has constituted the amplification step in most diagnostic protocols in PGT-M, although WGA-based protocols are becoming more widely adopted.

13.3.4 PCR-Based Protocols for Monogenic PGT

Polymerase chain reaction from minimal sample quantity is associated with certain limitations and pitfalls. These include the occurrence of complete amplification failure, the occurrence of ADO (when only one of the two target alleles amplifies to detectable levels), and finally the likelihood of sample contamination. Total PCR failure may preclude a diagnostic result, something that is undesirable but which will not lead to transfer of an affected embryo. However, the ADO and/or contamination may lead to an unacceptable misdiagnosis. It is paramount that PCR-based PGT-M protocols address and monitor these issues [9].

Both PCR failure and ADO can be minimized if protocols are stringently optimized prior to clinical application. To this end, for direct multiplex-based PGT-M protocols it is general practice that work-up involves the prior design of primer sets, with potential for robust and efficient co-amplification in multiplex PCR formats (e.g. similar annealing temperatures, minimal cross-hybridization). Preclinical experiments are carried out to optimize the efficiency of amplification and minimize ADO for all amplicons included in the multiplex PCR protocol. To "model" the amounts of DNA in just a single cell (around 6 pg) it is general strategy to test genomic DNA diluted to a few genome equivalents, and then single cells such as single lymphocytes, fibroblasts, or buccal cells, preferably isolated from members of the family requesting the PGT-M. These model single cells are considered acceptable in the light that single blastomeres from research embryos are rarely available, plus they are of uncertain quality and genotype.

The high number of amplification cycles required for analysis of a single or only a few cells leaves PCR-based PGT-M protocols extremely vulnerable to the accidental introduction of contaminating DNA at any of the many stages in PGT. Measures taken to preclude contamination during embryo fertilization have been mentioned already (see Section 13.2). With respect to the genetic diagnosis, both operator contamination and amplicon carryover contamination are a possibility, but these can be effectively precluded if all stages of PCR set-up are performed using stringent "single-cell analysis" conditions. It is extremely important to physically separate the pre-PCR steps from those of the PCR itself, and then finally any post-PCR steps. These include the set-up of PCR in dedicated ultraviolet (UV)-treated hoods, using exclusive UV-treated equipment, entirely separated from any post-PCR processing, and with operators wearing appropriate attire and employing correct handling procedures.

As an additional precaution, contamination during any PGT-M procedure should be monitored by the inclusion of negative controls and blanks at all stages. Additionally, the inclusion of STRs supports the detection of likely contamination within each individual sample, such that identification of alleles not consistent with those expected from the parents implies the presence of contaminating DNA. The inclusion of several STRs linked to the disease-causing locus (gene) also monitors ADO and supports the accuracy of the genotyping.

The preclinical work-up required for PGT-M tests is challenging and time-consuming and is the major bottleneck within a clinical PGT-M service. The pioneering PGT-M cases for autosomal recessive disorders used nonfluorescent nested PCR and gel electrophoresis to identify the causative mutation, for example nested-PCR followed by restriction enzyme analysis or nested-PCR followed by denaturing gradient gel electrophoresis (DGGE). A major improvement in detecting PCR products from single cells came with the introduction of the labeling of PCR products by fluorescence. This led to the development of more sensitive sophisticated assays and instruments, such as capillary electrophoresis for amplicon sizing or mini-sequencing, or alternatively real-time PCR for genotyping single nucleotide variants. Improved DNA polymerases and in silico PCR-primer design means that PCR-based protocols perform efficiently in multiplex formats to analyze many loci/markers directly in a single cell. However, they continue to be technically exacting and labor intensive, and must be performed using highly stringent conditions [9].

13.3.5 Whole-Genome Amplification

To overcome the limitations of testing single cells directly with PCR-based protocols, WGA of the DNA in the single (or few) cells prior to the genotyping step represents an efficient option. Whole-genome amplification methods can be grouped into those based on PCR (involving thermal cycling) or isothermal amplification (see Chapter 2).

The early PCR-WGA methods employed the use of *Taq* polymerases and degenerate or semidegenerate primers, but they had limited success. They were usually very time-consuming, involving protocols that required around 16 hours to complete, and the multiple degenerate primers required were relatively costly. Current PCR-based methods for WGA consist of a DNA fragmentation step followed by a step to ligate adaptors to enable PCR amplification. Available as commercial kits, these protocols produce about 3–5 μg of DNA from a single cell, with fragment sizes up to 500 bp long.

Isothermal amplification techniques include multiple displacement amplification (MDA) which is based on the strand displacement amplification properties of phi29 DNA polymerase in the presence of random exonuclease-resistant primers.

Both PCR-based and MDA WGA methods have been used for single-cell WGA in the context of PGT-M. However, both approaches have limitations;

neither method usually amplifies the entire genome with equal efficiency, and a few regions may consistently fail to amplify. Both methods are vulnerable to high levels of ADO (up to around 30%), which means that the derivation of the genotype may require subsequent analysis of many disease-linked loci to support an accurate and reliable result.

Although the initial amplification step in WGA requires measures to preclude contamination in the same way as PGT-M protocols based on direct PCR methods, all subsequent PCR-based assays do not require "single-cell analysis conditions" and almost any PCR-based method can be used for genotyping. This means that work-up for PGT-M involving WGA does not require optimization of a multiplex-PCR protocol, but only work-up of the genetic phasing within each family. Furthermore, once the sample has been subjected to WGA, it can be analyzed numerous times if needed. One of the first successful strategies for PGT-M based on MDA WGA was termed "preimplantation genetic haplotyping" (PGH), which uses multiple STRs across the disease locus to derive the linked normal and affected haplotype.

WGA is also an essential step when performing PGT for structural chromosomal abnormalities using CGH or array-CGH (see Section 13.4.2).

13.3.6 SNP-Based Genome-Wide Haplotyping for PGT-M

In the last few years, new technologies have supported the development of genome-wide linkage-based analysis of SNPs, applicable to PGT-M. For each embryo, SNPs are phased to identify the parental origin of chromosomes at all SNP loci. In order to establish the high-risk versus low-risk haplotypes relative to the disease, the analysis of embryo-biopsy material has to be performed alongside genotyping of the parents and other first-degree relatives.

These preferably include either already-born children, the parents of the couple, and less commonly on genetic material from a previous abortion of an affected fetus, all with known status relative to the familial disease targeted for exclusion in the embryos. Assuming all these prerequisites are met, then this approach offers a generic and universal approach for PGT-M, essentially suitable for any monogenic disorder, precluding the need for time- and labor-consuming design of customized protocols and their stringent optimization [10].

For an SNP to be informative one member of the couple (potential parents) must be heterozygous for the alleles and the other homozygous (e.g. AB and AA or AB and BB). As already mentioned, only 20–40% of all SNPs are likely to be informative in a couple, and thus many SNPs across the disease-associated locus will need to be analyzed to support a robust diagnosis. Thus, besides massively parallel analysis of SNPs in each sample, genome-wide haplotyping has to be supported by customized software to allow visualization and interpretation of the complex data (see also Chapter 2).

Genome-wide SNP haplotyping does have some limitations and pitfalls. It is not suited to couples who do not have available family members, or for highly consanguineous couples or if the disease-causing variant in one member of the couple is de novo. Furthermore, the interpretation of results may be confounded by chromosome recombination events. The robust analysis of X chromosome disorders is enhanced if the gender of the embryo is also assessed. Finally, genome-wide SNP haplotyping may reveal certain incidental findings, including the presence of copy number variations, nonpaternity, couple consanguinity, and also embryo gender (when the latter is not part of the diagnostic protocol).

The genome-wide analysis across all chromosomes potentially supports the detection of chromosome aneuploidies within the same PGT analysis [11]. However, to be cost-effective the analysis of complex genomic regions should be avoided (e.g. repetitive regions), facilitating the potential to analyze relevant regions with twofold to threefold higher efficiency. Whether for PGT-M only, or the simultaneous detection of certain chromosomal abnormalities (see below), SNP-based genome-wide analysis usually requires support by commercially available protocols, as few laboratories are capable of setting up both the laboratory and software analysis in-house.

13.3.7 PGT-M Applications with Special Considerations

Some indications for PGT-M may require variation of the classic strategies described above and, in addition, very rarely the genetic transmission within a family may not follow expected norms, for example in cases with mosaicism.

Among disorders that require a slightly different analytical approach are the repeat expansion disorders. Here, the disease-causing mutation comprises expansion of a nucleotide repeat (most commonly a trinucleotide repeat) whereby the affected allele size may be refractory to amplification by PCR (although all normal alleles should be within the range that do amplify). Some repeat expansion disorders are autosomal dominant (e.g. myotonic dystrophy, HD), others are X-linked (e.g. fragile X syndrome), and a few are autosomal recessive (e.g. Friedreich ataxia). In autosomal forms, normal embryos will be characterized by amplification of the normal allele from the affected parent(s). In addition, modified PCR protocols have been described that support amplification of both the expanded and nonexpanded repeats and distinguish embryos with an expansion from those without. These protocols use a PCR with three primers in a semi-nested format, whereby one of the primers is complementary to the triplet-repeat sequence (tri-primer format), The tri-primer reaction can be incorporated into the multiplex reaction alongside the informative STRs. In X-linked disorders, notably fragile X syndrome, affected males will give no amplification product from the primers used to amplify the repeat region, necessitating use of STRs to prevent misdiagnosis caused by ADO.

For the expansion repeat disorder HD, special considerations are associated with ethics rather than technical issues. For individuals aware of their HD status, PGT can involve direct testing of the disease-causing expanded CAG repeat in exon 1 of the *HTT* gene (supported by linked STRs). Alternatively, for individuals at risk of developing HD who do not wish to know their risk status, two approaches for PGT exist: exclusion testing or nondisclosure testing. Exclusion testing is based on identifying the grandparental origin of the two *HTT* alleles and only embryos with an *HTT* allele from the nonaffected grandparent are considered transferable. Disadvantages of exclusion testing include the requirement of DNA from family members with known disease status to support the derivation of phase, and also that statistically half of the rejected embryos are unaffected. For the nondisclosure approach, PGT embryos are analyzed directly for a CAG repeat, without revealing details of PGT results to the prospective parents. However, nondisclosure PGT remains controversial and is not widely accepted. For both exclusion testing and nondisclosure testing, all couples have to undergo ART, even if they are not at risk.

For gender-linked disorders, PCR was the original method used for sexing embryos [2]. However, it soon became apparent that the amplification of Y-chromosome-specific regions was susceptible to ADO, and thus FISH, a more robust technique, became the preferred method to determine gender. However, following marked improvements in PCR technologies, PCR-based methods have again become the preferred method for X-linked PGT cases, with the added advantage that they also support a specific diagnosis for the X-linked disease. With this approach, more genetically transferable embryos are available, advantageous compared with gender analysis alone, whereby half of all male embryos were likely unaffected but not transferred.

For families affected by mtDNA mutations, reproductive choices include prenatal diagnosis of a spontaneous conception, mtDNA replacement, the use of donor oocytes, and more recently PGT. However, the complicated pathobiology of genetic disorders caused by mtDNA mutations may limit the reliability of prenatal diagnosis and PGT. Mitochondria are maternally inherited and all cells have thousands of mitochondria, each containing many copies of the mitochondrial genome (see Chapter 12). A characteristic of disorders caused by mtDNA mutations is the coexistence of normal and mutated mtDNA within a single cell, known as heteroplasmy. The onset of clinical symptoms, as well as the variability in phenotype and penetrance of mtDNA diseases, is influenced by factors including the threshold effect (whereby the mutational load in a cell has to exceed a certain level), mitotic segregation, clonal expansion, and the so-called bottleneck effect. Although the precise mechanism for the latter is still unclear, one prevalent theory is that there is a rapid reduction of mtDNA in early germ cell development followed by rapid expansion as the oocyte matures, causing some oocytes to have reduced mutation levels, which may sometimes be below the threshold level. The outcome of prenatal diagnosis when analyzing the proportion of mutant mtDNA in amniotic or trophoblast cells is unpredictable. However, it is possible to measure the ratio of mutated to normal mtDNA in an embryonic cell with a fair degree of accuracy and recent studies (for certain mtDNA mutations at least) have indicated that if the mutation load is low then there is a low risk that this will lead to the development of an affected child. However, carriers of mtDNA mutations must be appropriately counseled, and although PGT for mtDNA disorders represents a promising option, further investigation and evaluation are required before wider clinical application [12].

Preimplantation genetic testing for HLA typing aims to establish a pregnancy that is HLA compatible with a sibling who requires hematopoietic stem-cell transplantation. The HLA typing alone can be performed when the affected child requires transplantation for treatment of a noninherited disease, or simultaneously with PGT to exclude a familial inherited single-gene disorder, as each offspring is also at risk of being affected. Although initially considered ethically controversial, HLA typing is now widely accepted as a valuable treatment option since transplantation success is strongly associated with the degree of HLA match between the donor and recipient. The first combined PGT-HLA matching case was reported in 2001 for a case of Fanconi anemia [12], and since then the number of reported cases has been increasing every year [5]. According to data from the European Society of Human Reproduction and Embryology (ESHRE), the most common indication to date is for HLA matching combined with PGT for β-thalassemia and/or sickle cell syndromes [5]. Fanconi anemia, Gaucher disease, adrenoleukodystrophy, and osteopetrosis are among other single-gene disorders for which HLA typing combined with PGT have been reported, and there are many reports of cycles for HLA typing for acquired diseases [5].

Reliable tests for HLA typing of preimplantation embryos have to address several technical challenges, including the large size of the region (3.6 Mb), the large number of loci within the region, the high level of polymorphism, and the possibility of recombination within the HLA region. HLA haplotyping by linkage analysis of polymorphic STR markers located throughout the HLA region is the established strategy for PGT-HLA, matching the haplotypes between the embryos and the affected sibling.

Genetic chance presents another limitation on the clinical utility of PGT for HLA typing. The chance that two siblings will be HLA matched is 25%, among which only 75% will be unaffected for an autosomal recessive disorder (18.8%, or 3 of every 16 embryos) and 50% for an autosomal dominant or X-linked disorder, the latter if only sexing is applied (12.5% of all embryos fertilized in any cycle). As only approximately 25% of PGT cycles result in pregnancy and delivery, the overall success rate for HLA-PGT matching is predicted to rarely surpass about 10–15% for any cycle initiated, limiting the ultimate success (i.e. the birth of an unaffected histocompatible baby). In fact a recent multicenter study by 14 centers involving 716 cycles performed between August 2001 and September 2015 [13] observed that the diagnostic efficacy of PGT-HLA protocols was lower in comparison to published PGT-HLA protocols and to that reported for general PGT applications by ESHRE (78.5 vs. 94.1% and vs. 92.6%, respectively). The clinical efficacy proved very difficult to assess due to inadequate follow-up of both the ART/PGT and hemopoietic stem cell transplant procedure outcomes. When successful, PGT-HLA is a valuable procedure for treating an affected child, but families should also be made aware of the limitations before embarking on such a procedure.

13.4 PGT for Structural Rearrangements (PGT-SR)

13.4.1 Indications for PGT-SR

Indications for PGT-SR include balanced translocations or Robertsonian translocations. It is estimated that 1 in 625 (0.16%) individuals carries a balanced chromosomal rearrangement, with translocations (reciprocal or Robertsonian) and inversions as the most frequent (see Chapters 1 and 7). An enormous number of reciprocal translocations have been described, with breakpoints scattered throughout all chromosomal regions. With few exceptions, the majority of reciprocal translocations are family-specific. Robertsonian translocations originate through centric fusion of the long arms of any two of the five acrocentric chromosomes (13, 14, 15, 21, and 22), with loss of the short arms. An inversion is a two-break event involving just one chromosome. Carriers of confirmed balanced chromosomal rearrangement are, most of the time, phenotypically normal. The production of unbalanced gametes, due to unfavorable meiotic segregation when the derivative chromosomes pair up at meiosis I, puts them at risk for repeated miscarriages, subfertility, or infertility, and unbalanced offspring. This explains why the incidence of balanced chromosomal rearrangement rises to about 3% in couples with reproductive problems. The theoretical chance of producing normal or balanced gametes is 4 of 32 for reciprocal translocation carriers and 4 of 16 for Robertsonian translocation carriers. However, the actual percentage depends on several factors, including which chromosomes are involved, the location of the breakpoints, and the gender of the carrier [14]. For carriers of an inversion the chromosomal imbalance occurs if there is, within the inverted segment, a crossover at the first meiotic division, between the inversion chromosome and the normal homolog resulting in a recombinant chromosome, which may produce four different segregates.

13.4.2 Analytical Methods for Structural Chromosomal Abnormalities

13.4.2.1 FISH-Based Protocols for PGT-SR

Fluorescence in situ hybridization utilizes chromosome-specific DNA probes labeled with different colored fluorochromes that are hybridized to the target DNA (interphase or metaphase) fixed to a microscopic slide. As a result, FISH detects a chromosome or a part of a chromosome, thereby allowing determination of the copy number of that region in a particular sample. Microscopic evaluation and computerized imaging systems enable fluorescent probe signals to be identified and counted.

Initial applications using FISH for diagnosing chromosomal translocations involved painting probes for metaphase chromosomes of PBs. A major shortcoming of this method was that only translocations of the female could be examined. Subsequently, homemade probes isolated from cosmid, YAC (yeast artificial chromosome), or BAC (bacterial artificial chromosome) libraries spanning or flanking the translocation breakpoints, were applied to the interphase nucleus of a blastomere from cleavage-stage embryos. These probes gave very accurate results because they allowed the distinction between normal and balanced genotypes, but were very time-consuming to isolate. The commercial availability of centromeric, locus-specific, and subtelomeric probes opened completely new perspectives for PGT in translocation carriers. The PGT for Robertsonian translocations is relatively simple as dual-color FISH with one probe to enumerate each chromosome is sufficient. The situation is more complex for reciprocal translocations. Ideally the protocol should include a FISH probe for both centric segments and both the subtelomeric regions of the translocated segments. In this case two scoring errors would have to occur to misdiagnose an unbalanced product as normal/balanced. However, four suitable probes may not be available. Most centers performing PGT for reciprocal translocations use a combination of two subtelomeric probes distal to the breakpoint, along with a centromeric probe for one of the two chromosomes involved, in order to support the distinction between normal/balanced genotypes and unbalanced genotypes; all probes are labeled with different colors [14]. Carriers of reciprocal translocations and inversions usually have a unique rearrangement, necessitating the development, preclinical testing, and optimization of the individualized probe mixture for each couple.

The optimal outcome of the FISH method also depends on cell fixation onto a microscope slide, a critical step that requires skill and experience. Another drawback of FISH for PGT-SR includes failure of FISH probes to hybridize, and the accuracy of interpretation can be affected by signal overlap or split signals. Finally, the limited number of distinct fluorochromes available for labeling the DNA probes limits the number of chromosomes that can be assessed in a single analysis, although PGT-SR usually interrogates only the chromosomes involved in the translocation.

13.4.2.2 PCR-Based Protocols for PGT-SR

Polymerase chain reaction-based strategies have also been used to detect chromosome imbalances in embryos derived from carriers of chromosome rearrangements. PCR-based protocols that utilize informative STRs on both segments (flanking the breakpoint) of the translocated chromosomes have been reported [15]. This approach offers the advantage of determining the inheritance of individual chromosomes and allowing the detection of uniparental disomy. PCR-based PGT protocols for translocations have the potential to overcome several inherent limitations of FISH-based tests. They are not dependent on cell fixation and microscopic signal interpretation but, on the other hand, are subject to amplification failure, ADO, and contamination as discussed above. For the PCR-based methods also work-up is necessary prior to any clinical case in order to identify which STR polymorphisms are informative. Similar to FISH-based PGT for structural chromosome abnormalities, the number of chromosomes assessed remains limited, usually to the chromosomes involved in the translocation.

13.4.2.3 CGH-Based PGT-SR

By contrast, genome-wide aneuploidy screening of a single cell can be performed by CGH. However, the DNA content of a single cell (about 5–10 pg) is insufficient for direct use in CGH with consistent results, and therefore the method requires the use of WGA. The CGH technique employs a competitive hybridization of differentially labeled DNA samples (DNA from the sample is green; chromosomally normal reference DNA is red) to normal metaphase chromosomes on a microscope slide. With image-processing software, the metaphase chromosomes are evaluated. An excess of green fluorescence on a specific chromosome is indicative of a chromosomal gain, whereas an excess of red fluorescence is indicative of chromosome loss. This CGH technology has been applied with success in PGT cycles for aneuploidy [16]. The obvious advantage of CGH over FISH is that the copy number of all chromosomes can be determined. In addition, CGH provides a more detailed picture of the entire length of each chromosome, enabling the detection of imbalance of chromosomal segments. However, CGH is labor-intensive and time-consuming, which limits its diagnostic potential for PGT. For this reasons, this technology has been scarcely applied to clinical cases in favor of array-CGH.

13.4.2.4 Array CGH-Based PGT-SR

Array comparative genome hybridization (aCGH) is a technique based on the same principle as CGH, but the competitive hybridization of differentially labeled test and reference DNA samples does not take place on metaphase spreads but on DNA probes fixed to a microscope slide (microarray). Each probe is specific to a different chromosomal region and occupies a discrete spot on a slide. After the labeling and hybridization procedure, a laser scanner is used to excite the hybridized fluorochromes and read and store the resulting images of the hybridization. Chromosomal loss or gain is revealed by the color adopted by each spot after hybridization based on software packages (i.e. ratio of fluorescence intensity for the two colors). Similar to CGH, an amplification step is required to increase the initial quantity of DNA. Different WGA procedures are available (see Section 13.3.5) but the one commonly used for aCGH applications is a WGA procedure based on random fragmentation of genomic DNA, ligation of adaptors, and subsequent amplification by PCR (see Chapter 2).

Over the last few years, a variety of single cells, including PBs, blastomeres, or TE biopsies, have been successfully screened using aCGH on different platforms with the aim of aneuploidy testing. The commercial availability of different types of high-resolution microarrays opened new opportunities for clinical applications for translocation carriers. One of the first validation studies reported for PGT in couples with structural chromosome abnormalities [17] reported 28 cycles of PGT for chromosomal translocations. In this study paired comparison between aCGH and PCR-based PGT testing on 200 embryos from translocation carriers was performed. A high percentage of embryos (93%) were successfully diagnosed with the aCGH technique. Results were concordant between both diagnostic techniques for all embryos tested. Moreover, the study demonstrated that aCGH analysis can identify segmental aneusomy, as small as 2.5 Mb, from patients carrying a translocation. Only 30 embryos (16%) were found balanced/normal for the translocation as well as for the other chromosomes. A total of 22 embryos were transferred in 17 transfer procedures. A pregnancy was obtained in 12 patients, from which all were ongoing at the time of publication. This demonstrates that aCGH can detect chromosome imbalances in embryos, also providing the added benefit of simultaneous aneuploidy screening for all 24 chromosomes [17]. Array-CGH potentially provides a generic solution which should be applicable to all translocations, avoiding the need for patient individualized tests.

Apart from errors directly related to the chromosomes involved in the translocation, it has been postulated that the segregation patterns could have an influence on meiotic synapsis and disjunction of other bivalents, causing aneuploidy in chromosomes not involved in the translocation. This phenomenon has been named the interchromosomal effect (ICE) [18], although its existence is controversial and conflicting results have been reported in the literature. In light of these findings, and with the technical feasibility offered by aCGH, aneuploidy screening, simultaneous to the PGT of translocations, could be considered to detect other possible postzygotic chromosome errors. The number of patients involved in the study of Fiorentino et al. [17] is too small to speculate on the beneficial effect of testing for all the chromosomes. In a recently published study, Mateu-Brull et al. concluded that there were no significant differences for total ICE among the different types of rearrangements, with higher pure ICE only for Robertsonian translocations [19].

13.4.2.5 SNP Microarrays for PGT-SR

Originally SNP arrays were designed to genotype human DNA at thousands of SNPs across the genome simultaneously; however, applications have expanded to include the detection and characterization of copy number variations. Biallelic SNPs, in which one of two bases are present, referred to generically as A and B, are valuable markers, and hundreds of thousands of SNPs can be genotyped simultaneously with the use of SNP arrays. Furthermore, for molecular cytogenetics, analysis of the ratio of the intensity of the B to the A alleles at heterozygous loci enables detection of duplications and deletions from whole chromosomes to small regions with high resolution. In the duplications, the B-allele ratio at heterozygous loci splits into two bands representing loci that are either AAB or ABB. In deletions, loss of heterozygosity (LOH) is detected by the absence of the heterozygous band. This technology was applied to identify balanced embryos in carriers of structural chromosome abnormalities, and more recently new platforms have been developed to offer a genome-wide approach for PGT-M, PGT-SR, and PGT-A in human embryo biopsy samples [11,20].

13.4.2.6 NGS-Based PGT-SR

Several NGS platforms have been validated for detecting whole chromosome aneuploidies, and a few studies have also shown high concordance for segmental aneuploidies in carriers of translocations. A retrospective analysis of 287 blastomere biopsies and 55 TE biopsies showed that the concordance rate on the overall diagnosis between aCGH and NGS on abnormal samples was 100% irrespective of the type of biopsy. This study confirmed the applicability of a commercial NGS-based workflow for preimplantation testing for reciprocal translocations/inversions [21].

13.5 Clinical Outcomes of PGT for Monogenic Diseases and Structural Chromosome Abnormalities

The ultimate goal of PGT is the birth of a healthy baby. To this end the optimization of every stage of PGT is important. To a great extent, the success of the stages of ART and embryology tend to be case-dependent, especially since a large percentage of cases undergoing PGT are also infertile (almost 40% of patients who undergo PGT for monogenic disorders, and over 50% for those who have PGT for structural abnormalities). It is really only the stage of genetic diagnosis which can be optimized and this has been facilitated by the continuous research efforts of PGT centers, along with improvements in reagents and instruments available for genetic analysis in single cells.

According to annual ESHRE PGT data collections for cycles performed between 1997 and 2015 inclusive [5], there have been over 21 000 monogenic PGT cycles (19 340 with identification of the single-gene disorder and 1832 in sex-linked diseases with only gender selection). These data clearly demonstrate an upward trend in the proportion of embryos with a successful diagnosis following testing for monogenic PCR; in earlier datasets successful diagnosis was achieved in just over 80% of embryos analyzed by PCR, whereas in the most recent data collection for cycles performed up to 2015, diagnosis was achieved in about 89% of the biopsies. In cases with only gender selection for X-linked diseases, informativity was also achieved in 89% of the biopsies. The percentage of cases with embryo transfer was 71.5% and 59.8%, respectively. Pregnancy rates with positive heartbeat for single-gene disorders in the last data collection were 23.9% per oocyte retrieval (OR) and 33.5% per embryo transfer (ET) with PCR, and 17.1% and 29.5% respectively in X-linked diseases with gender selection only.

With respect to PGT for chromosomal abnormalities, strategies based on FISH were proven effective, resulting in thousands of successful cycles according to the data collected by the ESHRE PGT consortium. However, in the latest data collection published [5], aCGH, SNP microarrays, and NGS were incorporated as new technologies to analyze not only the chromosomes involved in the arrangement but also aneuploidy for the remaining chromosomes. From 2013 to 2015, the use of arrays for genetic testing increased from 36 to 52% in PGT-SR, from 0 to 2% in SNP microarrays, and from 0 to 3% in NGS. The uptake of biopsy at the blastocyst stage (from 22% up to 36%) was only observed in cycles for structural chromosomal abnormalities, alongside the application of aCGH, SNP microarrays, and NGS [5]. Of the embryos successfully biopsied, 92.7% gave a diagnostic result, and 61% of the cycles had an embryo transfer. Clinical pregnancy rates were growing for this indication and becoming similar to PGT-M, 18.6% per OR and 34.2% per ET. The data demonstrate that there was little difference in outcome between male and female translocation carriers. However, there was a difference in outcome according to the type of translocation being tested, such that Robertsonian translocations showed a higher number of normal/balanced embryos available for transfer leading to higher pregnancy rates compared with reciprocal translocations, irrespective of whether they were male or female carriers. Reciprocal translocations accounted for most of the cases (60%).

At this point it is important to mention that data collection and follow-up of PGT cycles is challenging due to the increasing number of cycles and centers applying PGT, and also due to the different methodologic strategies that can be combined in the ART laboratory and for genetic diagnosis, including the analysis of aneuploidies in both PGT-SR and PGT-M cases.

Several reports on children born after PGT have been published. In one such report that performed pediatric follow-up in a large series of children born after cleavage-stage biopsy and PGT, the data, collected on 581 post-PGT children, showed that term, birthweight, and major malformation rates were not statistically different from those of 2889 children born after fertility treatment involving ICSI [22]. The overall rates of major malformation among the post-PGT and ICSI children were 2.13 and 3.38%, respectively, and the authors concluded that embryo biopsy does not add risk factors to the health of singleton children born after PGT. More recently, the perinatal follow-up of 430 children born after PGT showed no evidence that PGT treatment increases the risk of congenital malformations or adverse perinatal outcome [23]. Moreover, 5-year-old children born after PGT show normal growth, health, and motor development when compared with children born after IVF/ICSI and natural conception children from families with a genetic disorder [24].

13.6 Accuracy and Quality Assurance in PGT

The absolute accuracy of PGT is difficult to estimate since it is impossible to confirm the diagnosis in every embryo. Pregnancy allows access for reanalysis (either during pregnancy or after birth), but most embryo transfers do not result in pregnancy and confirmatory testing is done on only a minority of untransferred embryos. Thus, misdiagnosis is likely underreported by PGT centers. Unacceptable misdiagnosis has serious consequences whereby the embryo transferred was characterized to be unaffected at PGT, but identified during pregnancy or after birth to be affected. Alternatively, benign misdiagnosis includes cases in which embryos diagnosed as negative for the hereditary mutation during PGT, result in an unaffected pregnancy or child who in fact carries a mutation (for a recessive disease). To date misdiagnosis rates have been estimated based on reporting (anonymously) to the ESHRE PGT consortium, and overall the rates are very low (0.16%). This includes 0.1% for FISH-based cycles, although rates are relatively higher (0.5%) for PCR-based cycles [25].

Audit through confirmatory testing on untransferred embryos (superfluous to reproductive needs of couples) is a good way to evaluate the diagnostic value of PGT protocols and highlight any pitfalls in methods and strategies used. Audit is also an important part of quality control, especially relevant for monogenic PGT since there are few universally standardized protocols, and most are case-specific "in-house" protocols, especially for rare monogenic diseases. The issues of quality control and quality assurance within PGT can be supported through some form of internationally or nationally recognized accreditation [26]. External quality assessment is also an important part of quality management and is an essential component of accreditation schemes [26]. External quality assessment provides an independent measure of service quality and a comparison against other laboratories. Assessment data gives service users (clinics, clinicians, and patients) a way to evaluate whether services reach the required standards. External quality assessment also has an educational element by highlighting variation between laboratories in testing protocols, and differences in interpretation and reporting of results. This information is important not only to support continual improvement of each laboratory but also for professional bodies responsible for providing the most relevant best practice guidelines.

13.7 Future Developments in PGT

Even for fertile couples undergoing PGT, pregnancy rates rarely surpass 30–35%. Thus, ideally, a key objective associated with PGT is also the ability to select the best-quality healthy embryo(s) most likely to implant. Traditionally embryos are selected by morphologic criteria that are, however, difficult to quantify. The analysis of the euploid status of an embryo, a procedure currently termed PGT-A, presents a means of predicting potential embryo implantation and live birth, although there is still considerable controversy over its true clinical benefit. However, NGS for genome analysis does offer generic approaches for simultaneous ploidy analysis (PGT-A) and PGT-M and PGT-SR [10,11 20,21] (see Chapter 2).

Regarding molecular diagnosis, NGS in PGT may facilitate multiple-gene testing, the detection of SNPs, copy number variants, and chromosomal aneuploidies, and even potentially epigenetic profiling. Besides providing a more holistic approach to PGT, NGS in patients undergoing ART may additionally support more "personalized" procedures to optimize success of ART and pregnancy outcomes. However, this technology awaits more knowledge to support interpretation of the hugely complex genetic information it generates. In addition, there are many ethical issues foreseen; for example, how to interpret and report all the numerous genetic variants detected (see Chapter 15). Regarding the impact of biopsy, other alternatives such as blastocentesis or a noninvasive approach based on analysis of the embryonic cell-free DNA secreted into the culture media could have several important advantages when compared with current invasive PGT, avoiding the biopsy and extending the accessibility of PGT to a wider population of patients by minimizing laboratory and personnel expenses.

Although in the future PGT will likely incorporate many new technologies, it is mandatory to establish the safety and clear clinical benefit before they are incorporated within clinical practice. The innate biology of human reproduction will always limit the positive outcome of ART and thus PGT, while additional complex issues such as the impact of epigenetics (discussed in Chapters 5 and 14) on embryos remain unresolved. Ideally all developments in PGT should be based on hypothesis-driven research, and the exchange of expertise and knowledge through international forums, ensuring that PGT is always applied with the highest standards of laboratory, clinical, and ethical conduct.

References

1. Zegers-Hochschild F, Adamson GD, Dyer S, et al. The International Glossary on Infertility and Fertility Care, 2017. *Hum Reprod* 2017; 32(9):1786–801.
2. Handyside AH, Kontogianni EH, Hardy K, Winston RML. Pregnancies from biopsied human preimplantation embryos sexed by Y-specific DNA amplification. *Nature* 1990; 244: 768–70.
3. Griffin DK, Wilton LJ, Handyside AH, et al. Dual fluorescent *in situ* hybridization for simultaneous detection of X and Y chromosome-specific probes for the sexing of human preimplantation embryonic nuclei. *Hum Genet* 1992;89:18–22.
4. Xu K, Montag M. New perspectives on embryo biopsy: not how, but when and why? *Semin Reprod Med* 2012;30:259–66.
5. Coonen E, van Montfoort A , Carvalho F, et al. ESHRE PGT Consortium data collection XVI–XVIII: cycles from 2013 to 2015. *Hum Reprod Open* 2020;2020(4): hoaa043.
6. Rubio C, Rienzi L, Navarro-Sánchez L, et al. Embryonic cell-free DNA versus trophectoderm biopsy for aneuploidy testing: concordance rate and clinical implications. *Fertil Steril* 2019; 112(3):510–19.
7. Kokkali G, Vrettou C, Traeger-Synodinos J, et al. Birth of a healthy infant following trophectoderm biopsy from blastocysts for PGT of beta-thalassaemia major. *Hum Reprod* 2005;20:1855–9.
8. Craven L, Tang MX, Gorman GS, De Sutter P, Heindryckx B. Novel reproductive technologies to prevent mitochondrial disease. *Hum Reprod Update* 2017; 23(5):501–19.
9. Carvalho F, Moutou C, Dimitriadou E, et al. ESHRE PGT Consortium good practice recommendations for the detection of monogenic disorders. *Hum Reprod Open* 2020;2020(3): hoaa018. DOI 10.1093/hropen/hoaa018
10. Handyside AH, Harton GL, Mariani B *et al.* Karyomapping: a universal method for genome-wide analysis of genetic disease based on mapping crossovers between parental haplotypes. *J Med Genet* 2010; 47: 651–8.
11. Masset H, Zamani Esteki M, Dimitriadou E, et al. Multi-centre evaluation of a comprehensive preimplantation genetic test through haplotyping-by-sequencing. *Hum Reprod* 2019; 34(8):1608–19.
12. Hellebrekers DM, Wolfe R, Hendrickx AT, et al. PGT and heteroplasmic mitochondrial DNA point mutations: a

systematic review estimating the chance of healthy offspring. *Hum Reprod Update* 2012;18:341–9.

13. Kakourou G, Kahraman S, Ekmekci GC, et al. The clinical utility of PGD with HLA matching: a collaborative multi-centre ESHRE study. *Hum Reprod* 2018;33(3):520–30.
14. Scriven PN, Handyside AH, Ogilvie CM. Chromosome translocations: segregation modes and strategies for preimplantation genetic diagnosis. *Prenat Diagn* 1998;18:1437–49.
15. Fiorentino F, Kokkali G, Biricik A, et al. Polymerase chain reaction-based detection of chromosomal imbalances on embryos: evolution of preimplantation genetic diagnosis for chromosomal translocations. *Fertil Steril* 2010;94:2001–11.
16. Wilton L. Preimplantation genetic diagnosis and chromosome analysis of blastomeres using comparative genomic hybridization. *Hum Reprod Update* 2005;11: 33–41.
17. Fiorentino F, Spizzichino L, Bono S, et al. PGT for reciprocal and Robertsonian translocations using array comparative genomic hybridization. *Hum Reprod* 2011;26:1925–35.
18. Gianaroli L, Magli MC, Ferraretti AP, et al. Possible interchromosomal effect in embryos generated by gametes from translocation carriers. *Hum Reprod* 2002;17:3201–7.
19. Mateu-Brull E, Rodrigo L, Peinado V, et al. Interchromosomal effect in carriers of translocations and inversions assessed by preimplantation genetic testing for structural rearrangements (PGT-SR). *J Assist Reprod Genet* 2019;36(12):2547–55.
20. Treff NR, Zimmerman R, Bechor E, et al. Validation of concurrent preimplantation genetic testing for polygenic and monogenic disorders, structural rearrangements, and whole and segmental chromosome aneuploidy with a single universal platform. *Eur J Med Genet* 2019;62(8):103647.
21. Chow JFC, Yeung WSB, Lee VCY, Lau EYL, Ng EHY. Evaluation of preimplantation genetic testing for chromosomal structural rearrangement by a commonly used next generation sequencing workflow. *Eur J Obstet Gynecol Reprod Biol* 2018;224:66–73.
22. Liebaers I, Desmyttere S, Verpoest W, et al. Report on a consecutive series of 581 children born after blastomere biopsy for preimplantation genetic diagnosis. *Hum Reprod* 2010;25:275–82.
23. Heijligers M, van Montfoort A, Meijer-Hoogeveen M, et al. Perinatal follow-up of children born after preimplantation genetic diagnosis between 1995 and 2014. *J Assist Reprod Genet* 2018;35 (11):1995–2002.
24. Heijligers M, Peeters A, van Montfoort A, et al. Growth, health, and motor development of 5-year-old children born after preimplantation genetic diagnosis. *Fertil Steril* 2019;111(6):1151–8.
25. Wilton L, Thornhill A, Traeger-Synodinos J, Sermon KD, Harper JC. The causes of misdiagnosis and adverse outcomes in PGT. *Hum Reprod* 2009;24:1221–8.
26. Harper JC, Sengupta S, Vesela K, et al. Accreditation of the PGT laboratory. *Hum Reprod* 2010;25:1051–65.

Chapter

14 Epigenetics and Assisted Reproductive Technology

Aafke P.A. van Montfoort

14.1 Introduction

Since the birth of the first baby via in vitro fertilization (IVF) in 1978, there has been concern about the safety of IVF and other assisted reproduction technology (ART) procedures for the health of ART-conceived children. Data show that ART singletons are at increased risk for adverse perinatal outcomes such as low birthweight and being small for gestational age, and congenital malformations [1]. The biological mechanism behind these risks is mainly unresolved. Since the publication of a few case reports on the incidence of rare imprinting disorders such as Angelman and Beckwith–Wiedemann syndromes in ART-conceived children, epigenetic deregulation has gained increasing attention as a possible common cause for the adverse outcomes. This led to an expansion of ART literature on epigenetic effects. In this chapter, I focus on the current knowledge of epigenetic disturbances in humans, reported after ART in general and in relation to specific ART components, and the difficulties encountered in these kinds of studies. When needed, animal studies will also be mentioned. The subfertility of the population as a possible cause for the epigenetic deregulation is also taken into consideration. Finally, I discuss whether epigenetic effects can be related to the reported health outcome in ART children and if these possible derangements can affect their health at adult age.

14.2 Epigenetics and Gene–Environment Interactions

14.2.1 What Is Epigenetics?

An individual's susceptibility to disease or other phenotypes is not only determined by the DNA sequence inherited from both parents; epigenetic information also plays a role. The Greek prefix *epi* in "epigenetics" implies modifications that are "on top of" the DNA without interference in the DNA sequence itself. These include chromatin modifications such as DNA methylation, histone modifications, histone variants, and RNA-mediated chromatin modifications (see also Chapter 5). An important function of these epigenetic changes is the regulation of gene expression. In this chapter, without denying the importance of the other modifications, mainly DNA methylation is discussed. Methylation is commonly investigated because of the availability of robust techniques to interrogate targeted as well as global changes in DNA methylation, even in a few cells.

Mammalian DNA can become methylated at the carbon-5 position of a cytosine in CpG dinucleotides. Most CpGs within the genome, including CpGs residing in gene bodies, intergenic regions, and repetitive elements such as transposons, are highly methylated, while CpG islands associated with regulatory regions such as gene promoters and enhancers display low or intermediate methylation (with the exception of imprinted regions; see below). Methylation is regulated by DNA methyltransferases (DNMTs). DNMT3a and DNMT3b, together with the cofactor DNMT3L, are responsible for de novo methylation of unmethylated CpG sites. DNMT1, responsible for maintenance methylation, uses the hemimethylated DNA strand to copy methylation patterns to the new strand after DNA replication. DNA methylation of promoter regions is generally associated with transcriptional repression, either directly by interfering with the binding of transcription factors or indirectly by recruitment of methyl CpG-binding domain (MBD) proteins and their associated repressive chromatin remodeling activities [2].

Epigenetic regulation is instrumental in development and cell differentiation. All cells within an organism have the same genotype, but they can differ in phenotype due to epigenetically regulated changes in gene expression that are transmitted to

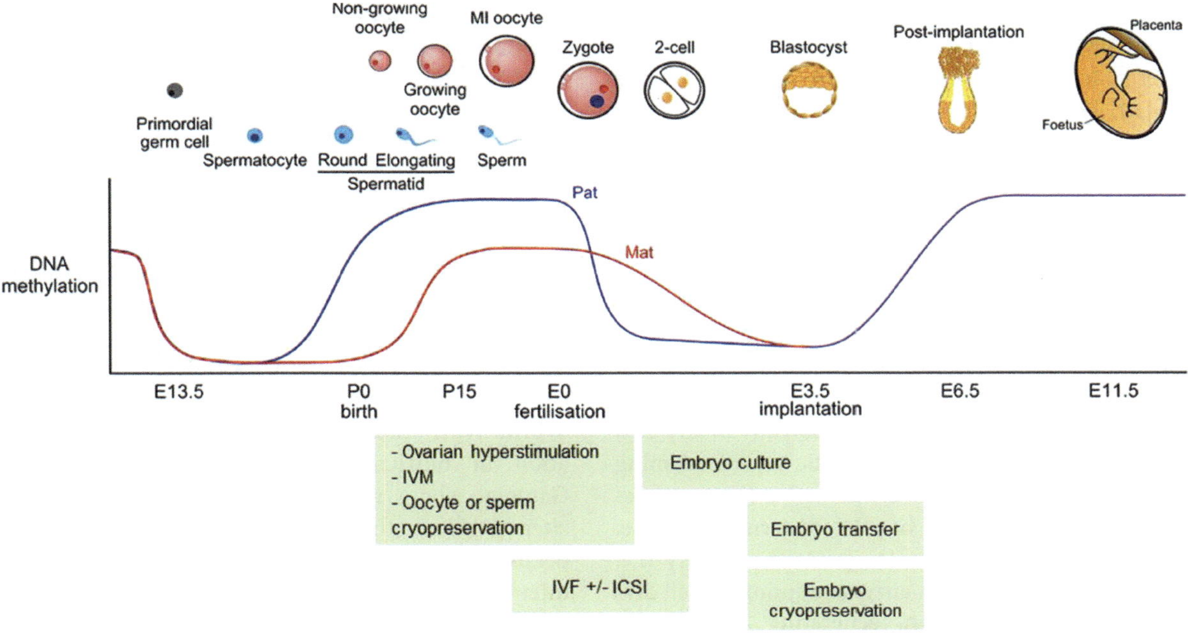

Figure 14.1 DNA methylation changes during developmental epigenetic reprogramming. The two phases of epigenetic reprogramming are schematically depicted. See text for an extensive description. The boxes below indicate the timing of ART procedures in relation to reprogramming events. Source: adapted with permission from Hanna et al. [2].

daughter cells. Also, chromosome stability, stable X chromosome inactivation in females, and sex-specific genomic imprinting are under epigenetic control (the latter is extensively described in Chapter 5). Briefly, in gametes, sex-specific epigenetic marks are deposited (mainly DNA methylation) that, after fertilization, result in parent-of-origin-specific gene expression of these genes. With imprinted genes playing important roles in brain function and fetal development, aberrant imprinting because of uniparental disomy (UPD) or defective methylation can lead to severe diseases characterized with growth problems and/or mental retardation. Given their role in placental and fetal growth and development and the increased risk for low birthweight in IVF children, these genes are often the targets of interest in ART-related epigenetic studies. In particular, *H19*, *IGF2*, and the genes regulated by the *KCNQ1OT1* imprinting control region (ICR; also called germline differentially methylated region or gDMR) draw attention since these are involved in Beckwith–Wiedemann and Silver–Russell syndromes that are characterized by overgrowth and growth retardation, respectively.

In order to acquire germ cell-specific imprints, to remove the parental epigenetic marks acquired during lifetime, and to restore the totipotent state of an early preimplantation embryo after fusion of two differentiated germ cells, DNA methylation marks are erased and remodeled by two phases of epigenetic reprogramming: first at the primordial germ cell (PGC) stage and second after fertilization at the preimplantation embryo stage (see Hanna et al. [2] for a review on this topic and Figure 14.1). Most of the knowledge on epigenetic reprogramming events comes from studies in mouse, but with the development of low-input and single-cell sequencing techniques information on the comparability between mouse and human has been gradually revealed. Early in PGC development, global erasure of preexisting DNA methylation marks in gDMRs of imprinted genes starts, together with the DNA demethylation of promoters, genic, intergenic, and transposon sequences. In human, this demethylation lasts until embryonic week 11, preserving less than 10% methylation (comparable to mouse). Now, the female and male germ cells enter meiotic and mitotic arrest, respectively. Remethylation in the male germ cells starts before birth. In mice it starts at the prospermatogonial stage and is completed in the late spermatogonial stage after birth. Next to sperm-specific methylation, the imprints are also established. At present, three human paternally imprinted gDMRs are known, all located at intergenic regions: between *H19* and *IGF2* (H19 DMR), between *DLK1* and *MEG3* (IG-DMR), and between *GPR1* and *ZDBF2*. In the female germline, around 35 gDMRs have been identified so far. These are generally located in promoter regions (e.g. of *SNRPN* and *KCNQ1OT1)* and are estimated to regulate 50–90 imprinted genes [2]. Remethylation in oocytes starts after birth, around resumption of meiosis, and is completed just before ovulation for some loci. Human oocytes have an average methylation of 54% mainly restricted to gene bodies, while sperm is highly (around 80%) and more uniformly methylated [3]. The second phase of epigenetic reprogramming starts after fertilization with the active (i.e. enzyme-driven) demethylation of the paternal genome. Shortly thereafter, the female genome passively (i.e. maintenance methylation enzyme dilution by cell divisions) loses methylation, generating a totipotent embryo. This continues until the blastocyst stage. The residual methylation of the maternal genome is considerably higher than that of the paternal genome (around 23% vs. 15%, respectively), indicating that more oocyte-derived methylation is transmitted to the embryo than paternal methylation. Among these, but not exclusively, are the paternally and maternally imprinted gDMRs and some repetitive elements. These methylation marks are in most cases maintained throughout the rest of life and any perturbation is therefore likely to have major implications. After implantation, remethylation starts and the inner cell mass becomes differentially methylated from the extraembryonic lineages (Figure 14.1).

14.2.2 Gene–Environment Interactions

Besides the developmentally induced epigenetic changes, the epigenetic marks are also responsive to environmental cues. When this occurs during gametogenesis or the periconceptional and preimplantation embryo stages, epigenetic perturbations will be mitotically amplified and transmitted to the next cell generations and therefore affect all or at least a large number of cells within an individual. This is in contrast to perturbations arising in more differentiated cells that will only affect (part) of a particular organ or tissue. Epigenetic perturbations (or are they adaptations?) around the time of conception have been proposed to be one of the mechanisms linking parental lifestyle (diet, smoking, stress) or toxic exposures to increased morbidity risks in offspring [2,4]. A good example of a periconceptional environmental effect is the rodent model where the mother is given a low-protein diet in the days around conception only. This intervention has been shown to result in a metabolic syndrome-like phenotype with altered growth trajectories, elevated blood pressure, and reduced arterial vasodilation capacity. Indications for similar relations in human stem from the increased risk for cardiometabolic consequences in offspring that were in utero during the Dutch famine, a period of severe food restriction during the Second World War. Another example is the seasonal variation in nutrient availability in rural areas of Gambia, where nutrient availability at conception alters DNA methylation patterns in offspring that persist into childhood [2,4]. Real periconceptional intervention studies in human are difficult to conduct and often unethical. ART is therefore a good "natural" model to study the action of environmental effects around conception on the epigenome in gametes and embryos and on the so-called developmental programming of the offspring.

14.3 Imprinting Disorders in ART Offspring

Concern about epigenetic risks of ART in humans started with two case reports of children with the rare imprinting disorder Angelman syndrome (AS) that were born after intracytoplasmic sperm injection (ICSI) (see van Montfoort et al. [5] and DeAngelis et al. [6] for reviews and Table 14.1). AS has an incidence of 1 in 15 000 births and is characterized by severe mental retardation, behavioral problems, and absence of speech (OMIM #105830). The defect lies in an imprinted region (*SNRPN* cluster) on the maternally inherited chromosome 15 that leads to loss of maternal *UBE3A* expression (Figure 14.2). In more than 95% of cases, the defect is genetic: a deletion of the critical region on the maternally inherited chromosome, maternally inherited *UBE3A* mutations, or paternal UPD of the chromosomal region. Less than 5% arises from a loss of maternal DNA methylation at *SNRPN*. At least eight cases of AS have been reported after ICSI (Table 14.1). Of these, the unexpectedly high number of four cases showed a methylation defect. In one case there was a complete loss of methylation of the *SNRPN* region, while in

Table 14.1. The incidence of IVF/ICSI among cases with imprinting disorder and the general population (see van Montfoort et al. [5] and DeAngelis et al. [6] for more extensive reviews of these studies)

Reference	Type of study	No. of cases	No. of cases with IVF or ICSI	Per cent IVF or ICSI in cases	Per cent IVF or ICSI in population	RR (IVF or ICSI vs. population)	Molecular defect
Beckwith–Wiedemann syndrome							
DeBaun et al. (2003)	Case series	65	IVF (n = 2) and ICSI (n = 5)	4.6[a]	0.76	6.2	5/7 LOM *KCNQ1OT1* gDMR 1/7 GOM *H19* DMR 1/7 no imprint defect 1/7 not known
Maher et al. (2003)	Case series	149	IVF (n = 3) and ICSI (n = 3)	4.0	0.997	4.2[b]	2/6 LOM *KCNQ1OT1* gDMR 4/6 not analyzed
Gicquel et al. (2003)	Case series	149	IVF (n = 4) and ICSI (n = 2)	4.0	1.3	3.2[b]	6/6 LOM *KCNQ1OT1* gDMR
Halliday et al. (2004)	Case control	37	IVF (n = 3) and ICSI (n = 1)	10.8	0.67	4.4[b]	3/4 LOM *KCNQ1OT1* gDMR 1/4 not analyzed
Sutcliffe et al. (2006)	Case series	79	IVF (n = 1) and ICSI (n = 5)	2.9–7.6[c]	0.8	3.7–10.0[b,c]	6/6 LOM *KCNQ1OT1* gDMR
Doornbos et al. (2007)	Case series	71	IVF (n = 4)	5.6	0.92	2.8[b]	4/4 LOM *KCNQ1OT1* gDMR
Hiura et al. (2012)	Survey	70	Not specified (n = 6)	8.6	0.98	NC[d]	3/5 DNA methylation error
Mussa et al. (2017)	Case series	38	ICSI (n = 6)	16.2	2.1	9.1	
Angelman syndrome							
Cox et al. (2002)	Case series	2	ICSI (n = 2)	–	–		2/2 LOM *SNRPN*
Orstavik et al. (2003)	Case report	1	ICSI (n = 1)	–	–		1/1 LOM *SNRPN*

Table 14.1. *(cont.)*

Reference	Type of study	No. of cases	No. of cases with IVF or ICSI	Per cent IVF or ICSI in cases	Per cent IVF or ICSI in population	RR (IVF or ICSI vs. population)	Molecular defect
Ludwig et al. (2005)	Survey	79	ICSI (n = 3)	3.8	–		1/3 LOM *SNRPN* 2/3 maternal deletion 15q11
Sutcliffe et al. (2006)	Case series	75	0	0	0.8	–	–
Doornbos et al. (2007)	Case series	63	0	0	0.92	–	–
Hiura et al. (2012)	Survey	123	Not specified (n = 2)	1.6	0.98	NC[d]	0/2 DNA methylation error

LOM, loss of methylation; GOM, gain of methylation; –, not analyzed; gDMR, germline differentially methylated region; NC, not calculable.
[a] Seven ART-BWS cases were identified in a database. Only three patients were from after 2001, when use of ART was systematically assessed. Only this period is included in the risk assessment.
[b] Risk is significantly increased in IVF compared with non-IVF pregnancies.
[c] Range takes into account the large number lost to follow-up by assuming that all nonresponders conceived naturally.
[d] Not calculable as population data are not specified.

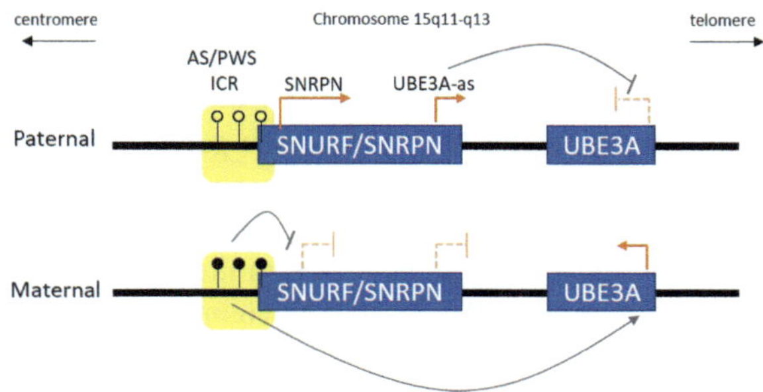

Figure 14.2 Simplified representation of the Prader–Willi syndrome (PWS)/Angelman syndrome (AS) locus on chromosome 15q11–q13. PWS arises from loss of expression of the paternal genes, in most cases caused by a deletion within this region or maternal UPD and rarely by methylation of the paternal ICR. AS arises from loss of maternal *UBE3A* expression as a result of genetic (deletion within the region, *UBE3A* mutation or paternal UPD of the region) or epigenetic (loss of methylation on the maternal ICR) defects. Open circle denotes unmethylated, black circle methylated. Several transcripts are generated from the *SNURF/SNRPN* locus among which the long *SNRPN* transcript and a smaller antisense (as) RNA transcript that is thought to inhibit the expression of *UBE3A*. *UBE3A* is only imprinted in brain.

another case only 10% of the alleles were methylated. This suggests a lack of (re)methylation during oogenesis, or an error in imprint maintenance early in embryonic development. In three case-series the incidence of ART in AS cases was not increased; 2 of 261 AS cases were conceived through IVF or ICSI. This is what would be expected based on a 0.8–0.9% IVF/ICSI prevalence in the general population. Another seven AS cases originated from ovulation induction and/or intrauterine insemination, of which one had a methylation defect at *SNRPN* [5,6].

Beckwith–Wiedeman syndrome (BWS, OMIM #130650), which is also a maternal imprinting disorder, has an incidence of 1 in 13 700. It is a fetal and postnatal overgrowth disorder characterized by a predisposition to tumor development and a variety of complications. The five most common features are macrosomia, macroglossia, abdominal wall defects, ear creases, and neonatal hypoglycemia, but some cases even lack these. BWS is caused by a gain of methylation at the *H19* DMR (5–10% of cases), loss of methylation at *KCNQ1OT1* (50% of cases), paternal UPD of the 11p15.5 region (20–25% of cases), and a mutation in the *CDKN1C* gene (5% of cases). The relative risk (RR) for BWS after IVF (with or without ICSI) versus the risk for BWS in the general population is estimated at between 2.8 and 10.0 (Table 14.1). BWS occurs irrespective of the cause of subfertility and ART characteristics, such as level of hormonal stimulation, mode of fertilization, fresh or frozen embryo transfer, or day of embryo transfer. BWS has also been reported after intrauterine inseminations with or without ovulation induction, or after the use of fertility drugs alone. It should be kept in mind that the variability in clinical features makes it sometimes hard to identify BWS cases, especially since BWS children grow up as adults of normal size with normal intelligence. This might introduce a bias of overidentification of BWS in ART offspring since these tend to be followed up more thoroughly. Nevertheless, it appears that after IVF almost all cases are related to a loss of maternal methylation at *KCNQ1OT1*, while in the general BWS population only around 50% is caused by an epigenetic defect. In six of the ART-BWS cases, the actual level of methylation at the *KCNQ1OT1* locus was measured and ranged from 0 to 28%. Since most of the cases had some maternal methylation left, either an incomplete gain of methylation during oogenesis or a postfertilization imprint maintenance effect is likely to be the cause [5,6].

Prader–Willi syndrome (PWS) is a paternal imprinting disorder on the same region as AS. The molecular defects are mostly deletions of the paternal chromosome (Figure 14.2), just as in the reported ART-PWS cases with molecular data available [6]. Furthermore, four PWS case-series show no or a slightly increased risk of being conceived via ART.

In the other paternal imprinting disorder Silver–Russell syndrome (SRS), hypomethylation of *H19* on paternal chromosome 11 (BWS region) is the most common molecular mechanism, although loci on several other chromosomes are implicated as well. Two case series reported an increased risk for SRS after ART with RRs of 6.6 (95% CI 2.3–19.1) and 12.6 (95% CI 5.0–32.2) [6]. In spite of this, an association between ART and SRS should be interpreted with caution as the complex molecular origin and the wide variation in phenotypes make it difficult to estimate a true prevalence of the disease. The presence of hypomethylation of *H19* in nearly all ART-SRS cases, and the higher rate of methylation errors at other regions in ART-SRS cases, remains nevertheless remarkable [6].

In summary, there is more evidence for an ART-induced increased risk of maternal imprinting disorders than for paternal imprinting disorders, although an association with the latter cannot be excluded. One explanation for an increased incidence could be that some imprints escape erasure or are aberrantly reestablished during gametogenesis, either because of ART-related manipulations or related to the parental subfertility. The cellular mosaicism in methylation errors, as reported in some BWS, AS and SRS cases, is however more suggestive of a postfertilization effect, just as the combination of abnormal maternally and paternally methylated imprints found in some BWS and SRS cases [6,7]. The likelihood of both mechanisms is discussed below. It is important to stress that besides the retrospective nature of all studies and the limited number of cases due to the rarity of the disorders, one other major limitation of most studies on ART and imprinting disorders is the lack of control for confounding factors such as parental age. Even if the increased risk for imprinting disorders after ART appears to be true, the absolute risk is still very small.

14.4 Epigenetic Effects Transmitted by the Subfertile Parents

Before describing the de novo-induced epigenetic effects by ART, the possibility of transmittance of epigenetic alterations from the subfertile patient to their offspring is discussed. There are two types of epigenetic inheritance: intergenerational and transgenerational. The former is used to describe "inheritance" of epigenetic changes by the inherent exposure of the germline (leading to F1), and the germline of the fetus in pregnant females (leading to F2), when an individual (F0) is exposed. Transgenerational epigenetic inheritance refers to transfer of chromosome/chromatin modifications without direct exposure. When F0 is pregnant during exposure, evidence for transgenerational epigenetic inheritance can first be seen in the F3 generation. When F0 is not pregnant during exposure or male, the transgenerational effect can be first seen in the F2 generation.

Evidence for the existence of intergenerational and transgenerational epigenetic inheritance comes mainly from animal studies. After exposure of midgestation rats to the antiandrogenic compound vinclozolin, persistent CpG methylation changes in promotors of F3 sperm, and decreased spermatogenic capacity up to F4 were found. Epigenetic variation can also be transmitted via the male germline. Premating fasting or exposure to a high-fat or low-protein diet in rodents let to phenotypic alterations in glucose and fat metabolism, next to changes in methylation and gene expression of several lipid metabolism-related genes. One mechanism proposed for epigenetic inheritance is resistance to erasure of DNA methylation and histone modifications during reprogramming in the PGCs and the embryo. Epigenetically labile regions (or metastable alleles), such as the Agouti viable yellow (A^{vy}) allele originating from the insertion of a retrotransposon just before the promotor region of the Agouti locus, are good candidates for this resistance. The fur color of mice with this A^{vy} allele differs according to the methylation status of the A^{vy} locus and is susceptible to diet of the midgestation mother for instance. The shift in fur color of offspring is maintained for multiple generations. Another possible mechanism of epigenetic inheritance is via the transmission of molecules other than DNA, such as small noncoding RNAs (ncRNAs). These RNAs pair with complementing mRNA transcripts and label them for degradation. The best-known example for this is a modified *Kit* gene in mouse. In *Kit* wild-type offspring from *Kit* heterozygous parents, the phenotypic characteristics of the mutant allele were still present for several consecutive generations. This could be ascribed to the ncRNAs derived from the mutant *Kit* allele that were transmitted via the male germline. The effect gradually fades away with every generation.

In human, the long generation course complicates studies on epigenetic inheritance. A first indication for at least intergenerational inheritance in human can be derived from a "natural" experiment involving Dutch pregnant women who were subjected to severe famine during the Second World War. In offspring exposed to starvation in the earliest stages of development, when epigenetic reprogramming occurs, the DMR of *IGF2* was significantly hypomethylated compared with nonexposed siblings. This was not the case in offspring exposed at a later gestational age.

Altogether, the existence of intergenerational inheritance has been proven in several models. Transgenerational inheritance is also possible, but probably occurs at a lower frequency. The phenomenon and examples of epigenetic inheritance are more extensively described in van Montfoort et al. [5] and Cavalli and Heard [8].

14.4.1 Maternal Aging and the Oocyte Epigenome

Several studies have shown that DNA methylation, histone modifications, and the expression of epigenetic modifiers are altered in oocytes of mice with advanced age. Transmittance of these methylation defects is theoretically possible. However, with aging, the quality and developmental potential of the oocytes is reduced, which might hinder transmittance of epigenetic defects to offspring. Furthermore, the reduced expression of epigenetic modifiers such as DNMTs with increased maternal age is associated with smaller litter sizes and increased postimplantation embryo loss as these modifying enzymes are needed for proper embryo development [9].

14.4.2 Male Subfertility and DNA Methylation in Spermatozoa

In the male germline, the paternal and maternal imprints are erased in the fetal prospermatogonia. The male imprints are reestablished after birth in the adult spermatogonial stage, before the spermatocytes enter meisosis I and hence are completely methylated in mature spermatozoa. A disturbance in spermatogenesis, whether in concentration, morphology or motility, is associated with incorrect imprinting. An overview of the numerous studies showing this is given by van Montfoort et al. [5]. In general, in spermatozoa from oligozoospermic men, the occurrence of hypermethylation of maternal imprints and the hypomethylation of paternal imprints such as *H19* and IG-DMR is increased, especially in semen samples with less than 10×10^6 spermatozoa per milliliter. In normozoospermia only a few CpGs within a DMR are aberrantly methylated, while in azoospermia this ranges from only a few to all CpGs. The latter occurs only in a minority of alleles and thus in a minority of the spermatozoa. Although less extensively investigated, reduced sperm motility and morphology are also associated with aberrant methylation at imprinted regions.

Global DNA methylation of nonimprinted sequences such as long and short interspersed nucleotide elements (LINE1 and SINE) is not affected in azoospermia.

Aging is related to alterations in the methylome of spermatozoa. Just as in somatic cells, the alterations occur in such a consistent manner that algorithms have been developed to predict the age of an individual based on the sperm methylome. However, analysis of spermatozoa from individuals with older and younger grandfathers showed that these alterations were not transmitted transgenerationally [10]. Nevertheless, it remains striking that part of the regions associated with age-associated methylation effects are located at genes previously associated with schizophrenia and bipolar disorder, and that the prevalence of neuropsychiatric disorders in offspring of older fathers is increased.

14.4.3 Is There Evidence for Transmission of Methylation Defects in Spermatozoa to IVF Offspring?

The numerous aberrations in DNA methylation in spermatozoa from the subfertile male might be worrying. However, transmission of these defects to viable offspring is not so evident. To the best of my knowledge this has never been reported. The reported reduced fertilization rate and increased embryo developmental arrest in ART after using this methylation-defective sperm suggest that spermatozoa with aberrant imprints have reduced capacity to fertilize the oocyte or to guide the embryo through the first stages of development. However, the embryos were not analyzed so paternal transmission is not confirmed.

In trophoblastic villi samples from ART miscarriages between 6 and 9 weeks of gestation, *H19* methylation defects were found. For part of these samples, similar defects were found in the spermatozoa from the father [11]. This suggests transmittance, but with no viability for the offspring.

14.5 Epigenetic Effects of ART

Considering the effect of the periconception environment on offspring, ART could be a risk for inducing epigenetic effects. Not only does ART lead to changes in environment during a sensitive period, but ART procedures coincide with the major epigenetic reprogramming events during gametogenesis and embryogenesis (Figure 14.1). During oogenesis, and coinciding with ovarian hyperstimulation, in vitro maturation (IVM) and cryopreservation, the maternal-specific imprints are established. During spermatogenesis paternal imprints are established and should be maintained during cryopreservation for example. After fertilization, nonimprinted genes are demethylated and imprinted ones should maintain their methylation, withstanding oocyte and sperm manipulation (IVF or ICSI), embryo culture, and embryo cryopreservation. The current knowledge of ART-induced epigenetic alterations at these developmental stages as well as in IVF progeny is described in the following sections.

14.5.1 Epigenetic Effects in Oocytes

In humans, because of the difficulty in obtaining fetal and adult oogonial tissue with immature oocytes, studies on epigenetic reprogramming of imprinted genes during oogenesis are very limited. Moreover, the analysis of hormonal priming or IVM on imprinting in oocytes might be confounded by maternal age and/or suboptimal oogenesis related to the subfertility and often lacks proper controls. Only one study has used immature oocytes from nonstimulated fertile patients obtained after laparoscopy and compared these with superovulated germinal vesicle (GV) and metaphase I (MI) oocytes from IVF treatments [12]. In oocytes with fully grown GVs, the methylation of *MEST* was almost complete, while after ovarian stimulation only 10 of 16 GV/MI oocytes were methylated. The demethylation of the paternal imprint of *H19* DMR was not yet completed at the antral follicle stage of unstimulated follicles, leaving a remnant of around 12.5%. In superovulated GV/MI oocytes, two of six were still erroneously methylated [12]. Since the cause of subfertility of the donating couples was male or tubal obstruction, the effects are likely to be attributed to superovulation.

From studies in mice it seems as if superovulation does not directly affect methylation at the oocyte stage, but that it exerts its effects indirectly resulting in a response at blastocyst stage or even later. Although no causal effects were reported in superovulated mouse oocytes, except for a gain of methylation at *H19*, in blastocysts a superovulation dose-dependent disturbance of CpG methylation in some imprinted loci (*Snrpn*, *Kcnq1ot1*, *Peg3*, and *H19*) was found by Market-Velker and coworkers (see review by Marshall and Rivera [9]). Within the blastocysts, methylation on the loci ranged from complete methylation to severe hypomethylation, confirming a post-fertilization imprint maintenance effect instead of an imprint acquisition effect in the oocyte. This would also explain the reduced methylation on the paternal *H19* allele. Either superovulation impacts the maternal gene products required for imprint maintenance or exerts its effect via a temporary change in the environment in the female reproductive tract. It should be noted that in mice the maturation of multiple oocytes per cycle is natural, while in human it is not. This might explain why in mice superovulation does not interfere with imprint establishment and maintenance in oocytes. In mice, at embryonic day 9.5 superovulation-induced methylation differences were no longer seen according to Fortier and colleagues, but *Snrpn* and *H19* (but not *Kcnq1ot1*) showed biallelic expression in the placenta (see review by Marshall and Rivera [9]). In the embryo the expression of the three genes was monoallelic. The placenta effect vanished for *Snrpn* and was reduced for *H19* when embryos were transplanted to the uterus of nonstimulated foster mothers [9].

By culturing GV and MI oocytes that remain after an IVF treatment with hormonal stimulation, the effect of IVM of human oocytes is investigated. After an overnight culture, all in vitro-matured metaphase II (MII) oocytes display correct methylation at maternal (*KCNQ1OT1* and *SNRPN*) and paternal (IG-DMR) imprinted genes. However, in a study comparing in vivo- and in vitro-derived MII oocytes, the methylation level of *KCNQ1OT1* was lower after IVM. *H19* is also vulnerable to the maturation environment. In two of six patients, the pool of MII oocytes derived from IVM of GV oocytes displayed aberrant *H19* methylation in all or part of the oocytes [5].

The demand for cryopreservation of oocytes to preserve fertility is growing and the vitrification technique is gaining ground as the method of choice. GV stage oocytes that were vitrified and, after thawing, matured in vitro to MII oocytes showed a similar methylated and demethylated pattern in *KCNQ1OT1* and *H19* DMR, respectively, as compared with non-cryopreserved in vitro-matured oocytes [13].

In conclusion, superovulation has the potential to affect imprint maintenance while IVM more likely affects erasure and establishment in oocytes. Until now, studies have focused on imprinted genes only, but with the single-cell omics techniques that are being developed, it is expected that a plethora of information on this topic will become available soon.

14.5.2 Epigenetic Effects in the Embryo

The inevitable connection of subfertile patients, superovulation, IVF, and embryo culture makes it difficult to investigate the separate contribution to epigenetic deregulation in human preimplantation embryos and blastocysts. Moreover, in vivo control embryos are lacking. The limited amount of material per embryo is no longer an issue with the generation of new sequencing techniques that allow the profiling of a few or even single cells. Guo and colleagues were one of the first to profile DNA methylation in human gametes and embryos at different developmental stages. Until the four-cell stage, the human embryos were created by IVF with gametes from healthy fertile donors. From the eight-cell stage onward, the embryos (cryopreserved) were donated by subfertile IVF couples. At all stages of preimplantation development and even post implantation, methylation at several imprinted genes fluctuated considerably around the expected 50% in these IVF embryos. However, without an in vivo control, no conclusions regarding an IVF effect on DNA methylation at imprinted regions can be drawn [14].

In two previous studies with a DMR-targeted approach and using human surplus embryos of low quality (low number of blastomeres and/or low morphologic grade) that were not suitable for transfer or cryopreservation, 19 and 38% of human embryos showed hypomethylation of *H19*, while the corresponding sperm samples were normal. Five cryopreserved good-quality blastocysts that were donated for research all showed normal methylation [5]. The hypomethylation within the low-quality embryos ranged from relaxation of methylation at a single CpG to loss of methylation in the whole region. This suggests a lack of imprint maintenance induced by in vitro manipulation as the main cause [5]. In a more recent study, 2 of 14 good-quality human blastocysts showed aberrant DNA methylation at *H19*; one displayed loss of paternal methylation and one a gain of maternal methylation. In total, 76% of good-quality, human, day 3 embryos and 50% of good-quality blastocysts displayed abnormal methylation (both gains and losses) in at least one of three tested imprinted regions (*H19*, *SNRPN* and *KCNQ1OT1*). Abnormal was defined as at least four standard deviations above or below the mean DNA methylation in buccal cells, a poor surrogate for in vivo control embryos [15]. The relation between embryo quality and abnormal methylation needs further investigation.

From these limited number of analyzed embryos there is no evidence that one of the two fertilization methods, IVF or ICSI, differently affects vulnerability for epigenetic defects. On the other hand, culture, and more specifically culture medium, does, even in human. From several mouse studies, where a parental subfertility factor plays no role, embryo culture results in biallelic expression of *H19*, and loss of methylation at the paternally imprinted *H19*, and maternally imprinted *Snrpn* and *Peg3* loci in blastocysts [16]. *H19* and *Snrpn* are involved in BWS and AS respectively, and therefore probably investigated the most. Other regions are found to be affected as well, but these regions are not verified by other studies. The culture-induced alterations persisted at least until embryonic day 9.5, where placental tissue appears to be more susceptible than the corresponding fetal tissue. The effects exaggerate when multiple manipulations are used (e.g. superovulation in combination with embryo culture, or embryo culture in combination with embryo transfer) or when an inferior culture medium is used. Market-Velker and colleagues compared six different culture media (Whitten's, KSOM supplemented with amino acids, Global, HTF, P1MB and G1/G2) and found that all media, albeit to a different extent, affected DNA methylation of imprinted genes (*H19*, *Peg3* and *Snrpn*) in mouse blastocysts, as compared with in vivo-derived blastocysts. All media systems also led to loss of imprinted gene expression of *H19*, but not of *Snrpn* and *Peg3*. In E9.5 concepti, biallelic expression of *Snrpn* and *Peg3* is reported, although to a lesser extent than *H19* [16,17]. In human, culture medium also gained attention as a multicenter randomized controlled trial showed that the birthweight of children, which as an embryo were cultured in one of two different commercially available culture media (Vitrolife or HTF), differed by 158 g after adjustment for several confounders. This resembles the results in animal models where several studies have shown that embryo culture in general, specific culture media of even specific components like serum affect the growth of the fetus, but also the expression of several imprinted genes [4,18]. Analysis of blastocysts left over from IVF cycles within the Vitrolife/HTF trial showed that the transcript abundance of genes involved in cell cycle regulation, apoptosis, protein processing, metabolism, and development was altered according to the medium. Similar altered processes were found by the Rinaudo group who compared gene expression in mouse embryos cultured in either Whitten's or KSOM medium supplemented with amino acids [17]. When the culture media were used in combination with high oxygen levels (20%) instead of 5%, the effect on gene expression was exaggerated [19].

An important observation from both human and mouse studies is that not all embryos are affected, not even within the same woman or mouse, and within an embryo not all imprinted loci are affected nor are the loci affected to the same degree. The epigenetic response within embryos is probably stochastic or related to (genetically induced?) interindividual variation in susceptibility. If true, this could lead to different phenotypes among ART offspring, rendering the correlation between ART and epigenetic effect even more complex to analyze.

14.5.3 Epigenetic Effects in Placenta and Umbilical Cord Blood

Initially, mainly DNA methylation at targeted, often imprinted, genes in placental tissue or umbilical cord blood from ART and naturally conceived pregnancies was investigated (see Mani et al. [20] for an extensive overview of studies). Many studies, but not all, have reported significant differences. Regions often found to be affected are the *H19/IGF2* region, *KCNQ1OT1*, *MEST*, and the LINE-1 repeat elements (see Chapter 6). Since both paternally and maternally methylated regions are involved, and since the loss or gain in methylation concerns only a minority of the alleles (i.e. only a few percent of hypomethylation or hypermethylation is recorded), the epigenetic deregulation is expected to occur at a time point after fertilization. However, the results are not consistent among studies, which is possibly attributable to small sample sizes, different ART protocols, different methylation measurement techniques, and improper correction for potential confounding factors. Others argue that comparing mean DNA methylation is not appropriate as only a subset of ART offspring is susceptible for epigenetic disruptions or, in other words, that ART does not induce universal epigenetic changes in every offspring on the same locus but acts stochastically. This hypothesis is substantiated by the increased variation in DNA methylation seen in ART offspring as compared with naturally conceived offspring, and in the increase of so-called outliers (i.e. samples with an outlying methylation level at a certain region) in the ART group [21]. We recently substantiated this outlier hypothesis. In a comparison of placenta samples from IVF and natural conceptions, the mean methylation status of CpGs within 34 DMRs of well-defined imprinted genes was similar among the two groups, while the number of placenta samples exhibiting outlier levels in at least one of the CpGs was significantly higher in the IVF group [IVF 97/97 (100%) vs. NC 53/69 (77%)] [22].

With the advent of methylation arrays, a considerable number of CpG sites can be analyzed at once, also including those outside imprinted regions (see Table 14.2). In general, unsupervised clustering is unable to distinguish the ART and control groups based on DNA methylation, meaning that overall the DNA methylation is largely similar between the groups. However, when examining individual CpG sites many differences are found, both hypermethylation and hypomethylation. In three studies an enrichment of CpGs within imprinted regions was found, suggesting that these regions are more vulnerable for disruption. Three studies did not find this enrichment, but in two studies older array versions with a limited number of imprinted CpG sites were used (Table 14.2).

In an attempt to find consistently recurring genes within the studies using arrays, Mani et al. [20] compared the gene lists of significantly differentially methylated genes from five studies. Only four genes appeared in at least two of the lists. However, this low amount should not be a surprise as the studies are very heterogeneous with respect to sample size and therefore power to detect differentially methylated sites, ART technique, analysis method, selection of CpGs to be measured, the statistical analysis, and definition of a differentially methylated region. Comparing lists with designated "significant" genes only is therefore not appropriate; raw data are to be preferred. Moreover, these gene lists are based on group means in DNA methylation, which might overlook stochastic outlier effects as described above. The group of Melamed and Choufani reported a higher variation in DNA methylation and more outliers in the ART group, just as reported in the targeted gene studies. However, the group of Gentilini suggested that the variation induced by ART is smaller than the variation induced by other neonatal and parental characteristics (see Table 14.2).

DNA methylation effects are not only reported after IVF or ICSI in subfertile couples. In placental tissue from offspring conceived with ovulation induction alone, aberrations in DNA methylation are found as well [20]. This suggests that the superovulation primed uterine environment might play a role. The more natural level of methylation found in neonatal blood spots and placental tissue from frozen embryo transfers (Table 14.2) supports this [20,29]. Furthermore, the methylation differences found in placentas from pregnancies using donor oocytes suggest that the in vitro manipulations play a role, and not solely the parental subfertility or a hormonal primed uterus.

Table 14.2. Overview of studies comparing DNA methylation with arrays in placental tissue or umbilical cord blood in human ART versus naturally conceived offspring

Study	No. of NC	No. of ART	Treatment	Tissue	Analysis method	CpG sites	DNA methylation
Katari et al. (2009) [23]	12	10	IVF	UCB	Goldengate array Illumina	1536 (736 genes)	358 (23%) CpGs with altered methylation, of which 277 CpGs were hypermethylated in IVF group No overrepresentation of imprinted genes 2/7 genes with altered methylation had higher expression
	13	10	IVF	Placenta	Goldengate array Illumina	1536 (736 genes)	246 (16%) CpGs, of which 154 CpGs were hypomethylated in IVF group No overrepresentation of imprinted genes 2/4 genes with altered methylation had higher expression
Melamed et al. (2015) [24]	8	10	IVF	UCB	Infinium HumanMethylation27 array + pyrosequencing	26 486	733 differentially methylated CpG sites (63.2% hypomethylated in ART and 36.8% hypermethylated). No overrepresentation of imprinted genes Higher variation in DNA methylation in IVF group
Estill et al. (2016) [25]	43	94	IUI (n = 18) ICSI with fresh ET (n = 38) ICSI with frozen ET (n = 38)	Neonatal bloodspots	Infinium HumanMethylation450 BeadChips + pyrosequencing	394 454	Unsupervised clustering did not cluster NC, IUI and ICSI groups. Differential methylation was observed between NC and all ART groups, and considerable differences were observed between fresh ET and IUI and fresh ET and frozen ET. Only a few differences were seen between IUI and frozen ET An enrichment of imprinted genes was observed
El Hajj et al. (2017) [26]	46	48	ICSI	UCB	Infinium HumanMethylation450 BeadChips + pyrosequencing	428 227	4730 (0.11%) CpGs are differentially methylated, 2743 up and 1987 down ICSI children had a lower epigenetic clock age (half a week) CpGs within imprinting control regions are enriched among the differentially methylation CpGs. Effect size is small (<10%)
Gentilini et al. (2018) [27]	41	23	ICSI	UCB	Infinium HumanMethylation450 BeadChips	?	No significant difference in mean methylation per CpG between the groups. Differences are corrected for many neonatal, pregnancy, and parental factors No difference in the number of outliers. The stochastic variation is mainly associated with other variables (season of birth, neonate and parental characteristics). The variation induced by ART is smaller than variation induced by the other variables

Table 14.2. *(cont.)*

Study	No. of NC	No. of ART	Treatment	Tissue	Analysis method	CpG sites	DNA methylation
Choufani et al. (2019) [28]	44	44	IVF (n = 5) ICSI (n = 18) IUI (n = 11) Subfertile (n = 10)	Placenta	Infinium HumanMethylation450 BeadChips + pyrosequencingl	?	In a regression analysis on DNA methylation, ART was not a significant predictor. However being an outlier was, and 11/15 outliers were ART. This included four subfertile without ART, four ICSI, one IVF and two IUI. In the outlier group 84 270 CpGs were significantly different, with 56% hypermethylation and 44% hypomethylation. Differences up to 30% are noted. When comparing imprinted genes only, more or less the same (14/15) outliers appear, meaning that differential methylation (mainly hypomethylation) at imprinted genes contributes significantly to the outlier status. Only a subset of ART pregnancies is susceptible to induction of imprinting defects or other genes When comparing subfertile women conceiving naturally or with IUI (in vivo) with IVF/ICSI (in vitro) 436 CpG sites are different, mostly increased in the in vitro group. Male factor and increased paternal age within the in vitro group had the most deviant DNA methylation
Novakovic et al. (2019) [29]	58	149	IVF GIFT Unknown	Neonatal bloodspot	Infinium HumanMethylation EPIC BeadChips	724 897	2340/724 897 probes were differentially methylated; 79.1% were hypermethylated and 20.9% hypomethyled in ART No overrepresentation of imprinted genes The epigenetic variation at birth is largely resolved at adulthood

ET, embryo transfer; GIFT, gamete intrafallopian transfer; ICSI, intracytoplasmic sperm transfer; IUI, intrauterine insemination; UCB, umbilical cord blood.

Overall, the reported effect sizes in DNA methylation are relatively small, and, at birth, no phenotypic differences were noted between the groups. Nevertheless, based on these results, it cannot be ruled out that the overall genome-wide instability of the epigenome by ART may make some offspring more vulnerable for the (inevitable) additional epigenetic changes induced by lifetime exposures that are yet to come; the figurative threshold of epigenetic changes needed to result in a phenotype will be reached earlier.

14.5.4 Epigenetic Effects in ART Children

Although imprinting disorders in ART are rare, epigenetic errors at other (imprinted or nonimprinted) genes might account for a wider spectrum of ART-related complications. The investigation of organ-specific methylation alterations is not possible in human, which limits the analyses to blood or buccal cells. In four studies, in children aged between 0 and 7 years old, where a few imprinted genes were examined, no effect except for a difference in *SNRPN* methylation in one study was seen [20]. In one study that found no effect, the samples were collected from children who were also subjected to anthropometric measurements and serum screening for several metabolic and growth factors. IVF children were taller than the controls and also had a more favorable lipid profile [30].

Recently, Novakovic and colleagues used a longitudinal cohort of ART and non-ART individuals to profile DNA methylation at birth and at adulthood. Although they found 2340 out of more than 700 000 CpGs to be differentially methylated between the two groups in the neonatal blood, none was found at adult age (Table 14.2). However, some patterns in methylation were maintained from birth to adulthood (albeit not significant) and no reversal in methylation direction (hypomethylation or hypermethylation in ART compared with non-ART) was seen [29]. After a thorough medical examination, no adverse health outcomes associated with ART were found in this cohort. Although the results need to be confirmed by other studies, this report could be reassuring. Epigenetic effects recorded during preimplantation development and around birth might attenuate during life. Possibly, the epigenetic variation acquired during life exposures overrules the ART-induced variation.

14.6 Perinatal and Childhood Outcome of ART Children: Epigenetic Influence or Not?

ART singletons show an increased risk for very preterm birth (<32 weeks), preterm birth (<37 weeks, RR 1.4–2.0), very low birthweight (<1500 g), low birthweight (<2500 g, RR 1.6–1.7), small for gestational age (RR 1.5), and perinatal mortality (RR 1.7–2.0). In addition, an increased risk for congenital malformations such as cardiovascular defects, neural tube defects or genital malformations such as hypospadias is reported, with RR between 1.3 and 1.4 [1,5].

Recently, interest has been directed toward cardiovascular and metabolic outcome parameters. A meta-analysis on these outcomes showed that IVF/ICSI children had a slight but significant increased blood pressure (systolic +1.88 mmHg [95% CI 0.27, 3.49]; diastolic +1.51 mmHg [95% CI 0.33, 2.70]), higher aortic and carotid intima-media thickness, suboptimal cardiac diastolic function and lower low-density lipoprotein and higher fasting glucose levels [31].

When assuming a random distribution of genotypic variation among ART and non-ART offspring, a genetic component for these perinatal and childhood IVF outcomes is unlikely. Therefore, it is reasonable to assume that this is an effect of an epigenetic adaptive response to the preimplantation embryo environment. Further, for most of the congenital malformations (e.g. neural tube defects and heart defects) the incidence decreases when the mother takes the dietary supplement folic acid (a factor in the DNA methylation cycle) during the periconception phase. The physiologic outcomes suggest a predisposition for metabolic syndrome, just as seen in offspring that were in utero during the Dutch famine and in animal studies where the maternal diet was changed during conception phase (described in Section 14.2.2). Furthermore, in vitro culture of mouse embryos leads to similar outcomes, such as increased systolic blood pressure, changes in postnatal organ weight, and altered angiotensin-converting enzyme and phosphoenolpyruvate carboxykinase activity in serum and liver, respectively [4]. In these examples, altered gene expression and methylation levels are described. Altogether, these outcomes fit the Developmental Origin of Health and Disease (DOHaD) paradigm with the fetal programming concept. This concept states that the fetus is programmed in order to prepare itself for life after birth. Based on the fetal environment (i.e. nutritional and endocrine status) epigenetic adaptations in organs and tissues are made that permanently adjust the physiology of these organs and the metabolism of the fetus. This ensures optimal health if both prenatal and postnatal environments match. If not, an extra burden is put on the organs and tissues that might lead to chronic diseases in later life. Long-term follow-up of ART offspring is needed to verify whether the subclinical aberrant cardiometabolic outcomes aggravate into disease and whether the epigenetic adaptations persist in later life or attenuate as suggested by Novakovic et al. [29]. This will reveal whether programming for life starts only at the fetal stage or already before implantation.

14.7 Discussion and Conclusion

Studying the relation between ART-related aspects and the induction of epigenetic changes is not straightforward. The study of ART-specific effects in human oocytes and embryos is hindered by the scarcity and quality of the material (immature or unfertilized oocytes and often surplus embryos of low quality). Moreover, in vivo control oocytes and embryos are lacking. In offspring, only blood or epithelial cells are available, and overall the inevitable connection between subfertility, superovulation, IVF, and embryo culture is not beneficial either. Studies are often not comparable since different ART protocols and/or different techniques to analyze methylation are used and/or different regions are analyzed. The inconsistent and heterogeneous results within and between studies suggest that the effect of ART on epigenetic regulation occurs either completely at random or is related to interindividual variation in susceptibility; some individuals might be more susceptible for epigenetic alterations than others. The genetically heterogeneous population complicates the studies even more. Nevertheless, keeping these shortcomings in mind, important and relevant observations have been done.

Even though the absolute risk is still low, there is evidence that ART is related to an increased incidence of maternal and paternal imprinting disorders such as AS, BWS, and SRS. As loss of imprinting is the most prevalent cause of imprinting disorders in ART offspring, an incorrect imprint establishment during gametogenesis is a likely cause. In the few studies investigating the effect of ovarian stimulation, the number of oocytes with an unmethylated DMR was much higher than the prevalence of maternal imprinting disorders. This suggests that most of the embryos derived from these oocytes are not viable. The same applies to transmittance of sperm DNA methylation defects, which can also be prevented by selection of gametes for ART or the nonviability of the embryo.

In addition to affecting imprint establishment during oogenesis, superovulation can also exert an effect at two other levels: by impacting on the level of maternal gene products needed for imprint maintenance post fertilization or by changing the environment in the female reproductive tract. Further to the direct evidence for this in mouse, there is the indirect confirmation in human, where a more natural level of DNA methylation was found in neonatal blood spots and placental tissue from frozen embryo transfers as compared with fresh embryo transfers.

There is extensive evidence, mainly from animal studies, that embryo culture and more specifically culture medium can affect the methylation status of imprinted genes. In human surplus embryos with corresponding normal human sperm samples, paternal hypomethylation has been reported, implying that embryo culture and methylation defects are related in human as well. Moreover, the different transcriptome in human blastocysts cultured in two different media is indicative that the human embryo responds to its environment. In mouse, the culture (medium)-induced effects are maintained at least until embryonic day 9.5, with the placenta being more vulnerable for aberrant methylation at imprinted genes than the fetus. In human, this tissue-specific susceptibility is not evident.

Both superovulation- and in vitro manipulation-induced epigenetic effects seem to occur in a stochastic nature (in embryos, placenta, and umbilical cord blood), meaning that only a subset of ART individuals is susceptible for epigenetic changes and cellular mosaicism in methylation exists within an embryo. Methylation defects will not necessarily lead to an imprinting disorder, but hypothetically could account for the wide spectrum of ART-related health effects, depending on the location of the defect. The mechanism behind this intraindividual difference in susceptibility needs to be resolved but is probably genetic, epigenetic, environmental, or more likely the result of crosstalk between all three.

Studies in children and adults suggest that the ART-induced epigenetic variation seen at birth resolves during life. This would be a reassuring message, but needs confirmation in other and larger study populations. If true, probably either the epigenetically aberrant cells obtain a proliferative disadvantage over healthy cells after birth, through which the defects are gradually faded out, or the epigenetic variation acquired during life exposures overrules the ART-acquired variation. However, the observed cardiovascular and metabolic outcomes in ART children question that the ART-acquired epimutations blend with the ones acquired during life. It is clear that much research on the long-term health effects of ART offspring and the fate of ART-acquired epimutations needs to be done.

References

1. Berntsen S, Soderstrom-Anttila V, Wennerholm UB, et al. The health of children conceived by ART: "the chicken or the egg?" *Hum Reprod Update* 2019;25:137–58.
2. Hanna CW, Demond H, Kelsey G. Epigenetic regulation in development: is the mouse a good model for the human? *Hum Reprod Update* 2018;24:556–76.
3. Zhu P, Guo H, Ren Y, et al. Single-cell DNA methylome sequencing of human preimplantation embryos. *Nat Genet* 2018;50:12–19.
4. Fleming TP, Watkins AJ, Velazquez MA, et al. Origins of lifetime health around the time of conception: causes and consequences. *Lancet* 2018;391:1842–52.
5. van Montfoort AP, Hanssen LL, de Sutter P, et al. Assisted reproduction treatment and epigenetic inheritance. *Hum Reprod Update* 2012;18:171–97.
6. DeAngelis AM, Martini AE, Owen CM. Assisted reproductive technology and epigenetics. *Semin Reprod Med* 2018;36:221–32.
7. Hiura H, Okae H, Miyauchi N, et al. Characterization of DNA methylation errors in patients with imprinting disorders conceived by assisted reproduction technologies. *Hum Reprod* 2012;27:2541–8.
8. Cavalli G, Heard E. Advances in epigenetics link genetics to the environment and disease. *Nature* 2019;571:489–99.
9. Marshall KL, Rivera RM. The effects of superovulation and reproductive aging on the epigenome of the oocyte and embryo. *Mol Reprod Dev* 2018;85:90–105.
10. Jenkins TG, James ER, Aston KI, et al. Age-associated sperm DNA methylation patterns do not directly persist trans-generationally. *Epigenetics Chromatin* 2019;12:74.
11. Kobayashi H, Hiura H, John RM, et al. DNA methylation errors at imprinted loci after assisted conception originate in the parental sperm. *Eur J Hum Genet* 2009;17:1582–91.
12. Sato A, Otsu E, Negishi H, Utsunomiya T, Arima T. Aberrant DNA methylation of imprinted loci in superovulated oocytes. *Hum Reprod* 2007;22:26–35.
13. Al-Khtib M, Perret A, Khoueiry R, et al. Vitrification at the germinal vesicle stage does not affect the methylation profile of H19 and KCNQ1OT1 imprinting centers in human oocytes subsequently matured in vitro. *Fertil Steril* 2011;95:1955–60.
14. Guo H, Zhu P, Yan L, et al. The DNA methylation landscape of human early embryos. *Nature* 2014;511:606–10.
15. White CR, Denomme MM, Tekpetey FR, et al. High frequency of imprinted methylation errors in human preimplantation embryos. *Sci Rep* 2015;5:17311.
16. Velker BA, Denomme MM, Mann MRW. Embryo culture and epigenetics. *Methods Mol Biol* 2012;912:399–421.
17. Mani S, Mainigi M. Embryo culture conditions and the epigenome. *Semin Reprod Med* 2018;36:211–20.
18. Zandstra H, Van Montfoort AP, Dumoulin JC. Does the type of culture medium used influence birthweight of children born after IVF? *Hum Reprod* 2015;30:530–42.
19. Feuer S, Liu X, Donjacour A, et al. Transcriptional signatures throughout development: the effects of mouse embryo manipulation in vitro. *Reproduction* 2017;153(1):107–22.
20. Mani S, Ghosh J, Coutifaris C, Sapienza C, Mainigi M. Epigenetic changes and assisted reproductive technologies. *Epigenetics* 2020;15:12–25.
21. Ghosh J, Mainigi M, Coutifaris C, Sapienza C. Outlier DNA methylation levels as an indicator of environmental exposure and risk of undesirable birth outcome. *Hum Mol Genet* 2016;25:123–9.
22. Mulder CL, Wattimury TM, Jongejan A, et al. Comparison of DNA methylation patterns of parentally imprinted genes in placenta derived from IVF conceptions in two different culture media. *Hum Reprod* 2020;35:516–28.
23. Katari S, Turan N, Bibikova M, et al. DNA methylation and gene expression differences in children conceived in vitro or in vivo. *Hum Mol Genet* 2009;18:3769–78.
24. Melamed N, Choufani S, Wilkins-Haug LE, et al. Comparison of genome-wide and gene-specific DNA methylation between ART

and naturally conceived pregnancies. *Epigenetics* 2015;10:474–83.

25. Estill MS, Bolnick JM, Waterland RA, et al. Assisted reproductive technology alters deoxyribonucleic acid methylation profiles in bloodspots of newborn infants. *Fertil Steril* 2016;106:629–39.e10.

26. El Hajj N, Haertle L, Dittrich M, et al. DNA methylation signatures in cord blood of ICSI children. *Hum Reprod* 2017;32:1761–9.

27. Gentilini D, Somigliana E, Pagliardini L, et al. Multifactorial analysis of the stochastic epigenetic variability in cord blood confirmed an impact of common behavioral and environmental factors but not of in vitro conception. *Clin Epigenetics* 2018;10:77.

28. Choufani S, Turinsky AL, Melamed N, et al. Impact of assisted reproduction, infertility, sex and paternal factors on the placental DNA methylome. *Hum Mol Genet* 2019;28:372–85.

29. Novakovic B, Lewis S, Halliday J, et al. Assisted reproductive technologies are associated with limited epigenetic variation at birth that largely resolves by adulthood. *Nat Commun* 2019;10:3922.

30. Miles HL, Hofman PL, Peek J, et al. In vitro fertilization improves childhood growth and metabolism. *J Clin Endocrinol Metab* 2007;92:3441–5.

31. Guo XY, Liu XM, Jin L, et al. Cardiovascular and metabolic profiles of offspring conceived by assisted reproductive technologies: a systematic review and meta-analysis. *Fertil Steril* 2017;107:622–631.e5.

Chapter 15

Human Reproductive Genetics in Medically Assisted Reproduction: Ethical Considerations

Guido de Wert and Wybo Dondorp

15.1 Introduction

This chapter on the ethics of human reproductive genetics focuses on current debates in three corners of the field of medically assisted reproduction (MAR). First, we discuss preimplantation genetic testing (PGT). Until recently, ethical debates concentrated on PGT "on indication," formerly termed preimplantation genetic diagnosis (PGD), namely testing of embryos on behalf of couples mostly at high risk of transmitting a specific genetic disorder. As important, however, is the ethics of (the offer of) routine testing of in vitro fertilization (IVF) embryos, formerly termed preimplantation genetic screening (PGS). Although it is still unclear what the precise value of such screening may turn out be, the scenario of next generation sequencing (NGS)-based "comprehensive PGS" raises a number of challenging ethical issues. Secondly, we discuss carrier screening. In MAR, until recently this type of screening targeted gamete donors. Presently, however, a growing number of clinics have started offering carrier screening to applicants of MAR as well. As discussed in Chapter 10, new NGS technologies are expected to make broad-scope testing for many genetic conditions feasible and affordable. Accordingly, what should we test gamete donors and applicants of MAR for and why? Finally, we add a brief ethical reflection on what may become a (or the) real revolution in human reproduction: germline genome editing.

According to the European Society of Human Reproduction and Embryology (ESHRE), professionals working in MAR have a double responsibility, not just to the couples or individuals who need their help in order to fulfill their child wish, but also to the resulting children, whose very existence they causally and intentionally bring about. More specifically, this last point means that professionals should refrain from providing assisted reproduction if there is a high risk of a child with a seriously diminished quality of life [1]. Similar guidance is given by authorities and professional bodies in various European countries. In the literature, moral contraindications for MAR are mainly discussed with an eye to concerns about helping couples or individuals with less than sufficient parental competence, leading to children being born in families with a high risk of child abuse or neglect. However, specific welfare-of-the-child concerns may also arise where MAR could lead to children with genetic disorders. Here, the double responsibility of MAR professionals can take two different forms. It may constitute a further contraindication for providing MAR in cases where proceeding with a requested transfer would amount to a high risk of a child with a very serious disorder, or it may provide a reason for offering or perhaps insisting on testing as a means of risk reduction where doing so is reasonably possible and proportional. In each of the following sections, questions pertaining to this special responsibility for the welfare of the child will emerge.

15.2 Preimplantation Genetic Testing on Indication (PGT-M/SR)

We wish to start this section with a critical note on new terminology. Debates about embryo selection traditionally distinguished between indication-based PGD on the one hand and PGS as routine testing of IVF embryos on the other. As this is a morally relevant distinction, we regard it as unfortunate that it is no longer visible under the new nomenclature, where all embryo testing is now referred to as PGT (PGT-M for monogenic disorders, PGT-SR for structural rearrangements, PGT-A for aneuploidy). Not only does the new terminology obscure the difference between indication-based testing and screening, it may also contribute to a further blurring of this distinction in practice.

When the first papers about clinical PGT were published in 1990, ethical comments varied widely.

People in favor argued that PGT was a welcome reproductive alternative for people at high risk of having a child with a serious genetic condition, while critics protested, among other objections, that the deselection of affected embryos was at odds with the high moral value (or "status") of human life "right from conception," and that PGT was a problematic step on a slippery slope to selective reproduction for trivial reasons. The objections were not widely considered to be convincing. The dominant view in ethics and in jurisdictions worldwide holds that the moral status of preimplantation embryos is relatively low; though these embryos may be "*potential* persons," they do not share in the high moral status or dignity of actual persons. Furthermore, a possible slippery slope is a good reason for regulation, not for outright prohibition.

It should not be a surprise, then, that the principled issue about whether PGT can be ethically justified at all has become a rearguard action. This is not to say that the ethics of PGT has been solved. Paradoxically, in view of its broad acceptance, the ethics of PGT has become increasingly complex. A distinction can be made between what we refer to as "front door" and "back door" issues [2]. Front-door issues arise with the widening range of conditions for which PGT-M may be offered, whereas back-door issues pertain to increasingly difficult decisions about what embryos to select for transfer. As will be made clear, front-door and back-door issues are interrelated. Starting at the front, we first discuss different views of acceptable indications for PGT-M. In this connection, three different approaches or "models" can be distinguished [3]. A strict version of what we call the *medical model* holds that PGT can only be justified if it aims at avoiding the conception of a child affected with a serious disease or handicap. A wider version of this model also allows for PGT-M to avoid conditions that are *indirectly medical*. Thirdly, what we call the *autonomy model* leaves it to prospective parents to decide for what conditions PGT would be justified.

15.2.1 PGT within the Medical Model

PGT-M was developed to help prospective parents at a known high risk of transmitting a serious monogenic disorder to have healthy children. Soon after, PGT-SR was also offered to prospective parents carrying a chromosomal abnormality, such as a translocation, that may either lead to repeated pregnancy loss or to the birth of a severely handicapped child. To date, PGT has been performed for hundreds of genetic disorders of different levels of severity, and the numbers are steadily increasing (see Chapter 13). For many people, this illustrates the relevance and potential of PGT, but for some others this underscores the urgency of the question: how serious does a disorder/handicap have to be in order to qualify for PGT? In countries where PGT-M/SR is available, its lawful application is often limited to couples with a "significant" or "high" risk of transmitting a "serious" genetic disorder to their offspring. The reasoning behind this requirement often remains unspecified. However, ESHRE's Task Force Ethics and Law has suggested that the standard should be understood as reflecting the "proportionality" of PGT-M/SR [4]. This notion refers to the balance between the benefits that PGT may have for the applicants on the one hand and the various aspects that make it a morally sensitive technology on the other. Let us take a brief look at some particular categories of genetic disorders for which PGT-M is widely regarded as proportional.

15.2.1.1 PGT-M for Untreatable, Midlife-Onset Disorders Caused by Completely Penetrant Genetic Variants

In the category of PGT for untreatable midlife-onset disorders, Huntington's disease (HD) is the paradigm case. Critics object that PGT-M for midlife (let alone late)-onset disorders is unwarranted as a carrier will have (many) decades of good and unimpaired living. However, this view seems to disregard that the prospect of HD – a progressive, highly invalidating, ultimately lethal disorder – imposes an extremely severe burden on both the carrier and the members of affected families. It should be no surprise, then, that HD is one of the main, and widely accepted, indications for PGT-M in practice.

Nevertheless, in more recent debates at least three ethical questions arise. A first issue arising with PGT-M (as well as with other forms of MAR) for HD carriers concerns the future loss of parental competence. Can reproductive physicians involved in IVF accept this risk, taking into account their professional responsibility for the welfare of the future child, more particularly their responsibility to avoid high risks of serious harm? And what if the prospective father or mother is already symptomatic? Black-and-white approaches seem to be inadequate. Although the development of HD symptoms in a parent is burdensome for children, many children are able to cope reasonably well. Relevant variables include the coping skills of the partner not affected with HD, communication about HD among family members, and the quality of the network of the family. Further discussion is needed to see how these variables would allow decisions to be made case by case. Second, what about exclusion PGT-M? Though the procedure is "unnecessary" in 50% of the cases, the weighing of other relevant factors, including the wish not to know one's genetic status and the burdens involved in IVF/PGT-M, is rather personal. Apparently, applicants consider this option to be valuable. Possible unnecessary embryo loss cannot be regarded as a major moral obstacle given the dominant view regarding the lower status of the preimplantation embryo. And third, what about applicants who carry a reduced penetrance allele (RPA), entailing 35–39 CAG repeats, for HD? Is their risk serious enough for a PGT-M indication? PGT-M may reveal that the embryo carries an RPA or (after expansion) a full penetrance allele (FPA). It is estimated that some two-thirds of the future children carrying an RPA will have HD in (late) adulthood, before the age of 75. Furthermore, these children are at significant risk that their own children will carry an FPA (after expansion). PGT-M aimed at the nontransfer of embryos with an FPA or an RPA is morally justified in view of this combination of risks.

15.2.1.2 PGT-M for Preventable/Treatable Conditions Caused by Mutations with Incomplete Penetrance

A second category is PGT-M for preventable or treatable conditions that are caused by mutations with incomplete penetrance. A good example of this category is hereditary breast and ovarian cancer (HBOC). PGT-M for relevant mutations of *BRCA1* or *BRCA2* is somewhat more controversial than PGT-M for HD, because the penetrance of these mutations is incomplete and (future) carriers have preventive and/or therapeutic options, including preventive mastectomy and ovariectomy. Clearly, however, although the penetrance is incomplete it is still very high – the cumulative risk for breast and ovarian cancer may be even higher than 90%. And preventive surgery has major implications for women's welfare, and is not 100% effective. A thought experiment might be illuminative here: Would you tell the prospective parents of a boy at high genetic risk of future testicular cancer not to worry too much because they may simply opt for preventive castration? PGT-M for mutations in *BRCA1*/*BRCA2* is widely – and, we think, rightly – considered to be morally justified. Most demands are from carriers with highly penetrant mutations and a dramatic family history. Likewise, PGT-M may be justified for other hereditary cancer syndromes and for some of the cardiogenetic conditions. The view that the availability of treatment makes PGT-M obsolete is one-dimensional and too restrictive; if children's and families' quality of life is seriously (adversely) affected even though treatment is available, PGT-M may still be a reasonable option.

15.2.1.3 PGT for Mitochondrial Disorders

There seems to be growing interest in PGT for mitochondrial disorders caused by a mutation in the mitochondrial (mt)DNA, such as Leigh syndrome and MELAS (mitochondrial encephalomyopathy, lactic acidosis, and stroke-like episodes). Ideally, one would like to transfer embryos without a (detectable) mutant load. Sometimes, however, these embryos are not available after PGT, so one has to consider transfer of the embryo with the lowest mutant load, so with the highest probability of leading to a healthy child. Obviously, the aim of PGT then changes from risk elimination to risk reduction. Possible objections include that this is at odds with the proper aim of PGT and with the doctor's responsibility to take the welfare of the future child thus conceived into account. Even though these objections do not provide compelling arguments to regard risk-reducing PGT as a priori unacceptable, in view of the responsibility to avoid a high risk of serious harm, a cutoff point (a threshold of mutant load) should be determined below which embryos are considered to be eligible for transfer. If only embryos above the threshold are found, options are limited to either engage in a new IVF/PGT cycle or to stop trying PGT. As it is the professional's responsibility to not only avoid a high risk of serious harm, but to also try to further reduce any risks that are not per se prohibitive, it should be considered whether it would be possible and proportional to start another cycle, possibly resulting in embryos with a lower or even zero mutant load. Looking for better embryos should not be seen as morally required if an embryo below the cutoff point is available. The number of cycles should be determined on a case-by-case basis, also depending on, for example, the specific mutation, the wishes of couples (especially the women involved), and the number of cycles allowed and reimbursed in a country. Taking account of the complexities and uncertainties involved, especially in the case of variable (unstable) mtDNA mutations with an unpredictable outcome, PGT for mitochondrial disorders should ideally be embedded in a scientific research protocol involving the follow-up of children conceived with this technology. When setting up such studies, it is important to be aware that, however valuable scientifically, they may also raise further ethical issues themselves. This is especially the case where children are involved in genetic testing for reasons other than their own best interest. While this should not be a reason for refraining from follow-up, it is certainly relevant for how such studies can be conducted in an ethically responsible way.

15.2.1.4 Contextualized Proportionality

Assuming that the moral acceptability of PGT-M depends on whether (1) the efforts, burdens and possible risks of IVF for the women involved, (2) the possible risks of IVF and the (so far theoretical) risks of PGT-M (especially the embryo biopsy) for future children thus conceived, (3) the inherent embryo loss, and (4) the costs are in proportion to the benefit of avoiding the conception of an affected child, it seems logical and justified to acknowledge that at least two contextual factors may influence this "proportionality" balance [4].

First, the fertility status of applicants. In the most common situation, fertile people undergo IVF/PGT simply in order to avoid the birth of an affected child. Regularly, however, a couple opts for IVF/intracytoplasmic sperm injection (ICSI) because of subfertility, and wants to add PGT in order to avoid the transmission of a particular genetic disorder. Second, patients with an accepted indication for PGT may want to avoid transmitting an additional disorder for which they are at a reproductive risk through so-called "combination PGT" [5]. In both situations, a significant part of the burdens and (moral) costs have already been taken into account, either with respect to IVF or ICSI as fertility treatment, or for performing PGT for the primary disorder. Given the different proportionality balance that this entails, PGT may in these situations be considered for conditions that would not be regarded as high risk and serious. As the decision to engage in IVF or ICSI as fertility treatment or in ICSI/PGT for the primary indication has already been taken and considered proportional, a more permissive policy regarding additional or secondary testing seems to be justified (2,5).

Although PGT-M/SR for couples who also happen to have a fertility problem is far from rare, PGT-M for more than one indication still is. However, this is about to change. Several factors may have a role in this development, including expanded possibilities for diagnosis of genetic disorders on the single-cell level, an increased familiarity with the role of genetics in disease, and a greater awareness of personal reproductive risks also as a result of more frequent genomic testing in families. The offer of expanded preconception carrier screening for recessive disorders (see below) may further add to this effect.

15.2.2 PGT for Cases That Are Indirectly Medical

Some applications of PGT do not fit the medical model sensu stricto, as (at least part of) the testing is not linked with possible health problems of the future child, whereas there is still a link to the medical model in the wider sense, in that the testing may be relevant for the health of a "third party" [6]. These applications are therefore "indirectly medical" or "intermediate" between medical and nonmedical.

A first example is PGT/HLA typing. The main ethical condition is that the future child (the term "donor" is a misnomer, of course) should be truly welcome – it should not be valued just as a cell bank. Regulations in Europe differ. Some countries, like the Netherlands, allow PGD/HLA typing only in the context of (in addition to) PGT-M for a particular disease, and prohibit the procedure if aimed to treat a child affected with a *non*-genetic disorder. From an ethical perspective this policy is debatable, as the future child may be welcomed "for itself" in the latter context as well. The take-home baby rate is, unfortunately, relatively low. This should not come as a surprise, as the double selection involved regularly results in no suitable embryo available for transfer. This should be clearly communicated to the applicants beforehand in order to avoid undue optimism.

A second example is PGT in situations where there is no concern about the health of the child to be conceived, but where the aim is to avoid reproductive dilemmas arising in the next generation. Think of the case of a male patient suffering from hemophilia. Some of these patients (and their partners) prefer to conceive male progeny only, because sons will not carry the mutation, whereas all daughters will. PGT aimed at avoiding this substantial transgenerational risk may well be morally justified.

15.2.3 PGT within the Autonomy Model

According to this view, prospective parents are free to use PGT in order to select embryos on the basis of any characteristic they prefer. Opponents argue that selecting for nonmedical characteristics violates the autonomy of the future child as the child is reduced to an object of parental ambitions and ideals. However, would embryo selection on the basis of characteristics that do not limit the possible life plans of the future child or that are useful in carrying out almost any life plan ("general purpose means") really violate the future child's autonomy? Should one not say that prospective parents undermine the ethical standard only when they deliberately try to mold the child's development into a predetermined life? Although the technical possibilities to use embryo selection for "superbabies," whatever that may be, are often exaggerated in the mass media, this is an area for further analysis and debate.

A paradigm case for the autonomy model in the current context is PGT/sex selection for nonmedical reasons. Sex selection for nonmedical reasons is prohibited in many countries. From an ethical point of view, however, this is not as evident as many seem to think [7]. The objection that such selection is inherently sexist is a matter for debate. The fear that it will result in a distortion of the sex ratio does not seem to apply to Western countries, where a preference for boys is generally weak or absent and where, apart from certain minority groups, sex-selective technologies are used for "family balancing," if at all. Moreover, the suggestion that sex selection for nonmedical reasons will reinforce gender stereotypes to the detriment of children's development and women's position in society are speculative at best. Since the conclusion must be that arguments against allowing sex selection for nonmedical reasons are weak, some have argued that categorically banning all forms of the practice would amount to an unjustified infringement of reproductive freedom. However, even if sex selection may under some conditions (such as family balancing) be acceptable, a further question still concerns the proportionality of the means.

A second instance of the autonomy model is PGT-M for "dysgenic" reasons. The paradigm case involves a deaf couple's request for PGT-M in order to selectively transfer embryos *affected* with (nonsyndromic) deafness. The couple may point to the psychosocial and developmental risks of *hearing* children growing up with (two) deaf parents. Concerns include that (young) hearing children will have difficulties in understanding the implications of their parents' disability and related behavior, that deaf parents will have only limited access to the experiences of hearing children, and that there is a risk of role inversion. Furthermore, applicants may argue that "deafness is not a handicap, but just a variant on the spectrum of normalcy." After all, deaf people have their own rich culture and their own (nonverbal) language. One can reasonably doubt, however, whether the "just a variant" view is tenable; after all, outside the microcosmos of the deaf subculture deafness is a disability that causes a variety of serious and lifetime challenges. Though deaf people still can (and usually do) live a reasonable happy life, selection for deafness is at odds with the professional responsibility of the reproductive doctor [4,8]. The couple's relational concerns should be tackled by educational support and advice, not by "dysgenic" PGT. Interestingly, ongoing technology development may contribute to solving the current moral puzzle. Until recently, cochlear implants have been controversial, among other things because their success is patchy. However, when the perfect version of the cochlear implant becomes available in the future, parents will harm a child they leave deaf. To select for a deaf child then becomes self-defeating [9].

15.2.4 The "Back Door": Ethics of the Transfer Policy

According to a classical rule of PGT practice, "affected embryos" should not be transferred to the womb. When PGT is inconclusive or when all embryos in a given IVF/PGT cycle prove to be affected, doctors will thus normally start a new cycle rather than transfer an affected embryo. An important argument for this is the responsibility of MAR professionals to take the welfare of the child into account. However, for this to lead to the "do-not-transfer" rule, it must be the case that PGT-M/SR is only done for conditions that clearly fall in the range of high risk and serious [4]. Given that in the past decennium the scope of accepted indications has widened beyond the limited range of classical disorders to also include conditions marked by less than complete penetrance, a later time of onset, and at least some treatment or surveillance options (e.g. HBOC or hypertrophic cardiomyopathy), it has become less obvious that the do-not-transfer rule would not allow for making exceptions. Here we see how what we call the front and the back door are linked: a further widening at the front inevitably leads to pressure at the back [2]. Even so, it is quite understandable that many PGT professionals remain reluctant to make exceptions to the rule to never transfer affected embryos. Confronted with such requests, many argue that if the applicants are ready to accept a child with the very condition that PGT was meant to avoid, they should take their further chances through natural conception, rather than implicating PGT professionals in responsibility for the result [10].

However, this is not helpful advice for all PGT applicants. Precisely for those in the two categories where the proportionality balance would allow doing PGT-M for milder conditions, either because they also have a fertility problem giving them an independent IVF/ICSI indication or because they already have a primary PGT-M indication for a serious condition, natural conception is not available as an alternative. For applicants with a fertility problem, this is quite obvious: after several unsuccessful PGT-M/SR cycles, those affected embryos constitute their last chance of having their own genetically related child. If they ask for a transfer of those "last-chance embryos," this should not be seen as a lack of seriousness on their part, but rather as an adjustment of priorities in the light of a reassessment of what is realistically feasible for them. While having tried for a healthy child, the bottom line is that they want a child. However, the notion of last-chance embryos also applies to cases where normally fertile couples who have PGT-M for more than one condition end up (after several cycles) with all embryos affected for both conditions. These are last-chance embryos in the wider sense of enabling the couple to start a pregnancy with the confidence that the resulting child will at least not be affected by the disorder that they want to avoid most.

As, at the front, combination PGT becomes less rare, it seems inevitable that, at the back, professionals will more often be confronted with requests to transfer last-chance embryos affected by what the applicants regard as the least serious of the two (or more) conditions [2].

Decision-making about such requests should be guided by the responsibility of professionals to take account of the welfare of the child-to-be. Thinking about these scenarios as a spectrum, on the one hand there will be cases where, notwithstanding a possible categorization by the applicants in terms of primary and secondary, both conditions are evidently highly serious. Clearly, professionals should refrain from transferring embryos leading to disorders in this category. At the opposite end are cases where the secondary condition is clearly not high risk and serious, whereas the primary condition is. There seems no good reason in principle why professionals should categorically refuse such requests. As often, the most difficult cases fall in the middle of the spectrum. These consist of transfer requests where the secondary condition, although an accepted indication, is in the gray area where it can be a matter of debate if the condition qualifies as high risk and serious. As an example, one may think here of *BRCA* mutations in female embryos. We propose shared decision-making about such gray-area cases, in which the particular views of the applicants and the history and context of their experiences with the condition (the story behind the request) are taken into account [10].

The importance of this discussion will only increase, also given that with the introduction of generic genome-analysis methods for PGT, incidental findings can be expected that will lead to embryos known to be affected with mutations or abnormalities not related to the condition or conditions for which PGT was done, leading to difficult last-resort transfer decisions. Clearly, it is paramount that applicants should be informed at the pretest stage about the center's policy with regard to the ethically laden choices that may emerge at the transfer stage [5].

15.3 Routine PGT

PGT to check the number of pronuclei (PGT-PN) is the classical type of routine PGT. Zygotes with three pronuclei (tri-pronuclear zygotes) are not transferred as these are not able to develop into a viable child. Such screening is so self-evident that it is not even mentioned in debates on routine PGT. The nontransfer of tri-pronuclear zygotes is, so we assume, even accepted by adherents to the sanctity-of-life doctrine; even they will consider it an absurdity to presume that there would be a moral duty to transfer an embryo lacking the potential to become a viable child.

15.3.1 PGT for Aneuploidy

PGT-A (formerly known as PGS) was introduced almost three decades ago to screen out aneuploid embryos and to transfer only euploid embryos, based on the hypothesis that this would improve IVF's pregnancy outcomes, especially its live birth rate (LBR) or take-home baby rate (THBR). While first-generation PGT-A (PGS 1.0) was based on either polar body (PB) biopsy or, mostly, blastomere biopsy followed by fluorescence in situ hybridization (FISH), this screening method proved to be ineffective if not counterproductive [11], and was replaced by PGT-A 2.0 and more recently PGT-A 3.0, characterized by among others a later, blastocyst-stage, biopsy and a comprehensive chromosomal analysis (see Chapter 13). This brief ethical comment focuses, first, on one principled point, and secondly on a rather practical point, with clear moral implications.

With regard to the principled debate, we argue that there are no valid categorical moral objections to PGT-A. After all, its primary aim, namely to increase the success rate of IVF, is commendable, while the means, namely to exclude embryos affected with serious chromosomal aberrations, which often lack viability, from transfer is clearly morally acceptable. Also in view of the strong consensus that preimplantation embryos have a lower moral status than fetuses, it would be difficult to consider prenatal screening for Down syndrome justified (as most of us do) and at the same time a priori reject PGT-A for principled reasons.

Some proponents of PGT-A on the basis of PBs claim that this strategy has a moral advantage in comparison with PGT-A on the basis of a blastomere biopsy, as the PB approach would regard, so they argue, just oocytes, thus avoiding *embryo* selection. This view is, however, not evident. After all, the PB approach involves both the first and the second PBs. As the second PB is generated only *after* the penetration of the oocyte by the sperm, the underlying issue regards the status of the pre-syngamy zygote: is this still an oocyte or already an embryo? Views differ: in German law, the pre-syngamy zygote is not qualified as an embryo, but in Dutch law it is. However, we doubt as to whether it is justified to make a fundamental ethical difference between intra- and post-fertilization PGT.

The practical issue that makes PGT-A still controversial regards the lack of evidence of its efficacy [12–14]. It is disquieting that some proponents of PGT-A uncritically claim that its efficacy has already been proved years ago and that PGT-A is just a matter of good clinical practice. For example, under the auspices of the virtual Academy of Genetics, COGEN (Controversies in Preconception, Preimplantation and Prenatal Genetic Diagnosis) issued a so-called Consensus Statement, which reads: "PGS should no longer be considered an experimental procedure ... and, where possible, should be made available for routine practice. ...We therefore believe that PGS should be part of the discussion with all patients considering/undergoing IVF treatment" [15]. Clearly, to present PGT-A as post-experimental, evidence-based reproductive medicine while many experts and some of the major professional organizations consider the evidence to be of low quality and the "supportive" studies to be flawed and biased is at odds not only with the interests and rights of IVF patients, especially their right to be adequately informed, but also with professionals' responsibility to avoid futile interventions and to enable society to distribute scarce resources available for healthcare in a just and well-considered way. Even if some studies seem to show a modest advantage of PGT-A in terms of a somewhat shorter time to pregnancy for at least some subsets of women (a possible benefit not to be confused with the primary aim of PGT-A's proponents), the socio-ethical question remains whether this advantage is sufficiently impressive to justify the substantial costs of PGT-A. The message that PGT-A should only be offered in the context of rigid randomized controlled clinical trials, aimed at a systematic evaluation of its presumed advantages and possible disadvantages, is still valid.

15.3.2 PGT for Common Complex Disorders Based on Polygenic Risk Scores

Some commercial companies, like Genomic Prediction Inc. in the USA, recently started to offer screening of all IVF embryos using polygenic risk scores (PRS), often abbreviated as PGT-P (PGT-P may also be offered more selectively; see below). PGT-P aims at *ranking* embryos for transfer, based on a calculation of the cumulative effects of a (large) set of genetic variants to assess future children's susceptibility for multifactorial disorders, such as type 2 diabetes, stroke and schizophrenia, and their chance of having particular nonmedical traits, like a high (or at least a higher than average) IQ.

The ethics of this new type of screening is similar to the ethics of future "comprehensive" PGT, which was proactively addressed in the ethical literature some time ago (see below). While companies involved market such PGT-P as just a rational way to select the best embryo, critics have two clusters of objections and concerns. From a scientific perspective [16], it is argued that this method is flawed and the information provided to the public is misleading because, among other things, the genetics knowledge base is largely European-ancestry centered, there is less variation in PRS among a set of embryos produced by the same two biological (prospective) parents, the penetrance of the genetic variants screened for is (very) incomplete, the predictive value of such PGT-P is suboptimal, and the ranking aimed at may have adverse effects linked with antagonistic pleiotropy; after all, by selecting for a beneficial complex trait (like high intelligence) one may inadvertently (and unknowingly) select the embryo with one or more harmful traits associated with the same genes (like autism). From an ethical perspective, at least the following principles need to be taken into account. First, the principle of proportionality, requesting that the benefits of any screening should clearly outweigh its disadvantages. Obviously, the results of PGT-P seem to be a priori less important for embryo ranking than traditional morphologic scores used for viability assessment. Apart from that, should possible PGT-P be restricted to "predicting" medical conditions or would it be sound to include genes associated with nonmedical traits? Would the latter constitute a risk or a benefit of PGT-P? Second, respect for reproductive autonomy. Relevant points to consider include the quality and adequacy of the information provided (a precondition for valid informed consent) and decision-making authority regarding the ranking of embryos for transfer. For example, how to deal with the ranking request of a couple that considers PRS for medical traits less important than scores "predicting" a higher than average IQ. And finally, justice. If unequal access to PGT-P for only those who can afford to pay would be at odds with so-called *formal* justice, should PGT-P then be collectively funded? Or would this be at odds with the just use of scarce resources for healthcare, also in view of the debatable proportionality of PGT-P?

15.3.3 Comprehensive PGT

In the future, NGS technology may allow PGT to test for chromosomal aberrations, all (more common) Mendelian disorders, many susceptibilities for complex disorders and genetic codeterminants for nonmedical traits, including personality traits, *simultaneously*. This approach may seem to be ideal, as one could at the same time select the most viable embryo for transfer (increasing the take-home baby rate), optimally reduce the risk of having an affected child, and select the "best" embryo for transfer. However, an unmitigated enthousiasm for such comprehensive PGT is naive, and at best premature, as this screening raises a series of complex ethical and ethically relevant questions and issues [3,17], including the following (this sketch is not meant to be exhaustive):

To begin with, what, precisely, would be the proper aim of comprehensive PGT? Basically, there seem to be at least two possibilities, at least in theory. The first would be to exclude genetic risks for progeny. Clearly, however, this aim would be self-defeating and illusory, as we are all "fellow mutants"; each of us, and each embryo carries various mutations and predispositions for disorders. Most of these will involve complex multifactorial disorders, and some will relate to Mendelian diseases. Besides, risk-free reproduction does not exist, neither in vivo nor in vitro; if anyone wanted to eliminate any genetic risk for future children, they should refrain from any transfer or, better, from reproduction altogether. A second, in theory more rational, possible aim of comprehensive PGT would be to select the "best" embryo, the one with the best prospects of a healthy (or, maybe broader, a flourishing) life. But what criteria are then most relevant (see below)?

A rather practical concern regards, again, the quality of the information generated by whole-genome sequencing and analysis on a single-cell basis. In more technical terms, what about the analytic and clinical validity of this information? Obviously, the more false-positive results, the lower the number of embryos available for transfer, and the lower the take-home baby rate. Furthermore, most of the information will be related to risk factors for common complex disorders (as is also the case in PGT-P). This information is by definition probabilistic, and often has a rather low predictive value. This will easily undermine the clinical utility of the information. The burden of proof regarding the often-claimed better quality of future screening panels and methods is, obviously, on the side of the proponents of such screening.

Next, considering the complexity of comprehensive PGT, traditional informed consent would simply be impossible. Could a so-called "generic consent" for comprehensive PGT be ethically (and legally) acceptable and, if so, on what conditions?

Another autonomy-related concern is that prospective parents involved would regularly, if not systematically, be confronted with rather complex, if not impossible, trade-offs. In theory, it may be simple to develop risk profiles that facilitate prospective parents' well-considered choice. However, what about clinical practice, taking into account the fact that all embryos will carry lots of predispositions for a great number of common disorders – constituting "profiles" that may have a different meaning for different stakeholders in view of their own family history and experience?

A related issue concerns, again, decision-making authority: who has the final say, the doctor or the prospective parents (see above)? Even if all parties agree that the best embryo should be transferred, there will, so we presume, regularly be disagreement about what this means if confronted with the complex risk profiles of, say, some ten embryos. Obviously, before implementing comprehensive PGT, a proactive reflection about material and procedural criteria and trade-offs to be made regarding embryo transfer is required.

Last, but not least, comprehensive PGT may provide unexpected (predictive) information about the genetic status of (one of) the prospective parents themselves, thereby undermining their right not to know, which is part of their right to informational self-determination. Likewise, it may involve an interference with the future child's right not to know, namely the right of the child to later decide for itself whether or not to be tested for genetic risks for future diseases. In theory, the latter could be prevented by not transferring embryos carrying such risk factors but, again, as we are all "fellow mutants," there may well be no embryo suitable for transfer. Or would it be morally acceptable to adopt a narrower (less extensive) interpretation of future children's right not to know, and accept comprehensive embryo profiling and related transfers of embryos with known genetic risk factors for later-onset preventable and/or treatable disorders?

PGT-P might be applied in the context of future comprehensive PGT, aimed at increasing the predictive value of the genomic profiling of embryos for complex disorders and (nonmedical) traits, by taking account of the *cumulative* effects of the relevant genetic variants.

Our conclusion is that comprehensive PGT, with or without PGT-P, would be fully disproportional at the moment. It meets neither basic technical criteria related to analytic and clinical validity nor the fundamental requirement of proportionality. Furthermore, some ethical issues urgently need further analysis.

15.3.4 Blurring the Boundary

While the literature suggests there is a neat dichotomy between PGT-M and PGT-A, there is in fact a gray area that indicates a blurring of the boundary between the two. This may be best illustrated by the recent introduction of genome-wide haplotyping and similar technologies such as karyomapping [18]. Its advantages are manifold and include (1) the technique is generic, so no (time-consuming, expensive) optimization of the protocol per family/locus is needed as the analysis makes use of huge numbers of single nucleotide polymorphisms (SNPs) spread across the genome; (2) in case of parental reciprocal translocations, chromosomally normal embryos and embryos carrying a balanced or unbalanced translocation can be distinguished; and (3) both monosomies and trisomies can be identified, differentiating between meiotic and mitotic origins. If haplotyping in the context of PGT-M is framed as just a technique to improve the performance of such PGT, any findings regarding the embryo's chromosomal constitution should be classified as unsolicited or incidental findings. However, such haplotyping could also be framed as a type of PGS; after all, additional information about the chromosomal constitution can be easily filtered out, so if one wants to obtain this information in addition to the information directly linked with the indication for PGT-M, one is in fact transgressing the line between testing on indication and screening. No doubt this conceptual issue requires further clarification. Linked with this "classification" issue is the question: who has decision-making authority about generating or filtering out information not linked with the original indication? And, again, how does one define clear, sound guidelines for the (prioritization for the) transfer of (possibly) affected embryos, especially when they carry genetic variations or anomalies not related to the disease locus linked with the original indication for PGT-M? Tentative guidance has recently been published and is a useful starting point for further discussion [19]. This includes the recommendation that embryos carrying trisomy 21 should never be transferred, even if this trisomy has a mitotic origin, lowering the risk of developing into a child with Down syndrome, and even if this precludes couples from having a genetically linked child at all.

If PGT-P were to be used in the context of PGT on indication, this would constitute another example of blurring. The ethics of such use of PGT-P is similar to the ethics of its alternative possible applications in PGT (see above).

15.4 Expanded Preconception Carrier Screening

For many decades, preconception carrier screening programs have targeted individuals or couples in specific ethnic groups with a higher frequency of autosomal recessive disorders associated with significant morbidity and reduced life expectancy. Well-known examples are β-thalassemia carrier screening in several high-risk populations in the Mediterranean region, and carrier screening in Ashkenazi Jewish populations for Tay–Sachs disease and other recessively inherited conditions with a higher frequency in those groups. When both partners are carriers of the same autosomal recessive disease, they have a one in four risk of having an affected child in each pregnancy. Because carriers of recessive disease are mostly healthy themselves, most carrier couples are unaware of their carrier status and most children with recessive disorders are the first affected child born in the family. Couples who know they are at risk can adjust their reproductive plans. If they want to avoid the birth of a child with the disorder, they can choose to refrain from having children, try to become parents through adoption, become pregnant and make use of prenatal diagnosis followed by termination of pregnancy in case of a positive result, or opt for MAR in the form of either reproduction with donor gametes or PGT. Although carrier screening can also be offered during pregnancy, the clear advantage of preconception screening is the much larger range of these options. However, as a target population, pregnant women and their partners are more easily reached than the diffuse category of couples or persons of reproductive age who may or may not have an imminent child wish.

With the advent of new genomic technologies it has now become possible to think of preconceptionally offered expanded carrier screening (ECS) (see also Chapter 10), an offer to all couples or persons of reproductive age, regardless of ethnic background, to have themselves tested for carrier status for up to several hundred, possibly more than 1000, recessive disorders simultaneously. Although these disorders are individually rare, the chances of being a carrier couple is estimated as 1–2 per 100 couples in the general population. This means that the reproductive risk is significant, or at least comparable to the risk for which prenatal screening is being offered in many countries for already quite some time.

For around a decade, commercial laboratories, especially in the USA, have been offering ECS to interested clients in the general population. The nonprofit part of the health sector is (hesitantly) following with initiatives in several countries. In Israel, a national ECS program for severe incurable disorders was introduced in 2013. Other initiatives are more recent. For instance, a government-funded trial started in Australia in 2019 to test the feasibility of a national ECS program. In Belgium, in 2017 the Superior Health Council provided recommendations on the basis of which a nationwide ECS offer is now being prepared. In the Netherlands, a pilot in the north of the country has led to a government initiative to explore the practical and ethical challenges of ECS. On a European level, the European Society of Human Genetics has issued a position statement with recommendations for the responsible implementation of this new type of reproductive screening in the general population [20]. For the purposes of this chapter an important development is that several European fertility clinics have started offering ECS on a routine basis to their patients, in addition to using expanded carrier testing in their donor programs.

A systematic review of the still limited evidence regarding reproductive decision-making by carrier couples identified through screening programs found that most of these couples chose to use preventive measures such as PGT or prenatal diagnosis followed by termination of pregnancy, especially when confronted with life-limiting disorders such as cystic fibrosis or hemoglobin disorders [21].

15.4.1 Aim of Offering ECS

As with most forms of reproductive screening, ECS does not lead to preventive measures other than by enabling couples to avoid having a child with a specific disorder. In the debate about prenatal screening for Down syndrome for example, concerns were expressed that making it a public health aim to reduce the birth incidence of children with these disorders through the offer of such screening might interfere with reproductive freedom (e.g. nudging pregnant women into terminating affected pregnancies) or send a morally problematic message with regard to the worth of people living with the relevant conditions. The wide consensus has since been that prenatal screening is aimed not at achieving health gains for society (the so-called "prevention paradigm") but at enabling would-be parents to make autonomous reproductive choices ("autonomy paradigm"). This is important, as it has direct implications for how such screening is to be conducted and evaluated. The autonomy aim sets higher requirements for the provision of balanced information and nondirective counseling, and entails different criteria for determining whether the screening was a success or failure, than if the aim were to bring down the number of children born with specific disorders.

Although enabling couples to make autonomous reproductive choices is also what most seem to regard as the aim of ECS [20], the consensus about this seems less complete than is the case with regard to prenatal screening [22]. This may reflect that historically the concept of preconception carrier screening has a strong connection with the idea of lessening the burden of disease in communities hard hit by severe disorders with a high frequency in the population. However, it may also have to do with the fact that preconception screening gives carrier couples a wider array of reproductive options as compared with screening during pregnancy. The idea that carrier couples can be expected to at least consider making use of these options in order to avoid a child with a severe disorder seems less contentious than issues concerning prenatal diagnosis and termination of pregnancy [23,24]. Without reverting to prevention as a public health aim, this may be argued for in terms of acknowledging that the reproductive options opened up by ECS, but then also ECS itself, are not entirely noncommittal from a perspective of parental responsibility. As this is not captured in the usual understanding of the autonomy paradigm, it is not obvious that the ethical framework developed for prenatal screening can simply be copied for ECS. For instance, it can be asked if professional neutrality should be as strict a norm in the preconception context as it is prenatally [24]. Clearly, these issues are a matter for further ethical reflection and debate.

15.4.2 ECS in the Context of MAR

Against the background of this discussion, different answers can be given as to why ECS may or should be offered in the context of MAR, as clarified in recent ESHRE guidance [25]. The offer can either be construed as an extra service for MAR applicants, or as an essential part of responsible MAR practice.

15.4.2.1 An Extra Service for MAR Applicants?

If the emphasis is on enabling couples to make autonomous reproductive decisions about which professionals should take a neutral stance, making ECS available as a form of selective screening in this specific context should be seen as nothing else but an extra service for MAR applicants. The arguments for providing this service may seem obvious. Whereas in the general population the target group for ECS (couples of reproductive age who want to have children in the near future) is difficult to reach, MAR patients are a self-presenting selection of precisely this target group to whom ECS can easily be offered. Moreover, whereas in the general population ECS does not seem to respond to a strongly perceived need, this may well be different in a context where couples (or individual reproducers) already come for medical help in order to have a child. The fact that commercial clinics have started offering ECS to their patients suggests that a significant number of them regard this as an offer they would indeed want to consider. As they are already undergoing IVF, adding PGT in case they turn out to be a carrier couple for a recessive disorder may not be such a big step. And for those already having PGT, testing for that disorder could simply be added to the procedure they will be having anyhow.

For couples who can easily afford procedures such as IVF and PGT, the extra costs of ECS (estimated at "less than 10%" in the study last quoted) may not be a barrier, but they can be prohibitive for those who already barely manage to find the money for MAR. This would not be a further justice issue if ECS concerned a luxury add-on without much added value, but that is not indeed how it is presented and perceived. Of course these "equity of access" issues (with regard to both MAR itself and ECS) can only be addressed on a societal level, with a role for health insurers or governments, depending on the healthcare system. In countries where MAR is (partially) paid for from public or collective funds, the selective nature of ECS in MAR may be regarded as another justice problem. Why offer this only to those who happen to receive reproductive assistance, given that all couples in the general population have the same a priori risk? However, given the challenges of implementing ECS in the general population, it can perhaps be argued that a selective screening offer in an easy-to-reach population should be seen as a useful learning step toward wider implementation, also with an eye to finding out under what conditions an ECS offer would be proportional in the sense of bringing couples more benefits than harms [25]. One of the more general objections to ECS, that it would amount to a problematic medicalization of reproductive decision-making, would at least not seem to apply in this context, where ECS is offered to those who already seek to reproduce with professional help.

15.4.2.2 An Essential Part of Responsible MAR Practice?

A different answer as to why offer ECS in MAR connects with what we have earlier referred to (see Section 15.1) as the double responsibly of MAR professionals: not just toward those seeking their assistance, but also toward the resulting children. As stated by ESHRE, this includes a prima facie duty to reduce reproductive risks to the extent that doing so is reasonably possible and proportional [26]. Whether a routine offer of ECS to MAR patients would fit this profile is at least not obvious. This not only depends on the expected level of risk reduction, which at less than 1% in fact seems rather small [20], but also on the efforts and costs needed to achieve this in a responsible way, taking account of information and counseling needs, and under conditions that would minimize the potential for generating anxiety and other adverse psychosocial effects in MAR patients. If this balance can be shown to be positive (which at present is a big if), then reasoning from a professional duty to try to reduce reproductive risks might lead to presenting an ECS offer as an essential element of responsible MAR practice.

The main difference with the service approach is that, on this reasoning, all MAR applicants should (not may) be offered ECS. Moreover, the offer comes with a different frame. By being presented as following from the center's responsibility to reduce reproductive risks, the offer invites the applicants to consider the offer in the same terms, from the perspective, that is, of their parental responsibility. Note that it does not follow that they should take the offer and have themselves tested, or that professionals should pressure them to do so, even if only by casting the offer in the form of an opt-out, as some US centers have started doing. Given the low a priori risk in the general population, that would amount to a disproportional infringement of the applicants' autonomy. Even in individual cases with a higher a priori risk, as in consanguineous couples, professionals should at most invite them to consider ECS [25].

15.4.2.3 Scope of ECS

It is a misunderstanding that the autonomy paradigm would boil down to promoting reproductive autonomy as an end in itself. As if any reproductive decision, whatever its content, should be facilitated simply because it is something that people happen to want. If that were so, any restriction of the scope of the screening offer should be regarded as problematic. Instead, as underlined among others by the Health Council of the Netherlands, the paradigm is about enabling autonomous choices that are meaningful in a reproductive context. Also referring to this distinction, the European Society of Human Genetics has stated that (pending further research and discussion) the scope of ECS should be limited to serious congenital or childhood-onset disorders [20]. As also with this restricted scope, ECS may target hundreds of disorders, and informed consent will inevitably be based on information with a more generic character. Avoiding a mixed bag of disorders with highly different health implications will then be of essence in order to allow for consent to still be sufficiently informed.

Of course, the precise delineation of "serious" is a matter for debate. A strict line cannot be drawn, because subjective judgments inevitably play a role in the ascription of these qualifications. In a survey among health professionals, Lazarin and colleagues found large intersubjective agreement with regard to a mild to severe classification in which the most serious conditions included shortened lifespan, intellectual disability, impaired mobility, and internal physical malformation [27]. As for ECS, it would be important to know what disorders carrier couples regard as serious enough for adjusting their reproductive plans, and this is an area for further research.

15.4.2.4 Disclosure: To Couples or Individuals?

A further issue is how results are to be presented, either in the form of couple-based disclosure, or individually. In the first case, a result is positive only if both members carry a mutation in the same gene. By contrast, disclosing the carrier status of each partner separately will reveal discordant couples in which one of the partners has a positive and the other a negative test. Apart from avoiding possible negative psychosocial effects on the partner found to be a carrier in discordant couples (where these are not offset by the gains of relevant reproductive options), the benefits of couple-based disclosure include lower costs and lower capacity demands both for the laboratory and for post-test counseling. However, with couple-based disclosure, a much lower number of carriers will be identified, leading to far more limited opportunities to inform family members who may also be carriers and initiate cascade screening in their families. Moreover, individual partners of noncarrier couples who break up have no idea about their carrier status when entering a relationship with a new reproductive partner. It could be argued that the autonomy paradigm does not simply require providing the most comprehensive result for both partners individually. That would certainly be the case had this meant the promotion of reproductive autonomy per se. However, things are less obvious with the qualified interpretation of the autonomy aim in terms of providing couples with options for meaningful reproductive choice. As knowledge of individual results has less immediate reproductive value, this would in fact seem to strengthen the case for couple-based instead of individual disclosure [28].

15.4.2.5 MAR for Carrier Couples

When ECS leads to finding a carrier couple at risk of having a child with a recessive disease, they will of course have to be helped in careful counseling to understand the implications as well as the options open to them. It can be expected that many will decide to adjust their reproductive plans to avoid a child with a serious disorder. This may for instance entail adding PGT or shifting to donor conception. While this is very much the same as for carrier couples found in a non-MAR population, there is one important difference. Carrier couples not dependent on reproductive assistance may decide to simply go ahead, get pregnant and by forgoing the option of prenatal diagnosis, take the one in four risk of having a child with, say, cystic fibrosis. While also in that setting counselors need not (or, in our view, should not) refrain from questioning the morality of that choice, it should be accepted that this is ultimately a matter of the couple's reproductive freedom. However, where MAR applicants are found to be a carrier couple for a clearly serious disorder, just to go ahead as planned with IVF for example, is not an option that professionals should be willing to facilitate. In cases of a high risk of a child with a seriously diminished quality of life, ESHRE guidance would require them to make their assistance conditional on the use of PGT or other preventive option [25].

This confirms that in a MAR context ECS is never a mere matter of providing couples with options for reproductive choice. Even when presented as an extra service under the autonomy paradigm, it may lead to information that invites both applicants and professionals to reappraise the options from a perspective of reproductive responsibility. At the same time, it is important to guard against a possible "hypertrophy" of professional responsibility [29]. The warning that "providers of assisted reproduction services may adopt a view whereby the conception of a child with any preventable genetic disorder, however mild, is considered an iatrogenic failure and should therefore be avoided through PGT" [21] should certainly be heeded.

15.4.3 ECS in MAR with Donor Gametes

For decades, recommendations for genetic screening of gamete donors have included testing for carrier status of autosomal recessive disorders with a high frequency in the donor's ethnic background. Together with karyotyping to avoid translocation carriers (recommended in some countries but not in others), that is still very much all the genetic testing recommended for gamete donors, in addition to taking an extensive medical and family history. For some time, however, the tendency among commercial clinics and gamete banks is to go beyond these guidelines and test their donors for a much wider range of genetic conditions, although with large differences between test panels. In recent years several clinics and banks have moved to using forms of ECS for this purpose. This should first of all be seen against the background of a competitive market in which donor gametes are seen as a product and screening as a way for providers to demonstrate better quality than competitors and to avoid legal liability. On the recipient side, the call for increased testing is driven by highly publicized cases in which a rare but serious genetic disorder was transmitted to several donor offspring.

The view that donor conception should be as safe as possible is contested. For instance, in a recent statement on sperm donation by a working group of the Dutch O&G Society together with the National Society for Clinical Embryology, it is argued that donor conception need not be safer than reproduction between partners [30]. Screening that is not offered to the general population should therefore not be done in donors. However, this seems to ignore a morally relevant difference between these contexts. Partners want to reproduce together, whereas people needing donor gametes do not (in most cases) want to reproduce only with this donor. And whereas partners cannot be replaced, donors are in fact replaceable. Even so, it does not follow that it would be wise to strive for a zero-risk policy in this context. First of all because it is simply impossible to rule out all reproductive risks and suggesting otherwise would send a message of false reassurance; but also because efforts aimed at further risk reduction must be proportional [31].

A specific problem arising here is that when testing leads to excluding more donors, this would undermine the very practice one tries to improve. However, precisely as with ECS for autosomal recessive disorders, this problem can in principle be solved. As reported rates of individuals found to be heterozygous for at least one autosomal recessive disorder are high, using ECS to test gamete donors would be a self-defeating strategy if all donors found to be carriers were indeed to be excluded from the program. Crucially, however, that need not be the case, as donors and recipients can be matched in order to avoid the formation of carrier couples. While this is sometimes promoted as the most rational approach to donor screening, it is still an open question how this matching approach should be implemented in practice. One obvious problem is that it requires recipients not only to accept this approach, but also to be tested themselves. Clearly, it would be disproportional to impose testing upon recipients in order to achieve what can at most be a small reduction of the risk of gamete donation [32].

More generally, it is a matter of concern that ECS is being proposed as a means of maximizing safety, regardless of the impact on all stakeholders, including gamete donors. This invites comprehensive donor testing proposals where, apart from ECS with an eye to matching, donors are also tested for a range of X-linked and later-onset dominant disorders. Those found to be carriers for such conditions will not be used as donor. While here the interest in their role and contribution seems to end, they themselves may be left to cope with highly burdensome health information pertaining to themselves and/or their relatives. One example of this is testing oocyte donors for carrier status of fragile X syndrome (FXS). Although not recommended in current donor screening guidelines, it is part and parcel of all ECS panels used for screening oocyte donors. However, this ignores the consensus that in view of uncertainty about the clinical importance of findings in a "gray zone" (intermediate alleles), fragile X syndrome carrier testing should not be offered to women in the general population. Ignoring this guidance where it concerns the oocyte donor amounts to treating her as a mere contributor of reproductive material; in effect to reduce her to her oocytes. From the point of view of medical ethics, this is clearly at odds with the principle of respect for persons.

15.5 Germline Genome Editing

Human germline genetic modification was considered to be a futuristic, far-fetched option until 2015, when Chinese researchers published their research on human germline genome editing (GGE) using CRISPR/Cas9. Since then, such editing has become a major topic on the international ethics and science policy agendas. We here summarize the joint recommendations and background paper of ESHRE and the European Society of Human Genetics, differentiating between nonreproductive GGE on the one hand – entailing both basic and preclinical research – and reproductive GGE on the other [33].

15.5.1 Nonreproductive GGE

Basic research in GGE focuses primarily on fundamental questions regarding human embryology and the methods applied in genome editing. The main ethical issue is the possible use of human embryos in such research, particularly the creation of "research" embryos. Although the latter is prohibited in many countries, it could be justified from an ethical perspective taking account of the dominant view that early (preimplantation) embryos have a relatively low moral status – they don't qualify as human individuals, let alone persons.

Preclinical GGE research would aim at exploring the safety and effectiveness of possible clinical applications. Such research would fit the normative framework for responsible innovation in assisted reproduction generally [34]. Clearly, the ethics of clinical applications codetermine the ethics of linked preclinical research. If one considers moral objections to reproductive GGE to be overriding (see below), preclinical GGE research would not only be a complete waste, but would also bring us dangerously close to the edge of the slope. And as long as it is unclear as to whether safe and effective reproductive GGE can be justified, preclinical research can or even should be dismissed as premature. Obviously, however, if only safety issues stand in the way of sound reproductive GGE, preclinical safety research would be morally justified.

15.5.2 Reproductive GGE

Reproductive GGE is categorically forbidden in many countries. From an ethical point of view, this is not self-evident. Deontologic objections such as "GGE is against nature," "GGE is at odds with human dignity," or "GGE would undermine the autonomy of the child" are unconvincing; they may apply to some, mostly rather theoretical, "enhancement"-like reproductive GGE aimed at creating super or designer babies with, for example, exceptional intellectual, athletic and/or artistic capacities, but are simply not applicable to reproductive GGE aimed at therapy for, or prevention of, serious disorders in future children.

So-called consequentialist objections, pointing to adverse consequences, may have a more direct relevance. In view of the many unknowns regarding safety, reproductive GGE would be premature at present, and can only be considered on the basis of further basic and preclinical research. If clinical GGE is considered to be justified in the future, adequate clinical trials and long-term follow-up studies are necessary. Although societal concerns are important – linked with, for example, the possible misuse of GGE for debatable nonmedical purposes (enhancement) – the distinction between therapy/prevention and enhancement is not always clear-cut. Experience with the regulation of PGT-M/SR should help build a sound strategy for regulating future clinical GGE, including a licensing system and obligatory regular reporting by licensed clinics of policies and practices.

Commentators often point to safe and effective alternatives for GGE, especially PGT-M, concluding that "there is no real need for GGE" [35]. This, however, seems to be a somewhat premature conclusion. While PGT-M may be a good option for many people at high risk of having an affected child, there will still be situations where GGE, if safe and effective, may well be preferred and morally justified depending on, among others, (1) the genetic make-up of prospective parents (such as couples who can only have affected children, in which case PGT-M is no option), (2) their experiences with clinical PGT (think of, for example, PGT-M cycles which show all embryos to be unsuitable for transfer), or (3) peoples' moral and religious preferences (couples may want to minimize embryo loss in IVF).

15.6 Final Remarks

While PGT-M was widely contested when it was introduced in the clinic some three decades ago, in many countries it has become an established reproductive option for couples at high risk. This is not to say that ethical reflection on such PGT has become obsolete. Issues for further debate include PGT-M's "contextualized proportionality" and possible tensions between respect for prospective parents' reproductive autonomy and professional responsibility, both at the front and the back door. The moral landscape with regard to routine PGT is rather diffuse; while the offer of PGT-PN is imperative and PGT-A does not raise any principled objections but should be questioned because of its unproven efficacy, PGT-P and "comprehensive" PGT are truly new steps in the direction of the rationalization of human reproduction and needs in-depth ethical scrutiny, with regard also to artificial intelligence-based ranking issues at its back door.

ECS in MAR may well have added value. Still, it is problematic to simply assume its proportionality – rigorous pilot and evaluation studies are crucially important. Points for further ethical reflection include the scope of ECS, the place of ECS in matching MAR applicants with gamete donors, and possible tensions between the autonomy aim of reproductive screening and the responsibility of MAR professionals to take account of the welfare of the future child.

In the years to come, (human) GGE can only be responsibly performed in the context of basic and preclinical research. If in the future GGE is considered to be sufficiently safe (the criteria for the safety of new reproductive technologies generally need further debate), one can only speculate about its clinical impact. It is not far-fetched to assume that it would be a real revolution in reproductive medicine and would also have a considerable impact on its clinical ethics. In some high-risk situations, offering (safe and effective) GGE as a condition for access to assisted reproduction might well be morally justified.

A final remark: assisted reproduction has become a highly lucrative medical industry. This may help explain the premature clinical introduction of experimental assisted reproductive technologies (without adequate preclinical safety research), the large-scale overtreatment of "subfertile" couples who do not need fertility treatment, the growing number of add-ons, and expanding the use of potentially risky technologies to applications for which its effectiveness has not been proven [36]. Professional organizations like ESHRE and the American Society for Reproductive Medicine should set ethical standards in order to protect patients and maintain the public's trust in reproductive medicine.

References

1. Pennings G, de Wert G, Shenfield F, et al. ESHRE Task Force on Ethics and Law 13: the welfare of the child in medically assisted reproduction. *Hum Reprod* 2007;22(10):2585–8.
2. Dondorp W, de Wert G. Refining the ethics of preimplantation genetic diagnosis: a plea for contextualized proportionality. *Bioethics* 2019;33(2):294–301.
3. De Wert G. Preimplantation genetic diagnosis: normative reflections. In: Harper J, ed. *Preimplantation Genetic Diagnosis*, 2nd ed. Cambridge: Cambridge University Press, 2009.
4. De Wert G, Dondorp W, Shenfield F, et al. ESHRE Task Force on Ethics and Law 22: preimplantation genetic diagnosis. *Hum Reprod* 2014;29(8):1610–17.
5. van der Schoot V, Dondorp W, Dreesen J, et al. Preimplantation genetic testing for more than one genetic condition: clinical and ethical considerations and dilemmas. *Hum Reprod* 2019;34 (6):1146–54.
6. de Wert G. Preimplantation genetic diagnosis: the ethics of intermediate cases. *Hum Reprod* 2005;20(12):3261–6.
7. Dondorp W, De Wert G, Pennings G, et al. ESHRE Task Force on Ethics and Law 20: sex selection for non-medical reasons. *Hum Reprod* 2013;28(6):1448–54.
8. Davis DS. Genetic Dilemmas. Reproductive Technology, Parental Choices, *and* Children's Futures, 2nd ed. New York: Oxford University Press, 2010.
9. Glover J. *Choosing Children. Genes, Disability and Design.* Oxford: Clarendon Press, 2006.
10. Soto-Lafontaine M, Dondorp W, Provoost V, de Wert G. Dealing with treatment and transfer requests: how PGD-professionals discuss ethical challenges arising in everyday practice. *Med Health Care Philos* 2018;21(3):375–86.
11. Mastenbroek S, Twisk M, van der Veen F, Repping S. Preimplantation genetic screening: a systematic review and meta-analysis of RCTs. *Hum Reprod Update* 2011;17(4):454–66.
12. Mastenbroek S, Repping S. Preimplantation genetic screening: back to the future. *Hum Reprod* 2014;29(9):1846–50.
13. Pagliardini L, Vigano P, Alteri A, et al. Shooting STAR: reinterpreting the data from the "Single Embryo TrAnsfeR of Euploid Embryo" randomized clinical trial. *Reprod Biomed Online* 2020;40(4):475–8.
14. Orvieto R, Gleicher N. Preimplantation genetic testing for aneuploidy (PGT-A)-finally revealed. *J Assist Reprod Genet* 2020;37(3):669–72.
15. CoGEN. *CoGEN Position Statement on Chromosomal Mosaicism Detected in Preimplantation Blastocyst Biopsies*, 2016. https://ivf-worldwide.com/cogen/oep/publications/cogen-position-statement-on-chromosomal-mosaicism-detected-in-preimplantation-blastocyst-biopsies.html
16. Turley P, Meyer MN, Wang N, et al. Problems with using polygenic scores to select embryos. *N Engl J Med* 2021;385(1):78–86.
17. Hens K, Dondorp W, Handyside AH, et al. Dynamics and ethics of comprehensive preimplantation genetic testing: a review of the challenges. *Hum Reprod Update* 2013;19(4):366–75.

18. Zamani Esteki M, Dimitriadou E, Mateiu L, et al. Concurrent whole-genome haplotyping and copy-number profiling of single cells. *Am J Hum Genet* 2015;96 (6):894–912.

19. Dimitriadou E, Melotte C, Debrock S, et al. Principles guiding embryo selection following genome-wide haplotyping of preimplantation embryos. *Hum Reprod* 2017;32 (3):687–97.

20. Henneman L, Borry P, Chokoshvili D, et al. Responsible implementation of expanded carrier screening. *Eur J Hum Genet* 2016;24(6):e1–e12.

21. Cannon J, Van Steijvoort E, Borry P, Chokoshvili D. How does carrier status for recessive disorders influence reproductive decisions? A systematic review of the literature. *Expert Rev Mol Diagn* 2019;19(12):1117–29.

22. Molster CM, Lister K, Metternick-Jones S, et al. Outcomes of an international workshop on preconception expanded carrier screening: some considerations for governments. *Front Public Health* 2017;5:25.

23. Bonte P, Pennings G, Sterckx S. Is there a moral obligation to conceive children under the best possible conditions? A preliminary framework for identifying the preconception responsibilities of potential parents. *BMC Med Ethics* 2014;15:5.

24. van der Hout S, Dondorp W, de Wert G. The aims of expanded universal carrier screening: autonomy, prevention, and responsible parenthood. *Bioethics* 2019;33(5):568–76.

25. de Wert G, van der Hout S, Goddijn M, et al. The ethics of preconception expanded carrier screening in patients seeking assisted reproduction. *Hum Reprod Open* 2021;2021(1): hoaa063.

26. Dondorp W, de Wert G, Pennings G, et al. ESHRE Task Force Ethics and Law 17. Lifestyle-related factors and access to medically assisted reproduction. *Hum Reprod* 2010;25(3):578–83.

27. Lazarin GA, Hawthorne F, Collins NS, et al. Systematic classification of disease severity for evaluation of expanded carrier screening panels. *PLoS One* 2014;9(12): e114391.

28. Schuurmans J, van den Heuvel LM, Plantinga M, et al. Feasibility of couple-based expanded preconception carrier screening offered by the general practitioner. *Eur J Hum Genet* 2018;26:808–9.

29. De Wert G. Ethics of assisted reproduction: the case of preimplantation genetic diagnosis. In Fauser BCJM, Rutherford AJ, Strauss JF, Van Steirteghem A, eds. *Molecular Biology in Reproductive Medicine*. New York/London: Parthenon Publishers, 1999: 433–48.

30. Nederlandse vereniging Obstetrie Gynaecologie (NVOG), Vereniging voor Klinische Embryologie (KLEM). *Landelijk Standpunt Spermadonatie. Specifieke Eisen voor Spermadonoren*. Utrecht, 2018.

31. Dondorp W, De Wert G, Pennings G, et al. ESHRE Task Force on Ethics and Law 21: genetic screening of gamete donors: ethical issues. *Hum Reprod* 2014;29(7):1353–9.

32. Mertes H, Lindheim SR, Pennings G. Ethical quandaries around expanded carrier screening in third-party reproduction. *Fertil Steril* 2018;109(2):190–4.

33. de Wert G, Pennings G, Clarke A, et al. Human germline gene editing: recommendations of ESHG and ESHRE. *Eur J Hum Genet* 2018;26(4):445–9.

34. Dondorp W, de Wert G. Innovative reproductive technologies: risks and responsibilities. *Hum Reprod* 2011;26(7):1604–8.

35. Mertes H, Pennings G. Modification of the embryo's genome: more useful in research than in the clinic. *Am J Bioeth* 2015;15(12):52–3.

36. Jans V, Dondorp W, De Wert G. Between innovation and precaution: the role of offspring safety considerations in strategies of introducing new reproductive techniques. *Hum Reprod* Open 2020;2020(2):hoaa003.

Index